MONOGRAPHS AND RESEARCH NOTES IN MATHEMATICS

Series Editors

John A. Burns
Thomas J. Tucker
Miklos Bona
Michael Ruzhansky

Published Titles

Application of Fuzzy Logic to Social Choice Theory, John N. Mordeson, Davender S. Malik and Terry D. Clark

Blow-up Patterns for Higher-Order: Nonlinear Parabolic, Hyperbolic Dispersion and Schrödinger Equations, Victor A. Galaktionov, Enzo L. Mitidieri, and Stanislav Pohozaev

Cremona Groups and Icosahedron, Ivan Cheltsov and Constantin Shramov

Difference Equations: Theory, Applications and Advanced Topics, Third Edition, Ronald E. Mickens

Dictionary of Inequalities, Second Edition, Peter Bullen

Iterative Optimization in Inverse Problems, Charles L. Byrne

Modeling and Inverse Problems in the Presence of Uncertainty, H. T. Banks, Shuhua Hu, and W. Clayton Thompson

Monomial Algebras, Second Edition, Rafael H. Villarreal

Nonlinear Functional Analysis in Banach Spaces and Banach Algebras: Fixed Point Theory under Weak Topology for Nonlinear Operators and Block Operator Matrices with Applications
Aref Jeribi and Bilel Krichen

Partial Differential Equations with Variable Exponents: Variational Methods and Qualitative Analysis, Vicenţiu D. Rădulescu and Dušan D. Repovš

A Practical Guide to Geometric Regulation for Distributed Parameter Systems, Eugenio Aulisa and David Gilliam

Signal Processing: A Mathematical Approach, Second Edition, Charles L. Byrne

Sinusoids: Theory and Technological Applications, Prem K. Kythe

Special Integrals of Gradshetyn and Ryzhik: the Proofs – Volume I, Victor H. Moll

Forthcoming Titles

Actions and Invariants of Algebraic Groups, Second Edition, Walter Ferrer Santos and Alvaro Rittatore

Analytical Methods for Kolmogorov Equations, Second Edition, Luca Lorenzi

Complex Analysis: Conformal Inequalities and the Bierbach Conjecture, Prem K. Kythe

Computational Aspects of Polynomial Identities: Volume I, Kemer's Theorems, 2nd Edition
Belov Alexey, Yaakov Karasik, Louis Halle Rowen

Forthcoming Titles (continued)

Geometric Modeling and Mesh Generation from Scanned Images, Yongjie Zhang

Groups, Designs, and Linear Algebra, Donald L. Kreher

Handbook of the Tutte Polynomial, Joanna Anthony Ellis-Monaghan and Iain Moffat

Lineability: The Search for Linearity in Mathematics, Juan B. Seoane Sepulveda, Richard W. Aron, Luis Bernal-Gonzalez, and Daniel M. Pellegrinao

Line Integral Methods and Their Applications, Luigi Brugnano and Felice Iaverno

Microlocal Analysis on R^n and on NonCompact Manifolds, Sandro Coriasco

Practical Guide to Geometric Regulation for Distributed Parameter Systems, Eugenio Aulisa and David S. Gilliam

Reconstructions from the Data of Integrals, Victor Palamodov

Special Integrals of Gradshetyn and Ryzhik: the Proofs – Volume II, Victor H. Moll

Stochastic Cauchy Problems in Infinite Dimensions: Generalized and Regularized Solutions, Irina V. Melnikova and Alexei Filinkov

Symmetry and Quantum Mechanics, Scott Corry

MONOGRAPHS AND RESEARCH NOTES IN MATHEMATICS

Nonlinear Functional Analysis in Banach Spaces and Banach Algebras

Fixed Point Theory under Weak Topology for Nonlinear Operators and Block Operator Matrices with Applications

Aref Jeribi
University of Sfax
Tunisia

Bilel Krichen
University of Sfax
Tunisia

CRC Press
Taylor & Francis Group
Boca Raton London New York

CRC Press is an imprint of the
Taylor & Francis Group, an **informa** business
A CHAPMAN & HALL BOOK

CRC Press
Taylor & Francis Group
6000 Broken Sound Parkway NW, Suite 300
Boca Raton, FL 33487-2742

ISBN 13: 978-1-4987-3388-5 (hbk)

Visit the Taylor & Francis Web site at
http://www.taylorandfrancis.com

and the CRC Press Web site at
http://www.crcpress.com

To
my mother Sania, my father Ali,
my wife Fadoua, my children Adam and Rahma,
my brothers Sofien and Mohamed Amin,
my sister Elhem,
my mother-in-law Zineb, my father-in-law Ridha, and
all members of my extended family ....

Aref Jeribi

To
the memory of my mother Jalila,
my father Hassan,
my wife Nozha, and my children Mohamed and Zaineb.

Bilel Krichen

Contents

Preface

This book focuses on fixed point theory for block operator matrices and its applications to a wide range of diverse equations such as, e.g., transport equations arising in the kinetic theory of gas (see [63]), stationary nonlinear biological models (see [138]), and in particular two-dimensional boundary value problems arising in growing cell populations and functional systems of integral equations. In all these topics, we are faced with problems such as the loss of compactness of mappings and/or missing appropriate geometric and topological structure of their underlying domain. Hence, it is convenient that we focus on fixed point results under the weak topology.

As a general rule, emphasis is generally put on some generic and recent results regarding the main topic, since it would be impossible to aim for complete coverage. Simultaneously throughout the chapters (5–7), we tried to illustrate the diversity of the theoretical results in the different settings (Banach spaces and Banach algebras).

In recent years, a number of excellent monographs and surveys presented by distinguished authors about fixed point theory have appeared such as, e.g., [2, 3, 45, 88]. Most of the above mentioned books deal with fixed point theory related to continuous mappings in topology and all its modern extensions. However, it is not always possible to show that a given operator between Banach spaces is weakly continuous. Quite often, its weak sequential continuity does not present any problem. As a first aim, this book is devoted to the study of several extensions of Schauder's and Krasnosel'skii's fixed point theorems to the class of weakly compact operators acting on Banach spaces as well as Banach algebras, in particular on spaces satisfying the Dunford–Pettis property. Notice that both of the above mentioned theorems can be used in order to resolve some open problems, seen in [99] and [115]. We first give some extension forms of Schauder's theorem by using some tools of the weak topology. Then, we present other results which are deduced by quite simple arguments. The notion of weak sequential continuity seems to be the most convenient one to use. It is not always possible to show that a map is weakly continuous. However, weakly sequentially continuous maps are shown to be the most convenient ones to use. That is why some new variants of fixed point theorems involving the measure of weak noncompactness and based on the notion of weak sequential continuity are presented in Banach spaces as well as Banach algebras. Some nonlinear alternatives for the sum of two weakly sequentially continuous mappings, and belonging to Leray–Schauder and Furi–Pera, are also presented.

The second objective of this book is dealing with the following question: Under which conditions will a 2×2 block operator matrix with nonlinear entries (and acting on Banach spaces and Banach algebras) have a fixed point? Based on the previous extension established under the weak topology setting, we are planning to extend these results by proving that, under certain hypotheses associated with its nonlinear entries, the 2×2 block operator matrix

$$\mathcal{L} := \begin{pmatrix} A & B.B' \\ C & D \end{pmatrix}$$

has a fixed point in $\Omega \times \Omega'$, where Ω and Ω' constitute two nonempty, closed, and convex subsets of Banach spaces or Banach algebras. This discussion is based on the presence or absence of the invertibility of the diagonal terms of $\mathcal{I} - \mathcal{L}$. Several fixed point theorems from Chapter 2 enable us to get new results for a particular 2×2 block operator matrix involving operators such as, e.g., $\mathcal{D}$-Lipschitzian, convex-power condensing, weakly sequentially continuous..., acting on a Banach algebra X. A regular case is considered when X is a commutative Banach algebra satisfying the so-called condition $(\mathcal{P})$; that is, for any sequences $\{x_n\}$ and $\{y_n\}$ in a Banach algebra X such that $x_n \rightharpoonup x$ and $y_n \rightharpoonup y$, then $x_n.y_n \rightharpoonup x.y$; here $\rightharpoonup$ denotes the weak convergence. That condition is very important and plays a key role in the investigations conducted in the proposed monograph. A new recent work is considered, when the entries are assumed multi-valued mappings.

Many applications to a wide range of diverse equations such as, e.g., transport equations arising in the kinetic theory of gas [97], stationary nonlinear biological models, in particular the two-dimensional boundary value problem arising in a growing cell population, a functional system of integral equations, and differential inclusions, are presented.

We should emphasize that this book is the first one dealing with the topological fixed point theory for block operator matrices with nonlinear entries in Banach spaces and Banach algebras. This book can also be regarded as a modest contribution to the fixed point theory in Banach spaces and Banach algebras. Researchers, as well as graduate students in applicable analysis, will find that this book constitutes a useful survey of the fundamental principles of the subject. Nevertheless, the reader is assumed to be, at least, familiar with some related sections concerning notions like the fixed point theorems of Schauder, Krasnosel'skii, Leray–Schauder and Furi–Pera, the basic tools of the weak topology, the concept of measures of weak noncompactness, and the transport equations, etc. Otherwise, the reader is urged to consult the recommended literature in order to benefit fully from this book.

There are some theorems which are based upon a number of hypotheses. These results are very recent. We should notice that, in a future and improved version of this book, the number of hypotheses in some theorems could be lowered and yet lead to the same conclusions. In other words, it may be possible to find an optimum number of hypotheses in the future.

The first author should mention that in the thesis work performed under his direction, by his former students and present colleagues Afif Ben Amar, Ines Feki, Bilel Krichen, Soufien Chouayekh, Rihab Moalla, Bilel Mefteh, Najib Kaddachi, and Wajdi Chaker, the obtained results have helped us in writing this monograph.

Finally, we would like to thank our friends and colleagues, particularly Pr. Ridha Damak whose encouragement and valuable remarks influenced the development of this monograph. We are grateful to Dr. Bilel Mefteh, Dr. Wajdi Chaker, Dr. Rihab Moalla, and Dr. Najib Kaddachi, who have read and commented upon the entire manuscript. Their constructive criticism has led to many improvements and has been a very great help. We encourage any comments or suggestions from any researcher. These comments or suggestions will certainly enable us to integrate some improvements for a new version of this book.

Aref Jeribi and Bilel Krichen

Sfax

Symbol Description

The most frequently used notations, symbols, and abbreviations are listed below.

$\mathbb{N}$	The set of natural numbers.
$\mathbb{R}$, $\mathbb{C}$	The fields of real and complex numbers, respectively.
$\mathbb{R}^n$	The n-dimensional real space.
$\inf(A)$	The infimum of the set A.
$\sup(A)$	The supremum of the set A.
∂A	The boundary of the set A
$\partial_\Omega A$	The boundary in Ω of the set A.
B_r	The ball with radius r.
B_X	The unit ball in X.
$d(x,y)$	The distance between x and y.
(X,d)	The metric space X.
X^*, X'	The dual of X.
$diam(A)$	The diameter of the set A, where A is a subset of a metric space X.
$dim(X)$	The dimension of the space X.
$\|x\|$	The norm of x.
$(X, \|.\|)$	The linear normed space X.
$\{x_n\}$	The sequence $\{x_n\}$.
$(\mathcal{P})$	The $(\mathcal{P})$ property.
BOM	Block Operator Matrix.
$\mathcal{P}(X)$	Set of subset of X.
$Gr(T)$	The graph of T.
Φ_T	The $\mathcal{D}$-function of the mapping T.
$\mathcal{P}_{cl}(S)$	The family of all closed subsets of S.
$\mathcal{P}_{cv}(S)$	The family of all convex subsets of S.
$\mathcal{P}_{cl,cv}(S)$	The family of all closed convex subsets of S.
l.s.c	Lower semi-continuous.
u.s.c	Upper semi-continuous.
$\mathcal{C}(\Omega,\mathbb{R})$	The space of real continuous functions on Ω.
$\mathcal{C}_b(I)$	The space of real continuous and bounded functions on I.
$\mathcal{C}(J,X)$	The Banach algebra of real continuous functions from J into X.
$\mathcal{C}(J,\mathbb{R})$	The Banach algebra of real continuous functions from J into $\mathbb{R}$.
$\mathcal{C}(J)$	The Banach algebra of real continuous functions from J into $\mathbb{R}$.
L_p	The L_p space.

$\mathcal{W}(X,Y)$	The set of weakly compact linear operators from X into Y.
$\mathcal{W}(X)$	The ideal of weakly compact linear operators on X.
$\mathcal{L}(X,Y)$	The space of linear operators from X into Y.
$\mathcal{L}(X)$	The space of linear operators from X into X.
$x_n \rightharpoonup x$	x_n converges weakly to x.
$x_n \to x$	x_n converges strongly to x.
$\overline{M}$	The closure of the set M.
$\overline{M^w}$	The weak closure of the set M.
$co(.)$	The convex hull.
$conv(.)$	The convex hull.
$\overline{co}(.)$	The closed convex hull.
$\overline{conv}(.)$	The closed convex hull.
(S,Σ,μ)	The measure space S.
MNWC	Measure of weak noncompactness.
DP	Dunford–Pettis property.
FIE	Functional Integral Equation.
FDE	Functional Differential Equation.
(m)	The (m) property.
T_H	The streaming operator.
$\mathcal{N}_f$	Nemytskii's operator generated by f.
IVP	Initial Value Problem.

Part I

Fixed Point Theory

Introduction

Over the past few decades, fixed point theory has been an active area of research with a wide range of applications in several fields. In fact, this theory constitutes an harmonious mixture of analysis (pure and applied), topology, and geometry. In particular, it has several important applications in various fields, such as physics, engineering, game theory, and biology (in which we are interested). Perhaps, the most well-known result in this theory is Banach's contraction principle. More precisely, in 1922, S. Banach formulated and proved a theorem which focused, under appropriate conditions, on the existence and uniqueness of a fixed point in a complete metric space (see [149]). This result leads to several powerful theorems such as inverse map theorem, Cauchy–Picard theorem for ordinary differential equations among others. In mathematics, some fixed point theorems in infinite-dimensional spaces generalize a well-known result proved by L. E. J. Brouwer [42] which states that every continuous map $A : B_1 \longrightarrow B_1$, where B_1 is the closed unit ball in $\mathbb{R}^n$ has, at least, a fixed point in B_1. These theorems have several applications. For example, we may refer to the proof of existence theorems for differential equations. The first result in this field was Schauder's fixed point theorem, proved in 1930 by J. Schauder and which asserts that every continuous and compact mapping from a closed, convex, and bounded subset $\mathcal{M}$ of a Banach space X into $\mathcal{M}$ has, at least, a fixed point [147].

From a mathematical point of view, many problems arising from diverse areas of natural sciences involve the existence of solutions of nonlinear equations having the following form

$$Ax + Bx = x, \quad x \in \mathcal{M},$$

where $\mathcal{M}$ is a nonempty, closed, and convex subset of a Banach space X, and where $A, B : \mathcal{M} \longrightarrow X$ are two nonlinear mappings. M. A. Krasnosel'skii was motivated by the observation that the inversion of a perturbed differential operator could lead to the sum of a contraction and a compact operator. That is why, M. A. Krasnosel'skii proved in 1958 a fixed point theorem (called

Krasnosel'skii's fixed point theorem (see [154])) which appeared as a prototype for solving equations of the above type. This theorem asserts that, if Ω is a nonempty, closed, and convex subset of a Banach space X, and if A and B are two mappings from Ω into X such that (i) A is compact, (ii) B is a contraction, and (iii) $A\Omega + B\Omega \subset \Omega$, then the sum $A + B$ has, at least, one fixed point in Ω. It should be noticed that this result was the first important mixed fixed point theorem which combined both Banach's contraction mapping principle and Schauder's fixed point theorem.

Many problems arising in physics and biology can be modeled using integro-differential equations in a way which allows us to find a fixed point for a continuous map defined on a functional space. This type of reasoning is equivalent to the fact of solving these equations using standard arguments, for example, transport equations arising in the kinetic theory of gas (see [63, 97]), stationary nonlinear biological models (see [138]), in particular boundary value problems arising in growing cell populations and functional integral equations. At this level, the book of D. O'Regan and M. Meehan [136], as well as the book of R. Dautray and J. L. Lions, [63] provided for us on the one hand a comprehensive existence theory for integral and integrodifferential equations, and presented also some specialized topics in integral equations, hence helping us to develop our applications. On the other hand, these books made a study of nonlinear and partial differential equations, not only those of classical physics, but also those that model transport in the kinetic theory of gas. Unfortunately, in all these topics, we are faced with problems such as the loss of compactness of mappings and/or missing appropriate geometric and topological structure of their underlying domain. That is why, it is convenient that we focus on fixed point results under the weak topology setting in order to investigate the problems of existence of solutions for different types of nonlinear integral equations and nonlinear differential equations in Banach spaces. At this level, A. Ben Amar, A. Jeribi, and M. Mnif in [32] have studied the existence of solutions for a model introduced by M. Rotenberg [142] in 1983, which describes the growth of cells population. The stationary model was presented in [14] on the space L_1 by the following equations :

$$v_3 \frac{\partial \psi}{\partial x}(x, v) + \sigma(x, v)\psi(x, v) - \lambda\psi(x, v) = \int_K r(x, v, v', \psi(x, v'))\, dv' \text{ in } D, \tag{0.1}$$

$$\psi_{|D^i} = H(\psi_{|D^o}). \tag{0.2}$$

The model (0.1)–(0.2) can be transformed into a fixed point problem having two types of equations. The first type involves a nonlinear weakly compact operator on L_1 spaces. The second type involves two nonlinear operators depending on the parameter λ, say, $\psi = A_1(\lambda)\psi + A_2(\lambda)\psi$, where $A_1(\lambda)$ is a weakly compact operator (i.e., it transforms bounded sets into relatively weakly compact sets) on L_1 spaces and $A_2(\lambda)$ is a (strict) contraction mapping for a large enough $Re\lambda$. Consequently, Schauder's (resp. Krasnosel'skii's) fixed point theorem cannot be used in the first (resp. second) type of equations. This is essentially due to the loss of compactness of the operator $(\lambda - S_K)^{-1}B$, where

$$S_K\psi(x,v) = -v_3\frac{\partial\psi}{\partial x}(x,v) - \sigma(x,v,\psi(x,v)),$$

and

$$B\psi(x,v) = \int_a^b r(x,v,v',\psi(x,v'))dv'.$$

More precisely, the authors of [32] gave an extension of the Schauder and Krasnosel'skii's fixed point theorems in Dunford–Pettis spaces to weakly compact operators. Since L_1 is a Dunford–Pettis space, this generalization gives a positive answer to some open problems encountered in [115] concerning the existence results in L_1 spaces for stationary transport equations arising in the kinetic theory of gas. The boundary conditions were assumed to be linear because, in contrast to biological models (0.1)–(0.2), it seems that, in rarefied gas dynamics context, nonlinear boundary conditions have no physical meaning (see [115] and the references therein). Due to the nonlinearity of the boundary conditions (0.2), the generalization of Schauder and Krasnosels'kii results established in [32] cannot be used to solve the problem (0.1)–(0.2). Consequently, it is useful to establish some fixed point theorems on general Banach spaces which can be applied directly for solving the biological problem (0.1)–(0.2). For this purpose, the authors of [33] gave some fixed point theorems based on the notion of weak sequential continuity and the well-known Arino–Gautier–Penot fixed point theorem.

In this direction, several other attempts have been made in the literature in order to prove the analogousness of the Krasnosel'skii fixed point theorem under the weak topology. In 2003, C. S. Barroso established a version of Krasnosel'skii's theorem using the weak topology of a Banach space [22]. His result required both the weak continuity and the weak compactness of A, whereas B had to be a linear operator satisfying the condition $\|B^p\| < 1$ for some integer $p \geq 1$. The proof was based on the Schauder–Tychonoff's fixed point theorem and on the weak continuity of $(I-B)^{-1}$. In a more recent paper [23],

C. S. Barroso and E. V. Teixeira established in 2005 a fixed point theorem for the sum $A + B$ of a weakly sequentially continuous mapping A and a weakly sequentially continuous strict contraction B.

We should notice that the strategy of the authors in the above-mentioned works consists in giving sufficient conditions which ensure the invertibility of $I - B$ in order to deal with the mapping $(I - B)^{-1}A$. Hence, it would be interesting to investigate, in the weak topology setting, the case when the mapping $I - B$ is not injective. We notice that this case was considered in 2006 by Y. Liu and Z. Li in [124] in the strong topology setting by looking for the multi-valued mapping $(I - B)^{-1}A$.

A wide class of problems (for instance in integral equations and stability theory) has been investigated by using the Krasnosel'skii fixed point principle (see, for examples, [22, 23, 48, 51, 131, 150]). However, in some applications, the verification of the hypothesis (iii) is quite hard to achieve. As a tentative approach to resolve such a difficulty, many attempts were made in the literature in the direction of weakening the hypothesis (iii). For example, in [48], T. A. Burton improved the Krasnosel'skii principle by requiring, instead of (iii), the more general following condition:

$$(x = Bx + Ay, \ y \in \mathcal{M}) \Longrightarrow x \in \mathcal{M}.$$

In [104], A. Jeribi, B. Krichen, and B. Mefteh have weakened the previous condition that is, $(A + B)(\mathcal{M}) \subset \mathcal{M}$ in order to prove the existence of a fixed point of $A+B$. The measures of noncompactness have proved to be a very useful and efficient tools in functional analysis, for instance in fixed point theory and in the theory of operator equations in Banach spaces. They were also used in ordinary and partial differential equations, integral and integrodifferential equations, and also in the characterization of compact operators between Banach spaces. The first measure of noncompactness, denoted by α, was defined and studied by K. Kuratowski [114] in 1930. In 1955, G. Darbo [60] used the function α in order to prove his fixed point theorem which generalized Schauder's fixed point theorem to the class of set-contractive operators, that is, operators T which satisfy $\alpha(T(A)) \leq k\alpha(A)$ with $k < 1$. For more details, the reader may refer to [8].

The measure of weak noncompactness is a very important tool used in this book. This measure was first introduced by F. S. De Blasi [64] in 1977, who proved the analogousness of Sadovskii's fixed point theorem for the weak topology. As stressed in [10], in many applications, it is not always possible to show the weak continuity of the involved mappings, whereas the weak sequential continuity does not involve any problem. So, O. Arino, S. Gautier, and J. P. Penot proved the analogousness of Schauder's fixed point theorem for weakly sequentially continuous mappings. Then, several fixed point theorems have been proved for weakly sequentially continuous mappings of Darbo's and Sadovskii's types [86, 112] and also Krasnosel'skii's type (see, for examples [4, 22, 29, 48, 51, 104] and many other references).

The Leray–Schauder principle [123], one of the most important theorems in nonlinear analysis, was first proved for a Banach space in the context of degree theory. Other variations of this principle were due to F. E. Browder [43], H. H. Schaefer [144], W. V. Petryshyn [139, 140], and A. J. B. Potter [141]. These variations, which are based on the compactness results, are useful in terms of providing solutions for nonlinear differential and integral equations in Banach spaces. The major problem is the loss of compactness in L_1-spaces, which represents the convenient and natural setting of some problems. At this level, this approach fails when we investigate some nonlinear stationary transport equations in L_1 context. However, these equations can be transformed into fixed point problems involving nonlinear weakly compact operators. Nevertheless, an infinite-dimensional Banach space, equipped with its weak topology, does not admit open bounded sets, which represents a major problem. For example, we may refer to nonlinear one-dimensional stationary transport equations arising in the kinetic theory of gas, where we must describe the interaction of gas molecules with solid walls bounding the region where the gas follows (see [34, 115, 116]).

Fixed point theory for weakly completely continuous multi-valued mappings plays an important role in the existence of solutions for operator inclusions, positive solutions of elliptic equations with discontinuous non-linearities, and periodic and boundary value problems for second-order differential inclusions (see, for examples the papers, [12, 37, 137], the monograph [79] by S. Djebali, L. Górniewicz and A. Ouahaband, and the book of L. Górniewicz [89]).

In [135], D. O'Regan has proved a number of fixed point theorems for multi-valued maps defined on bounded domains with weakly compact and convex values and which are weakly contractive with a weakly sequentially closed graph. Recently, A. Ben Amar and A. Sikorska-Nowak [36] improved and extended these theorems to the case of weakly condensing and 1-set weakly contractive multi-valued maps with a weakly sequentially closed graph and with unbounded domains. Moreover, they didn't assume that they went from a point into a weakly compact and convex set. Their results may be viewed as an extension of some relevant and recent ones (the reader may refer to [10, 33, 35, 134, 135]).

More recently, fixed point theory has been treated in Banach algebras. It was initiated in 1977 by R. W. Legget [122] who proved the existence theorems for the particular following equation:

$$x = x_0 + x.Bx, \quad (x_0, x) \in X \times \mathcal{M},$$

where $\mathcal{M}$ is a nonempty, bounded, closed, and convex subset of a Banach algebra X, and where $B : \mathcal{M} \to X$ represents a compact operator. The study of functional integral equations (FIEs) and differential equations is the main object of research in nonlinear functional analysis. These equations occur in physical, biological, and economic problems. Some of these equations can be formulated in suitable Banach algebras through nonlinear operator equations:

$$x = Ax.Bx + Cx. \tag{0.3}$$

In recent years, several authors have focused on the resolution of the equation (0.3) and have obtained many valuable results. We can cite for examples, J. Banas in [15, 16, 19, 20], J. Caballero, B. Lopez, and K. Sadarangani in [50], and B. C. Dhage in [66, 70, 71, 73]. These studies were mainly based on the convexity of the bounded domain, as well as the well-known Schauder's fixed point theorem, in order to guarantee the compactness of the operator $\left(\frac{I-C}{A}\right)^{-1}B$. Moreover, some properties of the operators A, B, and C (such as completely continuous, k-set contractive, condensing) and the potential tool of the axiomatic measures of noncompactness, were used. We should notice that if $\left(\frac{I-C}{A}\right)$ is not invertible, $\left(\frac{I-C}{A}\right)^{-1}$ could be seen as a multi-valued mapping. This case is not discussed in the results of Chapter 3. To our knowledge, this question is still open.

Since the weak topology is the practice setting and it is natural to investigate the problems of existence of solutions of different types of nonlinear integral equations and nonlinear differential equations in Banach algebras, it turns out that the above-mentioned results cannot be easily applied. One of the difficulties arising, when dealing with such situations, is that a bounded linear functional φ acting on a Banach algebra X does not necessarily satisfy the following inequality:

$$|\varphi(x.y)| \leq c\ |\varphi(x)|\ |\varphi(y)|, \quad \text{with } c \geq 0 \ \text{ and } \ x,\ y \in X.$$

That is why, A. Ben Amar, S. Chouayekh, and A. Jeribi introduced in the *Journal of Functional Analysis*, 2010 [26] a new class of Banach algebra satisfying a certain sequential condition called here the condition $(\mathcal{P})$, which can be presented as follows:

$$(\mathcal{P}) \quad \left\{ \begin{array}{l} \text{For any sequences } (x_n)_{n\in\mathbb{N}} \text{ and } (y_n)_{n\in\mathbb{N}} \text{ of } X \text{ such that } x_n \rightharpoonup x \\ \text{and } y_n \rightharpoonup y, \text{ then } x_n.y_n \rightharpoonup x.y, \text{ where } X \text{ is a Banach algebra.} \end{array} \right.$$

Their aim was to prove some new fixed point theorems in a nonempty, closed, and convex subset of any Banach algebra or of a Banach algebra satisfying the condition $(\mathcal{P})$ under the weak topology setting. Their conditions were formulated in terms of a weak sequential continuity for the three nonlinear operators A, B, and C which are involved in Eq. (0.3). Besides, no weak continuity conditions were required in their work. The theoretical results were applied in order to show the existence of solutions for the following nonlinear functional integral equations in the Banach algebra $\mathcal{C}(J, X)$ of real continuous functions defined on the interval $[0, 1]$:

$$x(t) = a(t) + (T_1x)(t)\left[\left(q(t) + \int_0^{\sigma(t)} p(t, s, x(s), x(\lambda s))ds\right).u\right]$$

and

$$x(t) = a(t)x(t) + (T_2x)(t)\left[\left(q(t) + \int_0^{\sigma(t)} p(t, s, x(s), x(\lambda s))ds\right).u\right],$$

where $0 < \lambda < 1$, J is the interval $[0, 1]$, and X represents any Banach algebra. The functions a, q, σ are continuous on J; T_1, T_2, $p(.,.,.,.)$ are nonlinear functions and u is a non-vanishing vector.

Notice that the study of weakly condensing operators and weakly 1-set-contractive ones has been one of the main research objects in nonlinear functional analysis (see, for examples, [16, 35, 122, 127]). Due to the loss of compactness, continuity, and weak continuity, the authors in [27] provided some new fixed point theorems in Banach algebras satisfying the condition $(\mathcal{P})$ under the weak topology setting. Their results were based on the class of weakly condensing, weakly 1-set-contractive, weak sequential continuity, and weakly compact via the concept of De Blasi measure of weak noncompactness [64] for the three nonlinear operators A, B, and C. Besides, no artificial conditions like compactness, continuity, and weak continuity were required for their work. However, the weak continuity condition for an operator is not easily satisfied. In general, it is not even verified.

Since the convex-power condensing operators generalize condensing operators, we thought that it would be interesting to continue, in a subsection of this book, some previous studies around this concept. For this purpose, we should notice that all the domains of the single-valued operators discussed here are not assumed to be bounded. The discussed results may be viewed as extensions of [5, 27, 93, 96]. For example, we quote that in [96], S. Hong-Bo gave a generalization of the famous Sadovskii's fixed point theorem on a bounded domain (fixed point theorem of Schauder's type) which asserts that if S is a nonempty closed, bounded, and convex subset of a locally convex Hausdorff space X and if $F : S \longrightarrow S$ is a convex-power condensing operator, then F has, at least, one fixed point in S. In 1996, D. Guo, V. Lakshmikantham, and X. Liu [93] obtained another extension of the fixed point theorem of S. Hong-Bo. Indeed, let S be a nonempty closed, bounded, and convex subset of a Banach space X, $x_0 \in X$, let n_0 be a positive integer, and also let $F : X \longrightarrow X$ be a convex-power condensing operator about x_0 and n_0. If $f(S) \subset S$ (the condition of Schauder's type), then F has, at least, a fixed point in S. In [5], R. P. Agarwal, D. O'Regan, and M. A. Taoudi gave an extension of this result on an unbounded domain of the weak topology setting. More precisely, if we assume that $F : S \longrightarrow S$ is weakly sequentially continuous and convex-power condensing operator, and if $F(S)$ is bounded, then F has, at least, one fixed point in S. In [4], the same authors assumed that the operator F is ws-compact instead of being weakly sequentially continuous and gave an example in the Banach space $L_1 = L_1(0, 1)$ for a ws-compact operator which is not weakly sequentially continuous. In fact, there is no relationship between the two notions: the ws-compactness and the weak sequential continuity. Recently, A. Ben Amar, S. Chouayekh, and A. Jeribi [28] controlled the topological structure of the set of fixed-points of F by proving that this set is weakly compact.

Knowing that the product $W.W'$ of two arbitrary weakly compact subsets W, W' of a Banach algebra X is not necessarily weakly compact, a recent paper by J. Banas and M. A. Taoudi [21] gave a generalization of some results established in [26] in the case of WC–Banach algebra, that is, a Banach algebra such that the product $W.W'$ of two arbitrary weakly compact subsets W, W'

is weakly compact. We mention that this property is valid in finite-dimensional algebras and that it is probably weaker than $(\mathcal{P})$. In this direction, another new and recent work [105] was established by A. Jeribi, B. Krichen, and B. Mefteh in the case of WC–Banach algebras, for mappings involving ω-contractions, where ω represents the measure of weak noncompactness, and also mappings A satisfying the following assumptions:

$$(\mathcal{H}1) \quad \begin{cases} \text{If } (x_n)_{n\in\mathbb{N}} \subseteq \mathcal{D}(A) \text{ is a weakly convergent sequence in } X, \text{ then} \\ (Ax_n)_{n\in\mathbb{N}} \text{ has a strongly convergent subsequence in } X, \end{cases}$$

and

$$(\mathcal{H}2) \quad \begin{cases} \text{if } (x_n)_{n\in\mathbb{N}} \subseteq \mathcal{D}(A) \text{ is a weakly convergent sequence in } X, \text{ then} \\ (Ax_n)_{n\in\mathbb{N}} \text{ has a weakly convergent subsequence in } X. \end{cases}$$

We should mention that these assumptions were already considered in the papers [1, 80, 87, 103, 150]. Regarding these two conditions, K. Latrach, M. A. Taoudi, and A. Zeghal noticed in [119] the following facts: (i) Every ω-contractive map satisfies $(\mathcal{H}2)$. (ii) A mapping satisfies $(\mathcal{H}2)$ if, and only if, it maps relatively weakly compact sets into relatively weakly compact ones. (iii) A mapping satisfies $(\mathcal{H}1)$ if, and only if, it maps relatively weakly compact sets into relatively compact ones. (iv) Operators satisfying $(\mathcal{H}1)$ or $(\mathcal{H}2)$ are not necessarily weakly continuous. (v) The condition $(\mathcal{H}2)$ holds true for every bounded linear operator.

Question 1:

Are there any examples of WC–Banach algebras which do not satisfy the condition $(\mathcal{P})$?

Many problems arising in mathematical physics, biology, etc., may be described in a first formulation, using systems of integral equations as well as systems of partial or ordinary differential equations. The theory of block operator matrices opens up a new line of attack of these problems. In recent years, several papers were devoted to the investigation of linear operator matrices defined by 2×2 block operator matrices

$$\mathcal{L} = \begin{pmatrix} A & B \\ C & D \end{pmatrix},$$

where the entries were not necessarily bounded in Banach spaces. Our book does not pretend to be complete in any respect, e.g., it does not deal with the vast literature concerning block operator matrices in spectral theory (see, e.g., the monograph of A. Jeribi [100], the book of C. Tretter [151], and the papers [11, 24, 25, 30, 55, 56, 59, 106, 107, 129]) or with the semigroup theory (see, e.g., [84]). However, the studies of nonlinear functional integral and differential systems in Banach spaces and Banach algebras have also been discussed for a long time in the literature. These studies were achieved via fixed point techniques (see, e.g., the book of B. Krichen [111] and the papers [54, 133]). It should be noticed that these systems may be transformed into the following fixed point problem of the 2×2 block operator matrix

$$\begin{pmatrix} A & B \cdot B' \\ C & D \end{pmatrix} \tag{0.4}$$

with nonlinear entries defined on Banach algebras. Our assumptions are as follows: A maps a nonempty, bounded, closed, and convex subset S of a Banach algebra X into X, B, B', and D acting from X into X and C from S into X.

One of the most important problems in the fixed point theory is related to the existence of solution for a two-dimensional equation. That is why A. Ben Amar, A. Jeribi, and B. Krichen have established in [31] Schauder's and Krasnoselskii's fixed point theorems for the operator (0.4), when $B' = 1$, and X is a Banach space. These authors have applied their results to a two-dimensional mixed boundary problem in $L_p \times L_p$, $p \in]1, +\infty[$. Due to the loss of compactness of the operator $C(\lambda - A)^{-1}$ in L_1 spaces, their analysis was carried out via the arguments of weak topology and particularly the measure of weak noncompactness. Hence, it was useful for the authors of [103] to extend it by establishing some new variants of the fixed point theorems for a 2×2 block operator matrix involving weakly compact operators. This problem can be formulated by:

$$\begin{pmatrix} -v\dfrac{\partial}{\partial \mu} - \sigma_1(\mu, v, .) & R_{12} \\ R_{21} & -v\dfrac{\partial}{\partial \mu} - \sigma_2(\mu, v, .) \end{pmatrix} \begin{pmatrix} \psi_1 \\ \psi_2 \end{pmatrix} = \lambda \begin{pmatrix} \psi_1 \\ \psi_2 \end{pmatrix},$$

and

$$\psi_{i}|_{\Gamma_0} = K_i\left(\psi_{i}|_{\Gamma_1}\right), \quad i = 1, 2,$$

where $R_{ij}\psi_j(\mu, v) = \displaystyle\int_a^b r_{ij}(\mu, v, v', \psi_j(\mu, v'))dv'$, $(i, j) \in \{(1, 2), (2, 1)\}$, $\mu \in$

$[0,1]$, $v, v' \in [a,b]$ with $0 \leq a < b < \infty$. The functions $\sigma_i(.,.,.)$ and $r_{ij}(.,.,.,.)$ are nonlinear and λ is a complex number, $\Gamma_0 = \{0\} \times [a,b]$ and $\Gamma_1 = \{1\} \times [a,b]$. We denote by $\psi_{i|_{\Gamma_0}}$ (resp. $\psi_{i|_{\Gamma_1}}$) the restriction of ψ_i to Γ_0 (resp. Γ_1) while K_i are nonlinear operators from a suitable function space on Γ_1 to a similar one on Γ_0. We should notice that in the case where σ_i depends on the density of the population, we have assumed that $K_i \in \mathcal{L}\left(L_p(\{1\} \times [a,b]; vdv), L_p(\{0\} \times [a,b]; vdv)\right)$.

Question 2:

What happens if the reproduction rules are not generated by a bounded linear operator K_i from $L_p(\{1\} \times [a,b]; vdv)$ to $L_p(\{0\} \times [a,b]; vdv)$? To our knowledge, this question is not yet developed.

In the context of Banach algebras, N. Kaddachi, A. Jeribi, and B. Krichen [108] have initiated in 2012 the study of the existence of a fixed point for the block operator matrix (0.4), where B' is a continuous operator on a Banach algebra X. The theoretical tools were based on the fact that the mapping N which is defined on the nonempty, closed, and convex subset S of X by:

$$\begin{cases} N : S \longrightarrow S \\ \qquad y \longrightarrow N(y) = z, \end{cases}$$

where z represents the unique solution for the operator equation

$$z = Az + B(I-D)^{-1}Cz.B'(I-D)^{-1}Cy,$$

is completely continuous. Some other conditions using the Lipschitzian [3] and $\mathcal{D}$-Lipschitzian maps (see the paper of B. C. Dhage [66]) were also used. An application to a coupled system of nonlinear equations defined on bounded domains in $\mathcal{C}([0,1],\mathbb{R})$ was considered. In [101], the same authors continued their studies in the weak topology setting. However, the arguments used in their applications were not valid when the operators are defined on unbounded domains. In fact, their proofs were based on Ascoli's theorem which is not generally applicable. To compensate the loss of compactness encountered, only few alternative tools were known, for example measures of noncompactness [8, 122]. Recently, a study of integral equations in the space $\mathcal{C}_b(I)$, constituted of real functions which are defined, continuous and bounded on an unbounded interval, was developed by many authors [17, 49, 61, 62, 127].

The fixed point theory for multi-valued mappings is an important topic of set-valued analysis. Several well-known fixed point theorems of single-valued mappings such as those of Banach and Schauder have been extended to multi-valued mappings in Banach spaces. Following the Banach contraction principle, H. Covitz and S. B. Nadler [58] introduced the concept of set-valued contractions and established the fact that a set-valued contraction possesses a fixed point in a complete metric space. Subsequently, many authors generalized Nadler's fixed point theorem in different ways. The theory has found applications in control theory, differential inclusions, and economics (see [2, 89]).

In 2001, R. P. Agarwal, M. Meehan, and D. O'Regan [2] concentrated on contractive, nonexpansive multi-valued mappings. The result due to Nadler dealing with the Banach contraction principle for contractive mappings with closed values, was first presented. They were also interested in a nonlinear alternative of Leray–Schauder's type, a Furi–Pera type result, and some coincidence type results which were just some of the fixed point theories presented for maps. An extension of the Schauder–Tychonoff theorem to multi-valued maps with a closed graph was given. In 2009, L. Gorniewicz [89] provided a systematic presentation of results and methods dealing with the fixed point theory of multi-valued mappings and some of its applications were given. In selecting the material he has restricted himself to the study of topological methods in the fixed point theory of multi-valued mappings and their applications, mainly to differential inclusions. By using some techniques of the above works, it was useful for us to develop some theoretical results and applications for block operator matrices (BOM) with multi-valued inputs.

Chapter 1 of this book is devoted to some definitions and mainly to the basic tools of nonlinear functional analysis needed in the sequel. The final section of this chapter deals with some classic results of fixed point theory in order to study nonlinear problems in both Banach spaces and Banach algebras.

It is unavoidable that any discussion on completely continuous and contractive maps in the Schauder and Krasnosel'skii fixed point theorems will lead to another one dealing with weakly sequentially continuous and weakly compact maps. That is why we choose it as the topic of Chapter 2. First, a generalization of the Schauder and Krasnosel'skii fixed point theorems in Dunford–Pettis spaces is presented. Both of these two theorems can be used to resolve some open problems encountered in [99] and [115]. It is not always possible to show that a map is weakly continuous. However, weakly sequentially continuous maps appear as being the most convenient to be used. That is why some new variants of fixed point theorems involving the measure of weak noncompactness and based on the notion of weak sequential continuity are given. We conclude Chapter 1 with nonlinear alternatives of Leray–Schauder and Furi–Pera for the sum of two weakly sequentially continuous mappings.

Following the discussion in Chapter 2 about the existence results of a fixed point for the sum of two weakly sequentially continuous mappings acting on Banach spaces, it becomes natural to consider continuation principles for these mappings in the case of Banach algebras. Some equations occurring in physical and biological problems can be formulated into the nonlinear operator equation $Ax.Bx + Cx = x$. In recent years, many authors have focused on the resolution of this equation and have obtained several valuable results (see, for examples, [15, 16, 19, 20, 50, 66, 70, 71, 73, 149] and the references therein). These studies were mainly based on the convexity of the bounded domain, the famous Schauder's fixed point theorem [149], as well as the properties of operators A, B, and C (cf. completely continuous, k-set contractive, condensing and the potential tool of the axiomatic measures of noncompactness).

Knowing that the product of two weakly sequentially continuous functions is not necessarily weakly sequentially continuous, we will introduce, in Chapter 3, a class of Banach algebras satisfying certain sequential conditions called here the condition $(\mathcal{P})$, that is, we will say that the Banach algebra X satisfies the condition $(\mathcal{P})$, if for any sequences $\{x_n\}$ and $\{y_n\}$ in X such that $x_n \rightharpoonup x$ and $y_n \rightharpoonup y$, then $x_n.y_n \rightharpoonup x.y$; here $\rightharpoonup$ denotes the weak convergence. The first objective of this chapter is to prove some new fixed point theorems in a nonempty, closed, and convex subset of any Banach algebras or Banach algebras satisfying the condition $(\mathcal{P})$ under the weak topology setting. Our main conditions are formulated in terms of a weak sequential continuity related to the three nonlinear operators A, B, and C involved in the previous equation. Besides, no weak continuity conditions are required for this work. A nonlinear alternative of Leray–Schauder in Banach algebra involving the three operators is presented. The second objective of this chapter is to extend some results due to B. C. Dhage to the case of $\mathcal{D}$-Lipschitzian maps and other results to ws-compact operators. We conclude Chapter 3 by improving and extending several earlier works which used the condition $(\mathcal{P})$ to mappings acting on WC–Banach algebras.

The fundamental aim of this book is to provide a unified and comprehensive survey of fixed point theory for block operator matrices (BOM) acting on

Banach spaces and Banach algebras and its various useful interactions with topological structures. An important feature of this fixed point theory (which is developed in detail in Chapter 4) is that it is based on two-dimensional variants of fixed point theorems for the 2×2 block operator matrix $\mathcal{L}$. There are two objectives of Chapter 4. The first one is to discuss the existence of fixed points for the block operator matrix $\mathcal{L}$ by laying down some conditions on the entries, which are generally nonlinear operators. This discussion is based on the presence or absence of invertibility of the diagonal terms of $\mathcal{I} - \mathcal{L}$. As a second objective, we first recall that, in Chapter 3, the fixed point theorems have enabled us to get new results for the 2×2 block operator matrix (0.4) involving operators such as, e.g., $\mathcal{D}$-Lipschitzian, convex-power condensing, weakly sequentially continuous..., acting on a Banach algebra X. A regular case is also considered when X is a commutative Banach algebra satisfying the condition $(\mathcal{P})$.

Chapter 5 deals with some open problems chosen from [99, 115, 116, 117, 118] concerning the existence of solutions on L_1 spaces for nonlinear boundary value problems derived from three models. The first one deals with nonlinear one-dimensional stationary transport equations arising in the kinetic theory of gas where we must describe the interaction of gas molecules with solid walls bounding the region where the gas follows (see [115, 116]). The second model was introduced by J. L. Lebowitz and S. I. Rubinow [121] in 1974 for modeling microbial populations by age and cycle length formalism. The third model, introduced by M. Rotenberg [142] in 1983, describes the growth of a cell population. These three models can be transformed into a fixed point problem including two types of equations. The first type involves a nonlinear weakly compact operator on L_1 spaces. The second type deals with two nonlinear operators depending on the parameter λ, say, $\psi = A_1(\lambda)\psi + A_2(\lambda)\psi$, where $A_1(\lambda)$ is a weakly compact operator (i.e., it transforms bounded sets into relatively weakly compact sets) on L_1 spaces and $A_2(\lambda)$ is a (strict) contraction mapping for a large enough $Re\lambda$. Consequently, the Schauder's (resp. Krasnosel'skii's fixed point theorem) [149] cannot be used in the first (resp. second) type of equations. This is essentially due to the lack of compactness.

In Chapter 6, we start by studying the existence of solutions for some variants of Hammerstein's integral equation. Next, we prove the existence of solutions for several nonlinear functional integral and differential equations, in Banach algebras $\mathcal{C}([0,T],X)$, where X is a Banach algebra satisfying the condition ($\mathcal{P}$). We finish by giving an application of a Leray–Schauder's type of fixed point theorem under the weak topology.

Chapter 7 is first devoted to the study of existence of solutions for a boundary coupled system arising in growing cell populations. We start the discussion in the product Banach space $L_p \times L_p$ for $p \in (1, \infty)$. Due to the loss of compactness on L_1 spaces, the analysis does not cover the case where $p = 1$. Next, we extend the previous results to the case where $p = 1$, by applying the new variants of fixed point theorems for a 2×2 block operator matrix involving weakly compact operators already presented in Chapter 4. The second purpose of this chapter is to study several coupled systems of nonlinear functional integral equations with bounded or unbounded domains in the Banach algebra $\mathcal{C}([0, 1], X)$. We conclude Chapter 7 by presenting some existence results for coupled systems of perturbed functional differential inclusions of initial and boundary value problems.

Chapter 1

Fundamentals

The study of fixed point requires more prerequisites from the general theories of topological notions and nonlinear operators. The aim of this chapter is to introduce the basic concepts, notations, and elementary results that are used throughout the book. Moreover, the results of this chapter may be found in most standard books dealing with functional analysis and fixed point theory (see for examples [41, 83, 128]).

1.1 Basic Tools in Banach Spaces

1.1.1 Normed vector spaces

Let X be a vector space over $\mathbb{K}$, ($\mathbb{K} = \mathbb{R}$ or $\mathbb{C}$), the field of real or complex numbers. A mapping $\|\cdot\| : X \longrightarrow \mathbb{R}_+$ is called a norm, provided that the following conditions hold:

(i) $\|x\| = 0$ implies $x = 0$,

(ii) $\|\alpha x\| = |\alpha|\|x\|, \quad \forall \alpha \in \mathbb{K}, \forall x \in X$, and

(iii) $\|x + y\| \leq \|x\| + \|y\|, \quad \forall x, \ y \in X$.

If X is a vector space and $\|\cdot\|$ is a norm on X, then the pair $(X, \|\cdot\|)$ is called a normed vector space. Usually we simply abbreviate this by saying that X is a normed vector space. If X is a vector space and $\|\cdot\|$ is a norm on X, then X becomes a metric space, if we define the metric $d(. \, , .)$ by:

$$d(x, y) := \|x - y\|, \quad \forall x, \ y \in X.$$

A normed vector space which is a complete metric space, with respect to the above-defined metric $d(. \, , .)$, is called a Banach space. Thus, a closed subset of a Banach space may always be regarded as a complete metric space. Hence, a closed subspace of a Banach space is also a Banach space. Now, let us try to put together, for future reference, a small catalogue of examples of Banach spaces.

Examples of Banach spaces

Example 1 $(\mathbb{K}, |\cdot|)$ is a simple example of a Banach space.

Example 2 Let us fix $n \in \mathbb{N}^*$ and let us denote the following set by $\mathbb{K}^n$:

$$\mathbb{K}^n := \Big\{x : x = (x_1, \ldots, x_n),\ x_i \in \mathbb{K},\ \ i = 1, \ldots, n\Big\}.$$

There are many useful norms with which we can equip $\mathbb{K}^n$, such as the following ones:

1. For $1 \leq p < \infty$, let us define $\|\cdot\|_p : \mathbb{K}^n \longrightarrow \mathbb{R}_+$ by:

$$\|x\|_p := \Big(\sum_{i=1}^{n} |x_i|^p\Big)^{1/p}, \qquad x \in \mathbb{K}^n.$$

   These spaces are finite-dimensional l^p-spaces. The frequently used norms are $\|\cdot\|_1$, and $\|\cdot\|_2$.

2. Let us define $\|\cdot\|_\infty : \mathbb{K}^n \longrightarrow \mathbb{R}_+$ by:

$$\|x\|_\infty := \max\{|x_i| : 1 \leq i \leq n\}.$$

   This norm is called the sup norm on $\mathbb{K}^n$.

The next example extends the just considered one to the infinite-dimensional setting.

Example 3 Let

$$\mathbb{K}^\infty := \Big\{x : x = \{x_i\}_{i=1}^{\infty},\ x_i \in \mathbb{K},\ ;\ i = 1, 2, \ldots\Big\}.$$

Then, $\mathbb{K}^\infty$ with a coordinate-wise addition and a scalar multiplication, is a vector space, of which certain subspaces can be equipped with norms, allowing them to be complete.

1. For $1 \leq p < \infty$, let us define

$$l^p := \Big\{x = \{x_i\} \in \mathbb{K}^\infty : \sum_{i=1}^{\infty} |x_i|^p < \infty\Big\}.$$

Then, l^p is a subspace of $\mathbb{K}^\infty$, and

$$\|x\|_p := \Big(\sum_{i=1}^{\infty} |x_i|^p\Big)^{1/p}$$

defines a specific norm allowing l^p to be complete.

2. Let us define l^∞ by:

$$l^\infty := \Big\{x = \{x_i\} \in \mathbb{K}^\infty : \sup_i |x_i| < \infty\Big\},$$

and

$$\|x\|_\infty := \sup_i\{|x_i|,\ x \in l^\infty\}.$$

With respect to this sup norm, l^∞ is complete.

Example 4 Let Ω be a compact subset of $\mathbb{K}^n$, and let $\mathbb{K}$ be as above; we define $\mathcal{C}(\Omega)$ by:

$$\mathcal{C}(\Omega) := \Big\{f : \Omega \longrightarrow \mathbb{K} \text{ such that } f \text{ is continuous on } \Omega\Big\}.$$

Let

$$\|f\|_\infty := \sup_{x\in\Omega} |f(x)|.$$

Since the uniform limit of a sequence of continuous functions is also continuous, we deduce that the space

$$E := \Big\{f \in \mathcal{C}(\Omega) \ \text{ such that } \ \|f\|_\infty < \infty\Big\}$$

is a Banach space.

1.2 Contraction Mappings

Definition 1.2.1 *Let (X, d) be a metric space and let $T : X \longrightarrow X$ be a mapping. T is called a Lipschitz mapping (or a Lipschitzian mapping) with a Lipschitzian constant $k \geq 0$ (or k-Lipschitzian), provided that*

$$d(T(x), T(y)) \leq kd(x, y), \quad \forall x,\ y \in X.$$

Let us notice that Lipschitz mappings are necessarily continuous mappings and that the product of two Lipschitz mappings (defined by a composition of mappings) is also a Lipschitzian mapping. Thus, for a Lipschitz mapping T, and for all positive integers n, the mapping $T^n = T \circ \cdots \circ T$, the mapping T composed with itself n times, is a Lipschitz mapping, as well.

Definition 1.2.2 *Let (X, d) be a metric space and let $T : X \longrightarrow X$ be a Lipschitz mapping with a Lipschitzian constant k. T is called a nonexpansive mapping, provided that the constant k may be chosen so that $k \leq 1$.*

Definition 1.2.3 *Let (X, d) be a metric space and let $T : X \longrightarrow X$ be a Lipschitz mapping with a Lipschitzian constant k. T is called a contraction mapping, (or a k-contraction mapping) provided that the Lipschitzian constant k may be chosen so that $0 \leq k < 1$. In this case, the Lipschitzian constant k is also called the contraction constant of T.*

1.2.1 The contraction mapping principle

In this section, we will discuss the contraction mapping principle, also called the Banach's fixed point theorem. We will also give some extensions and examples. We have the following theorem.

Theorem 1.2.1 *Let (X, d) be a complete metric space and let $T : X \longrightarrow X$ be a contraction mapping with a contraction constant k. Then, T has a unique fixed point $x \in X$. Moreover, if $y \in X$ is arbitrarily chosen, then the sequence $\{x_n\}_{n=0}^{\infty}$, given by:*

$$\begin{cases} x_0 = y, \\ x_n = T(x_{n-1}), \quad n \geq 1, \end{cases}$$

converges to x.

Proof. Let $y \in X$ be an arbitrary point of X and consider the sequence $\{x_n\}_{n=0}^{\infty}$ given by:

$$\begin{cases} x_0 = y, \\ x_n = T(x_{n-1}), \quad n \geq 1. \end{cases}$$

We will prove that $\{x_n\}_{n=0}^{\infty}$ is a Cauchy sequence in X. For $m < n$, we use the triangle inequality and we note that

$$d(x_m, x_n) \leq d(x_m, x_{m+1}) + d(x_{m+1}, x_{m+2}) + \cdots + d(x_{n-1}, x_n).$$

Since T is a contraction mapping, it follows that

$$d(x_p, x_{p+1}) = d(T(x_{p-1}), T(x_p)) \leq kd(x_{p-1}, x_p),$$

for any integer $p \geq 1$. By using this inequality repeatedly, we obtain

$$d(x_p, x_{p+1}) \leq k^p d(x_0, x_1).$$

Hence,

$$d(x_m, x_n) \leq \left(k^m + k^{m+1} + \cdots + k^{n-1}\right) d(x_0, x_1),$$

i.e.,

$$d(x_m, x_n) \leq \frac{k^m}{1-k} d(x_0, x_1),$$

whenever $m \leq n$. From this, we deduce that $\{x_n\}_{n=0}^{\infty}$ is a Cauchy sequence in X. Since X is complete, this sequence has a limit, say $x \in X$. Furthermore, since T is continuous, we deduce that

$$x = \lim_{n\to\infty} x_n = \lim_{n\to\infty} T(x_{n-1}) = T(\lim_{n\to\infty} x_{n-1}) = T(x),$$

and then, x is a fixed point of T.
If x and z are both fixed points of T, then we get

$$d(x, z) = d(T(x), T(z)) \leq kd(x, z).$$

Since $k < 1$, we must have $x = z$.

The following is an alternate proof stated in [41]. It follows (by induction) that, for any $x \in X$ and any natural number m, we have

$$d(T^{m+1}(x), T^m(x)) \leq k^m d(T(x), x).$$

Now, let

$$\delta := \inf_{x \in X} d(T(x), x).$$

Then, if $\delta > 0$, there exists $x \in X$ such that

$$d(T(x), x) < \frac{3}{2}\delta$$

and hence, for any m, we have

$$d(T^{m+1}(x), T^m(x)) \leq k^m \frac{3}{2}\delta.$$

Moreover,

$$\delta \leq d(T(T^m(x)), T^m(x)) = d(T^{m+1}(x), T^m(x))$$

and then, for any $m \geq 1$, we have

$$\delta \leq k^m \frac{3}{2}\delta,$$

which is impossible, since $k < 1$. Thus, $\delta = 0$.
Now, let us choose a minimizing sequence $\{x_n\}$ such that

$$\lim_{n \to \infty} d(T(x_n), x_n) = \delta = 0.$$

For any m and n, the triangle inequality implies that

$$d(x_n, x_m) \leq d(T(x_n), x_n) + d(T(x_m), x_m) + d(T(x_n), T(x_m)),$$

and hence,

$$(1-k)d(x_n, x_m) \leq d(T(x_n), x_n) + d(T(x_m), x_m),$$

which implies that $\{x_n\}$ is a Cauchy sequence and hence, has a limit x in X. Now, we may conclude that

$$d(T(x), x) = 0$$

and then, x is a fixed point of T. Q.E.D.

It may be the case that $T : X \longrightarrow X$ is not a contraction mapping on the whole space X, but rather a contraction on some neighborhood of a given point. In such a case, we have the following result:

Theorem 1.2.2 *Let (X, d) be a complete metric space, let*

$$B = \{x \in X \text{ such that } d(z, x) < \varepsilon\},$$

where $z \in X$ and $\varepsilon > 0$ is a positive number and let $T : B \longrightarrow X$ be a contraction mapping such that

$$d(T(x), T(y)) \leq k\mathrm{d}(x, y), \quad \forall\ x,\ y \in B,$$

with a contraction constant $k < 1$. Furthermore, let us assume that

$$d(z, T(z)) < \varepsilon(1-k).$$

Then, T has a unique fixed point $x \in B$.

Proof. While the hypotheses do not assume that T is defined on the closure $\overline{B}$ of B, the uniform continuity of T allows us to extend T to a mapping defined on $\overline{B}$ which is a contraction mapping having the same Lipschitzian constant as the previous mapping. We may also note that, for $x \in \overline{B}$, we have

$$d(z, T(x)) \leq d(z, T(z)) + d(T(z), T(x)) < \varepsilon(1-k) + k\varepsilon = \varepsilon,$$

and hence, $T : \overline{B} \longrightarrow B$. As a result, and according to Theorem 1.2.1, since $\overline{B}$ is a complete metric space, then T has a unique fixed point in $\overline{B}$ which must be in B, by using the above calculations. Q.E.D.

Now, let us recall the definition of the concept of $\mathcal{D}$-Lipschitzian mapping playing an important role in fixed point theory [71].

Definition 1.2.4 *Let X be a Banach space with a norm $\|.\|$. A mapping $T : X \longrightarrow X$ is called $\mathcal{D}$-Lipschitzian (or $\mathcal{D}$-Lipschitz), if there exists a continuous nondecreasing function $\Phi_T : \mathbb{R}^+ \longrightarrow \mathbb{R}^+$ satisfying*

$$\|Tx - Ty\| \leq \Phi_T(\|x-y\|)$$

for all x, $y \in X$ with $\Phi_T(0) = 0$. The function Φ_T is sometimes called a $\mathcal{D}$-function of T on X.

In the special case where $\Phi_T(r) = kr$ for some $k > 0$, T is a Lipschitz mapping with a Lipschitzian constant k. In particular, if $k < 1$, T is a contraction on X with a contraction constant k.

Remark 1.2.1 *Every Lipschitz mapping is automatically $\mathcal{D}$-Lipschitz, but the converse may not be true. If Φ_T is not necessarily nondecreasing and satisfies $\Phi_T(r) < r$ for $r > 0$, then T is called a nonlinear contraction on X.*

For example, take $T(x) = \sqrt{|x|}$, $x \in \mathbb{R}$ and consider $\Phi_T(r) = \sqrt{r}$, $r \geq 0$. Clearly, Φ_T is continuous and nondecreasing. First notice that T is subadditive. To see this, let x, $y \in \mathbb{R}$. Then,

$$(T(x+y))^2 = |x+y| \leq |x| + |y| \leq (\sqrt{|x|} + \sqrt{|y|})^2 = (T(x) + T(y))^2.$$

Thus, for all x, $y \in \mathbb{R}$ we have :

$$(T(x+y)) \leq T(x) + T(y).$$

Using the subadditivity of T we get

$$|T(x) - T(y)| \leq T(x-y) = \Phi_T(|x-y|), \text{ for all } x, \ y \in \mathbb{R}.$$

Thus, T is $\mathcal{D}$-Lipschitzian with $\mathcal{D}$-function Φ_T. Now, suppose that T is Lipschitzian with constant k. Then, for all $x \in \mathbb{R}$ we have $T(x) \leq k|x|$. Hence, for all $x \neq 0$ we have $k \geq \frac{1}{\sqrt{|x|}}$. Letting x go to zero, we obtain a contradiction. Consequently, T is not Lipschitzian.

Definition 1.2.5 *Let X be a Banach space with a norm $\|\cdot\|$. An operator $T : X \longrightarrow X$ is called compact, if $\overline{T(S)}$ is a compact subset of X for any $S \subset X$. Similarly, $T : X \longrightarrow X$ is called totally bounded, if T maps a bounded subset of X into the relatively compact subset of X. Finally, $T : X \longrightarrow X$ is called a completely continuous operator, if it is a continuous and totally bounded operator on X.*

It is clear that every compact operator is totally bounded, but the converse may not be true. Take $T : \mathbb{R} \longrightarrow \mathbb{R}$ and $T(x) = x$. The operator T is totally bounded but is not compact. Also take

$$F(x) = \begin{cases} x & x > 0 \\ 1 & x = 0. \end{cases}$$

Notice that F is a totally bounded mapping but is not completely continuous.

Definition 1.2.6 *The topological space X is completely regular, if for each closed set H of X and each x in $X \backslash H$, there exists a continuous real-valued function Θ such that $\Theta = 0$ throughout H, and $\Theta = 1$ at x.*

In other terms, this condition says that x and H can be separated by a continuous function.

Notice that every pseudo-metric space is completely regular. That is, a set X equipped with a non-negative real-valued function $d : X \times X \longrightarrow \mathbb{R}^+$ (called a pseudo-metric) such that, for every x, y, $z \in X$,

(i) $d(x, x) = 0$,

(ii) $d(x, y) = d(y, x)$, and

(iii) $d(x, z) \leq d(x, y) + d(y, z)$.

As an example, consider the space $\mathcal{F}(X)$ of real-valued functions $f : X \longrightarrow \mathbb{R}$ together with a special point $x_0 \in X$. This point then induces a pseudo-metric on the space of functions, given by $d(f, g) = |f(x_0) - g(x_0)|$ for f, $g \in \mathcal{F}(X)$.

Lemma 1.2.1 *Let X be a completely regular space. Let E and F be two* *disjoint subsets of X, with E being closed and F being compact. Then, there exists a continuous function Θ such that $\Theta = 0$ throughout E and $\Theta = 1$ throughout F.*

Definition 1.2.7 *Let (X, d) be a metric space and $T : X \longrightarrow X$ is called a separate contraction mapping, if there exist two functions φ and $\psi : \mathbb{R}^+ \longrightarrow \mathbb{R}^+$ satisfying the following conditions:*

(i) $\psi(0) = 0$, ψ is strictly increasing,

(ii) $d(T(x), T(y)) \leq \varphi(d(x, y))$, and

(iii) $\psi(r) \leq r - \varphi(r)$ for $r > 0$.

Remark 1.2.2 *It is easy to see that, if T is a contraction mapping, then T is also a separate contraction mapping.*

The following example, stated in [125], provides a separate contraction mapping, but not a contraction mapping.
We consider the function $f : \mathbb{R} \times [0, 1] \longrightarrow [0, 1]$ defined by:

$$f(t, x) = x - \frac{x^4}{4} + \frac{sin^2(t)}{4}.$$

Let $X = \mathcal{C}(\mathbb{R}, [0, 1])$ and $T : X \longrightarrow X$ be defined by $(Tx)(t) = f(t, x(t))$. It is easy to verify that the operator T is well-defined. Moreover, T is a separate contraction mapping, but not a contraction mapping.

Another example of a separate contraction mapping which is not a strict contraction is given in [47]. Anyway, if we consider the mapping

$$T : \left[0, \frac{1}{\sqrt{2}}\right] \longrightarrow \left[0, \frac{1}{\sqrt{2}}\right]$$

defined by:

$$T(x) = x - x^3,$$

it is not difficult to prove that T is not a strict contraction but it is a separate contraction, taking

$$\varphi(r) = \begin{cases} r(1 - \frac{r^2}{4}) & r \leq 1 \\ \frac{3}{4}r & r \geq 1, \end{cases}$$

and $\psi(r) = r - \varphi(r)$.

Lemma 1.2.2 *Let $(X, \|.\|)$ be a normed space. If $S \subset X$, and if T is a separate contraction mapping, then $(I - T)$ is a homeomorphism of S onto $(I - T)(S)$.*

Proof. Obviously, $(I-T)$ is continuous. Let us prove that $(I-T)$ is one-to-one. If $x \neq y$, then by using the notations of Definition 1.2.7 we have

$$\begin{aligned}
\|(I - T)(x) - (I - T)(y)\| &\geq \|x - y\| - \|T(x) - T(y)\| \\
&\geq \|x - y\| - \varphi(\|x - y\|) \\
&\geq \psi(\|x - y\|) > 0.
\end{aligned}$$

Then, $(I - T)$ is one-to-one, and $(I - T)^{-1}$ exists.
Let us suppose that $(I - T)^{-1}$ is not continuous. Then, there exist $(I - T)(y)$ and $(I - T)(x_n) \to (I - T)(y)$, but there exist $\varepsilon_0 > 0$ and $(x_{n_k})_k$ with $\|y - x_{n_k}\| \geq \varepsilon_0$. Now, for each $\varepsilon > 0$, there exists N such that $n_k > N$. We have

$$\begin{aligned}
\varepsilon &\geq \|(I - T)(x_{n_k}) - (I - T)(y)\| \\
&\geq \|x_{n_k} - y\| - \|T(x_{n_k}) - T(y)\| \\
&\geq \|x_{n_k} - y\| - \varphi(\|x_{n_k} - y\|) \\
&\geq \psi(\|x_{n_k} - y\|) \\
&\geq \psi(\varepsilon_0) > 0.
\end{aligned}$$

A contradiction occurs when $\varepsilon \to 0$. This contradiction implies that $(I - T)^{-1}$ is continuous. This completes the proof. Q.E.D.

Definition 1.2.8 *Let (X, d) be a metric space, and let M be a subset of X. The mapping $T : M \longrightarrow X$ is said to be expansive, if there exists a constant $h > 1$ such that $d(Tx, Ty) \geq hd(x, y) \ \ \forall\ x,\ y \in M$.*

Proposition 1.2.1 *Let $(X, \|.\|)$ be a linear normed space, with $M \subset X$. Let us assume that the mapping $T : M \longrightarrow X$ is expansive with a constant $h > 1$. Then, the inverse of the mapping $F := I - T : M \longrightarrow (I - T)(M)$ exists, and we have*

$$\|F^{-1}(x) - F^{-1}(y)\| \leq \frac{1}{h - 1}\|x - y\|, \quad x,\ y \in F(M).$$

Proof. For each $x,\ y \in M$, we have

$$\|F(x) - F(y)\| = \|(Tx - Ty) - (x - y)\| \geq (h-1)\|x - y\|, \tag{1.1}$$

which shows that F is one-to-one and hence, the inverse of $F : M \longrightarrow F(M)$ exists. Now by taking $x,\ y \in F(M)$, and by using Eq. (1.1), we have the following estimate:

$$\|F^{-1}(x) - F^{-1}(y)\| \leq \frac{1}{h-1}\|x - y\|.$$

Q.E.D.

Definition 1.2.9 *Let (X, d) be a metric space, and let M be a subset of X. The mapping $T : M \longrightarrow X$ is said to be semi-expansive, if there exists $\Phi : X \times X \longrightarrow \mathbb{R}_+$ with $d(T(x), T(y)) \geq \Phi(x, y)\ \ \forall x,\ y \in M$ and satisfying the following conditions:*

(i) $\Phi(x, y) = 0$ implies $x = y$,

(ii) $\Phi(x, y) = \Phi(y, x)$, and

(iii) $\Phi(x_n, x) \to 0$ implies $x_n \to x$.

Remark 1.2.3 *Obviously, every expansive mapping is semi-expansive.*

1.3 Weak Topology

Definition 1.3.1 *A topological vector space is a vector space X over the field $\mathbb{K}$ such that the space is a Hausdorff topological space, and the operations $+ : X \times X \longrightarrow X$ and $\cdot : \mathbb{K} \times X \longrightarrow X$ are continuous.*

Let $(X, \|.\|)$ be a Banach space and let τ be the family of semi-norms

$$\{\rho_\varphi(x) = |\varphi(x)| : \varphi \in X^* \text{ and } \|\varphi\|_{X^*} \leq 1\},$$

where X^* denotes the dual of X. The topology generated by τ is called the weak topology. One may call subsets of a topological vector space weakly closed (respectively, weakly compact, etc.) if they are closed (respectively, compact, etc.) with respect to the weak topology. In practical situations we often face the problem of type (partial differential equation) in the weak topology setting. All results for a Hausdorff locally convex topology induced by a separating vector space of linear functionals hold for the weak topology of a normed space X. In particular :

Proposition 1.3.1 *If $(x_n)_{n\in\mathbb{N}}$ is a sequence of elements of X and x is an element of X, then $x_n \rightharpoonup x$ if, and only if, $\varphi(x) \to \varphi(x)$ for each $\varphi \in X^*$. Here, $\rightharpoonup$ denotes the weak convergence and $\to$ denotes the strong convergence in X, respectively.*

Theorem 1.3.1 *[83, Theorem 13, p. 422] A convex subset of locally convex linear topological X is X^*-closed if, and only if, it is closed.*

Notation: Throughout this book, $\rightharpoonup$ will denote the weak convergence and $\to$ will denote the strong convergence in X, respectively.

Now, let us introduce some types of weak continuity:

Definition 1.3.2 *A function $A : X \longrightarrow Y$ is called weakly continuous if it is continuous with respect to the weak topologies of X and Y.*

Definition 1.3.3 *Let X be a Banach space. An operator $A : X \longrightarrow Y$ is said to be weakly sequentially continuous on X if, for every sequence $(x_n)_n$ with $x_n \rightharpoonup x$, we have $Ax_n \rightharpoonup Ax$.*

Definition 1.3.4 *[154] Let X be a Banach space. An operator $A : X \longrightarrow Y$ is said to be strongly continuous (or sometimes called weakly-strongly sequentially continuous) on X if, for every sequence $(x_n)_n$ with $x_n \rightharpoonup x$, we have $Ax_n \to Ax$.*

For linear operators, we have the following:

Theorem 1.3.2 *[128, Theorem 2.5.11] A linear operator A for a normed space X into a normed space Y is norm-to-norm continuous if, and only if, it is weak-to-weak continuous (i.e., weakly continuous).*

1.3.1 Weakly compact linear operators

Definition 1.3.5 *Suppose that X and Y are Banach spaces. A linear operator A from X into Y is weakly compact if $A(B)$ is a relatively weakly compact subset of Y whenever B is a bounded subset of X.*

The collection of all weakly compact linear operators from X into Y is denoted by $\mathcal{W}(X, Y)$ or just $\mathcal{W}(X)$ if $X = Y$.

Proposition 1.3.2 *Every compact linear operator from a Banach space into a Banach space is weakly compact.*

Proposition 1.3.3 *Every weakly compact linear operator from a Banach space into a Banach space is bounded.*

Remark 1.3.1 *(i) $\mathcal{W}(X,Y) \neq \emptyset$. For example if X or Y is reflexive, then every bounded linear operator is weakly compact.*

(ii) Every weakly compact linear operator on X is weakly sequentially continuous.

Proposition 1.3.4 *If X and Y are two Banach spaces, then $\mathcal{W}(X,Y)$ is a closed subspace of $\mathcal{L}(X,Y)$, the Banach space of all bounded linear operators from X into Y.*

Proposition 1.3.5 *Suppose that X, Y, and Z are Banach spaces, that $A \in \mathcal{L}(X,Y)$ and that $B \in \mathcal{L}(Y,Z)$. If either A or B is weakly compact, then BA is weakly compact.*

Proposition 1.3.6 *If X is a Banach space, then $\mathcal{W}(X)$ is a closed two-sided ideal of X.*

Now, let us recall the following definitions:

Definition 1.3.6 *(i) A subset $\mathcal{M}$ of a topological space is called countably compact if every countable open covering of $\mathcal{M}$ contains a finite sub-covering. A subset $\mathcal{M}$ is called relatively countably compact if the closure is countably compact.*

(ii) A subset $\mathcal{M}$ of a topological space is called limit-point compact if every infinite subset of $\mathcal{M}$ has at least one accumulation point that belongs to $\mathcal{M}$. A subset $\mathcal{M}$ is called relatively limit-point compact if every infinite subset of $\mathcal{M}$ has, at least, one accumulation point.

(ii) A subset $\mathcal{M}$ of a topological space is called sequentially compact if every sequence in $\mathcal{M}$ has converging subsequence whose limit belongs to $\mathcal{M}$. A subset $\mathcal{M}$ is called relatively sequentially compact if every sequence in $\mathcal{M}$ has convergent subsequence.

One can verify that all just presented notions of compactness and their relative counterparts coincide in metrizable topologies. However there are examples of non-metrizable topologies where some types of compactness are equivalent.

The most known one is the weak topology of Banach spaces. One of the advantages of this special locally convex topology is the fact that if a set $\mathcal{M}$ is weakly compact, then every weakly sequentially continuous map $A : \mathcal{M} \longrightarrow X$ is weakly continuous. This is an immediate consequence of the fact that weak sequential compactness is equivalent to weak compactness (Eberlein–Šmulian's theorem). The Eberlein–Šmulian theorem states:

Theorem 1.3.3 *(W. F Eberlein 1947, V.L. Šmulian 1940)[128, Theorem 2.8.6] Let X be a normed space and let $\mathcal{M}$ be a subset of X. Then, the following assertions are equivalent:*

(i) The set $\mathcal{M}$ is relatively weakly compact.

(ii) The set $\mathcal{M}$ is relatively weakly countably compact.

(iii) The set $\mathcal{M}$ is relatively weakly limit point compact.

(iv) The set $\mathcal{M}$ is relatively weakly sequentially compact.

The following is a consequence of the Eberlein–Šmulian theorem.

Corollary 1.3.1 *If $\mathcal{M}$ is a relatively weakly compact subset of a normed space and $x_0 \in \overline{\mathcal{M}^w}$ (the weak closure of $\mathcal{M}$), then there is a sequence in $\mathcal{M}$ that converges weakly to x_0.*

The equivalence of the following characterization of weak compactness for linear operators is easily proved by using elementary arguments and the Eberlein–Šmulian theorem (see Theorem 1.3.3).

Proposition 1.3.7 *Suppose that A is a linear operator from a Banach space X into a Banach space Y. Then, the following assertions are equivalent:*

(i) The operator A is weakly compact.

(ii) The subset $A(B_X)$ is a relatively weakly compact subset of Y, where B_X denotes the closed ball in X centered at 0_X with radius 1.

(iii) Every bounded sequence $(x_n)_n$ in X has a subsequence (x_{n_j}), such that the sequence (Tx_{n_j}) converges weakly.

Theorem 1.3.4 *(S. Mazur, 1933)[128, Theorem 2.5.16] The closure and the weak closure of a convex subset of a normed space are the same. In particular, a convex subset of a normed space is closed if, and only if, it is weakly closed.*

Corollary 1.3.2 *If $\mathcal{M}$ is a subset of a normed space, then $\overline{co}(\mathcal{M}) = \overline{co^w}(\mathcal{M})$.*

We also need Krein–Šmulian's weak compactness theorem. A proof can be found in the book by N. Dunford and J. T. Schwartz [83, p. 434].

Theorem 1.3.5 *The closed convex hull of a weakly compact subset of a Banach space is itself weakly compact.*

The following two theorems will be important in the Part II of this book. For a proof, the reader can also see the book by N. Dunford and J. T. Schwartz [83, p. 434].

Theorem 1.3.6 *If* $\{f(.)\}$ *is a family of functions in* $L_1(S, \Sigma, \mu)$ *which is weakly sequentially compact, then the family* $\{|f(.)|\}$ *is also weakly sequentially compact.*

Theorem 1.3.7 *Let* (S, Σ, μ) *be a positive measure space. If a set* K *in* $L_1(S, \Sigma, \mu)$ *is weakly sequentially compact, then*

$$\lim_{\mu(E)\to 0} \int_E f(s)\mu(ds) = 0$$

uniformly for $f \in K$. *If* $\mu(S) < \infty$, *then conversely this condition is sufficient for a bounded set* K *to be weakly sequentially compact.*

Let us recall Dunford's theorem, which we will state for convenience.

Theorem 1.3.8 *(Dunford) Let* (Ω, Σ, μ) *be a finite measure space and let* X *be a Banach space such that both* X *and* X^* *have the Radon–Nikodym property. A subset* K *of* $L_1(\Omega, X)$ *is relatively weakly compact if, and only if,*

(i) K *is bounded,*

(ii) K *is uniformly integrable, and*

(iii) for each $B \in \Sigma$, *the set* $\{\int_B f d\mu,$ *such that* $f \in K\}$ *is relatively weakly compact.*

The Arzelà–Ascoli theorem (see [153]) plays a crucial role in the proof of existence of solutions for functional integral equations.

Theorem 1.3.9 *(Arzelà–Ascoli's theorem) Let* (X, d) *be a compact space. A subset* $\mathcal{F}$ *of the vector space of all real, continuous functions on* X, $\mathcal{C}(X)$ *is relatively compact if, and only if,* $\mathcal{F}$ *is:*

(i) Equibounded: there is some $L > 0$ *such that* $|\varphi(x)| \leq L$ *for all* $x \in X$ *and all* $\varphi \in \mathcal{F}$, *and*

(ii) Equicontinuous: for every $\varepsilon > 0$ *there is a* $\delta > 0$ *such that* $|\varphi(x) - \varphi(y)| < \varepsilon$ *for all* $\varphi \in \mathcal{F}$ *whenever* $|x - y| < \delta$, *and* x, $y \in X$.

1.3.2 The Dunford–Pettis property (DP property)

As properties for linear operators, both weak compactness and complete continuity lie between compactness and boundedness, which suggests that the two properties might be related. In general, none of them implies the other. However, it does happen that some common Banach spaces have the property that every weakly compact linear operator whose domain is that space is completely continuous. This property has been given a name by Grothendieck in honor of N. Dunford and B. J. Pettis, who proved in 1940, that $L_1(S, \Sigma, \lambda)$ has the property when λ is the Lebesgue measure on the σ-algebra Σ of Lebesgue measurable subsets of finite or infinite interval S in $\mathbb{R}^n$, where n is a positive integer.

Definition 1.3.7 *[91] A Banach space X has the Dunford–Pettis property (in short DP property) if, for every Banach space Y, each weakly compact linear operator from X into Y maps weakly compact sets in X to norm compact sets in Y.*

The space $L_1(S, \Sigma, \lambda)$, where (S, Σ, λ) is a finite measure space has the Dunford–Pettis property. This DP property is also valid for the space $C(K)$, where K is a compact Hausdorff space. For further examples, we may refer to [78] or [83], p. 494, 479, 508, and 511. Note that the DP property is not preserved under conjugation. However, if X is a Banach space whose dual has the DP property, then X has the DP property (see, e.g., [91]). For more information, we refer to the paper of J. Diestel [78] which contains a survey of the Dunford–Pettis property and related topics. The following theorem gives a characterization of the Dunford–Pettis property [128, Theorem 3.5.18].

Theorem 1.3.10 *Suppose that X is a Banach space. The following assertions are equivalent :*

(i) The space X has the Dunford–Pettis property.

(ii) For every sequence $(x_n)_n$ in X converging weakly to 0, and every $(\varphi_n)_n$ in X^* converging weakly to 0, the sequence $(\varphi_n(x_n))_n$ converges to 0.

(iii) For every sequence $(x_n)_n$ in X converging weakly to some x and every $(\varphi_n)_n$ in X^* converging weakly to some φ, the sequence $(\varphi_n(x_n))_n$ converges to $\varphi(x)$.

An important property for weakly compact operators on Dunford–Pettis spaces is:

Theorem 1.3.11 *Suppose that X is a Banach space, has the Dunford–Pettis property, and $A \in \mathcal{W}(X)$. Then, A^2 is a compact operator.*

We close this section with the following result.

Proposition 1.3.8 *Let X be a Dunford–Pettis space and let A be a weakly compact linear operator on X. Then, A is strongly continuous.*

Proof. Let $(x_n)_n$ be a weakly convergent sequence to x in X. We have $Ax_n \rightharpoonup Ax$ in X. Since X is a Dunford–Pettis space and A is weakly compact, then the sequence $(Ax_n)_n$ has a convergent subsequence to Ax, say $(Ax_{\rho(n)})_n$. We claim that $Ax_n \to Ax$ in X. Suppose that this is not the case; then we can find $\varepsilon_0 > 0$ and a subsequence $(Ax_{\rho_1(n)})_n$ such that $\|Ax_{\rho_1(n)} - Ax\| > \varepsilon_0$, for any $n \in \mathbb{N}$. But $x_{\rho_1(n)} \rightharpoonup x$ in X, so $Ax_{\rho_1(n)} \rightharpoonup Ax$ in X. Since X is a Dunford–Pettis space and A is weakly compact, then the sequence $(Ax_{\rho_1(n)})_n$ has a convergent subsequence to Ax in X, say $(Ax_{\rho_1(\rho_2(n))})_n$. Then for ε_0, there exist $n_2 \in \mathbb{N}$ such that for $n \geq n_2$, $\|Ax_{\rho_1(\rho_2(n))} - Ax\| \leq \varepsilon_0$ which is a contradiction. This proves the claim and completes the proof. Q.E.D.

1.4 Measure of Weak Noncompactness (MNWC)

Throughout this section, X denotes a Banach space. For any $r > 0$, B_r denotes the closed ball in X centered at 0_X with radius r, and B_X denotes the closed ball in X centered at 0_X with radius 1. Ω_X is the collection of all nonempty bounded subsets of X, and $\mathcal{K}^w$ is the subset of Ω_X consisting of all weakly compact subsets of X. Recall that the notion of the measure of weak noncompactness was introduced by De Blasi [64]; it is the map $\omega : \Omega_X \longrightarrow [0, +\infty)$ defined in the following way:

$$\omega(\mathcal{M}) = \inf\{r > 0 : \text{ there exists K} \in \mathcal{K}^w \text{ such that } \mathcal{M} \subset \text{K} + B_r\},$$

for all $\mathcal{M} \in \Omega_X$. For more convenience, let us recall some basic properties of $\omega(.)$ needed below (see, for example, [9, 64]) (see also [18], where an axiomatic approach to the notion of a measure of weak noncompactness is presented).

Lemma 1.4.1 *Let $\mathcal{M}_1$ and $\mathcal{M}_2$ be two elements of Ω_X. Then, the following conditions are satisfied:*

(1) $\mathcal{M}_1 \subset \mathcal{M}_2$ *implies* $\omega(\mathcal{M}_1) \leq \omega(\mathcal{M}_2)$.

(2) $\omega(\mathcal{M}_1) = 0$ *if, and only if,* $\overline{\mathcal{M}_1}^w \in \mathcal{K}^w$, *where* $\overline{\mathcal{M}_1}^w$ *is the weak closure of the subset* $\mathcal{M}_1$.

(3) $\omega(\overline{\mathcal{M}_1}^w) = \omega(\mathcal{M}_1)$.

(4) $\omega(\mathcal{M}_1 \cup \mathcal{M}_2) = \max\{\omega(\mathcal{M}_1), \omega(\mathcal{M}_2)\}$.

(5) $\omega(\lambda\mathcal{M}_1) = |\lambda|\omega(\mathcal{M}_1)$ *for all* $\lambda \in \mathbb{R}$.

(6) $\omega(co(\mathcal{M}_1)) = \omega(\mathcal{M}_1)$, *where* $co(\mathcal{M}_1)$ *is the convex hull of* $\mathcal{M}_1$.

(7) $\omega(\mathcal{M}_1 + \mathcal{M}_2) \leq \omega(\mathcal{M}_1) + \omega(\mathcal{M}_2)$.

(8) *if* $(\mathcal{M}_n)_{n\geq 1}$ *is a decreasing sequence of nonempty, bounded, and weakly closed subsets of* X *with* $\lim_{n\to\infty} \omega(\mathcal{M}_n) = 0$, *then* $\mathcal{M}_\infty := \cap_{n=1}^{\infty} \mathcal{M}_n$ *is nonempty and* $\omega(\mathcal{M}_\infty) = 0$ *i.e.,* $\mathcal{M}_\infty$ *is relatively weakly compact.*

Remark 1.4.1 $\omega(B_X) \in \{0, 1\}$. *Indeed, it is obvious that* $\omega(B_X) \leq 1$. *Let* $r > 0$ *be given such that there is a weakly compact* K *of* X *satisfying* $B_X \subset K + rB_X$. *Hence,* $\omega(B_X) \leq r\omega(B_X)$. *If* $\omega(B_X) \neq 0$, *then* $r \geq 1$. *Thus,* $\omega(B_X) \geq 1$.

In the following definitions, S is a nonempty, closed, and convex subset of X.

First, let us introduce the notion of ws-compact operator:

Definition 1.4.1 *[90] An operator* $F : S \longrightarrow X$ *is said to be* ws*-compact, if it is continuous on* S *and, for any weakly convergent sequence* $(x_n)_{n\geq 0}$ *of* S, *the sequence* $(Fx_n)_{n\geq 0}$ *has a strongly convergent subsequence in* X.

Remark 1.4.2 *(i) Obviously, every compact operator is* ws*-compact. If* X *is reflexive, then the notions of compactness and* ws*-compactness are equivalent. However, if* X *is not reflexive, then this equivalence does not hold.*

(ii) An operator F *is* ws*-compact if, and only if, it is continuous and maps relatively weakly compact sets into relatively compact ones.*

Remark 1.4.3 *We should notice that every strongly continuous operator is* ws*-compact. However, the converse of the previous proposition is not true in general (even if* X *is reflexive). Indeed, let* $X = L^2(0,1)$ *and let* $F : X \longrightarrow X$ *be defined by:*

$$(Fx)(s) := \int_0^1 x^2(t)dt = ||x||_2^2.$$

Clearly, since $\forall x,\ y \in L^2(0,1)$, $||Fx - Fy||_2 \leq ||x - y||_2\ ||x + y||_2$, *it follows* *that F is $||.||_2$-continuous. Moreover, since the range of F is homeomorphic to $\mathbb{R}$, we infer that F is ws-compact. Finally, if we take the sequence defined by:*

$$x_n(t) := \cos(n\pi t) \quad n \in \mathbb{N}^*,$$

then by using the density of staged functions in $L^2(0,1)$ and the Riemann–Lebesgue Lemma, we obtain $x_n \rightharpoonup \theta$ in $L^2(0,1)$. Moreover, we have:

$$\begin{aligned} ||Fx_n||_2 &= \int_0^1 (Fx_n)^2(t)dt \\ &= \int_0^1 (x_n)^2(t)dt \\ &= \frac{1}{2}\int_0^1 \cos(2\pi nt)dt + \frac{1}{2} \\ &= \frac{1}{2}. \end{aligned}$$

Definition 1.4.2 *[35] Let us fix $k > 0$. An operator $F : S \longrightarrow X$ is said to be a:*

(i) k-set-contraction with respect to the measure of weak noncompactness ω, if F is bounded, and for any bounded subset V of S, $\omega(F(V)) \leq k\omega(V)$.

(ii) Strict set-contraction, if F is a k-set-contraction with $k < 1$.

(iii) Condensing map with respect to the measure of weak noncompactness ω, if F is a 1-set-contraction and $\omega(F(V)) < \omega(V)$, for all bounded subsets V of S with $\omega(V) > 0$.

Remark 1.4.4 *We can easily notice that F is a weakly compact operator if, and only if, F is a 0-set-contraction with respect to the measure of weak noncompactness ω. If F is a strict set-contraction, then F is condensing. However, the converse is not true. From this, we may deduce that the strict set-contraction is an extension of the weakly compact operator.*

Now, let us give an example of a condensing operator which is not a strict set-contraction:

Example 5 We suppose that the Banach space X is not reflexive. Let

$$\begin{cases} \varphi : [0,1] \longrightarrow [0,1] \\ \qquad x \longrightarrow \varphi(x) = 1 - x \end{cases}$$

and let us define F by:

$$\begin{cases} F : B_X \longrightarrow B_X \\ \qquad x \longrightarrow F(x) = \varphi(\|x\|)x. \end{cases}$$

Let $B \subset B_X$. Since $FB \subset \overline{\text{co}}\,\{B \cup \{0\}\}$, it follows that $\omega(FB) \leq \omega(B)$. Now, let us show that

$$\partial(r\varphi(r)B_X) \subset F(rB_X) \text{ for all } r \in [0,1].$$

Let $x \in \partial(r\varphi(r)B_X)$. Then, $x = F\left(\frac{x}{\varphi(r)}\right)$, $r \neq 1$ and if $r = 1$, then $x = 0 = F(0)$. Therefore, (since X is not reflexive), we have

$$\begin{aligned} \omega(F(rB_X)) &\geq \omega(\partial(r\varphi(r)B_X)) \\ &= r\varphi(r) \\ &= \omega(rB_X)\varphi(r). \end{aligned}$$

Suppose that F is a strict set-contraction. Then, there exists $k \in (0,1)$ such that $\omega(rB_X)\varphi(r) \leq k\omega(rB_X)$. Hence, $\varphi(r) \leq k$, which is a contradiction if we pass to the limit as $r \to 0$. Moreover, let $B \subset B_X$ such that $\omega(B) = d > 0$, $0 < r < d$, $B_1 = B \cap rB_X$ and $B_2 = B \backslash rB_X$. Then, we have

$$\omega(FB_1) \leq \omega(B_1) \leq \omega(rB_X) = r < d = \omega(B).$$

Moreover, we can write

$$\begin{aligned} \omega(FB_2) &\leq \omega\left(\{\alpha x : 0 \leq \alpha \leq \varphi(r) \text{ and } x \in B_2\}\right) \\ &\leq \omega\left(\overline{\text{co}}\,\{\varphi(r)B_2 \cup \{0\}\}\right) \\ &\leq \varphi(r)\omega(B) \\ &< \omega(B). \end{aligned}$$

Notice that $B = B_1 \cup B_2$. Then, we have

$$\omega(FB) = \max\{\omega(FB_1), \omega(FB_2)\} < \omega(B).$$

Q.E.D.

Definition 1.4.3 *A map $A : \mathcal{M} \subset X \longrightarrow X$ is said to be ω-contractive (or a ω-contraction), if it maps bounded sets into bounded ones, and there exists a scalar $\alpha \in [0,1)$ such that $\omega(A\mathcal{N}) \leq \alpha\omega(\mathcal{N})$, for all bounded subsets $\mathcal{N} \subset \mathcal{M}$.*

Definition 1.4.4 *[96] An operator $F : S \longrightarrow S$ is said to be a convex-power condensing with respect to the measure of weak noncompactness ω, if F is bounded and there exist $x_0 \in S$ and a positive integer n_0 ($n_0 \geq 1$) such that, given any bounded subset V of S with $\omega(V) > 0$, we have*

$$\omega\left(F^{(n_0,x_0)}(V)\right) < \omega(V),$$

where

$$F^{(1,x_0)}(V) = F(V), \text{ and } F^{(n_0,x_0)}(V) = F\left(\overline{co}\left\{F^{(n_0-1,x_0)}(V), x_0\right\}\right).$$

F is also called convex-power condensing about x_0 and n_0.

Remark 1.4.5 *If V is bounded and $V \subset \left(F^{(n_0,x_0)}(V)\right)$, then V is relatively weakly compact. It is easy to see that a condensing operator is a convex-power condensing operator (since $n_0 = 1$). Therefore, the definition of the convex-power condensing operator is the generalization of the condensing operator.*

Definition 1.4.5 *(see [72]) A function $T : [a,b] \times \mathbb{R} \longrightarrow \mathbb{R}$ is said to be generalized Lipschitz, if there exists a function $\varphi \in L^1([a,b],\mathbb{R})$, such that*

$$|T(s,x) - T(s,y)| \leq \varphi(s)|x - y|$$

a.e., $s \in [a,b]$ for all x, $y \in \mathbb{R}$. The function φ is called the Lipschitz function of T.

The following result can be found in [5].

Lemma 1.4.2 *Let $S \subset X$ be closed and convex. Suppose that $F : S \longrightarrow S$ is weakly sequentially continuous and convex-power condensing with respect to ω. If $F(S)$ is bounded, then F has, at least, one fixed point in S.*

Proposition 1.4.1 *Let X be a Banach space and let $B : X \longrightarrow X$ be a weakly sequentially continuous map. Then, for any weakly compact subset K of X, $B(K)$ is weakly compact.*

Proof. By using Theorem 1.3.3, it is sufficient to show that $B(K)$ is weakly sequentially compact. For this purpose, let us take a bounded sequence $(x_n)_n$ in K. Since K is weakly sequentially compact, there exists a subsequence $(x_{n_k})_k$ of $(x_n)_n$ converging weakly for some x in K. Since B is weakly sequentially continuous, $(B(x_{n_k}))_k$ converges weakly to Bx. Q.E.D.

The following theorem is a fundamental tool for the proofs of the existence of solutions for several functional integral equations (see Chapters 6 and 7). A proof of this theorem can be found in [81, Dobrakov, p. 26].

Theorem 1.4.1 *Let K be a compact Hausdorff space and let X be a Banach space. Let $(f_n)_n$ be a bounded sequence in $\mathcal{C}(K,X)$, and $f \in \mathcal{C}(K,X)$. Then, $(f_n)_n$ is weakly convergent to f if, and only if, $(f_n(t))_n$ is weakly convergent to $f(t)$ for each $t \in K$.*

1.5 Basic Tools in Banach Algebras

Definition 1.5.1 *An algebra X is a vector space endowed with an inner composition law denoted by:*

$$\begin{cases} (.) : X \times X \longrightarrow X \\ \qquad (x,y) \longrightarrow x.y, \end{cases}$$

which is associative and bilinear. A normed algebra is an algebra endowed with a norm satisfying the following property, for all x, $y \in X$

$$\|x.y\| \leq \|x\| \|y\|.$$

A complete normed algebra is called a Banach algebra.

Examples of Banach algebras

Example 6 The set of real (or complex) numbers is a Banach algebra with a norm given by the absolute value.

Example 7 The prototypical example of a Banach algebra is $C^0(X)$, the space of (complex-valued) continuous functions on a locally compact (Hausdorff) space that vanishes at infinity. $C^0(X)$ is unital if, and only if, X is compact.

Example 8 The set of all real or complex n-by-n matrices becomes a unital Banach algebra, if we equip it with a sub-multiplicative matrix norm.

Example 9 Take the Banach space $\mathbb{K}^n$ with the sup-norm and define the multiplication componentwise: $x.y = (x_1, ..., x_n).(y_1, ..., y_n) = (x_1y_1, ..., x_ny_n)$.

Example 10 The algebra of all bounded real- or complex-valued functions defined on some sets (with pointwise multiplication and the sup norm) is a unital Banach algebra.

Example 11 The algebra of all bounded continuous real- or complex-valued functions on some locally compact spaces (again with pointwise operations and sup norm) is a Banach algebra.

Example 12 The algebra of all continuous linear operators on a Banach space E (with functional composition as multiplication and the operator norm as norm) is a unital Banach algebra. The set of all compact operators on E is a closed ideal in this algebra.

It is important to mention that the product of two weakly sequentially continuous functions is not necessarily weakly sequentially continuous.

Definition 1.5.2 *We will say that the Banach algebra X satisfies the condition $(\mathcal{P})$ if:*

$$(\mathcal{P}) \begin{cases} \textit{For any sequences } \{x_n\} \textit{ and } \{y_n\} \textit{ in } X \textit{ such that } x_n \rightharpoonup x \textit{ and } y_n \rightharpoonup y, \\ \textit{then } x_n.y_n \rightharpoonup x.y. \end{cases}$$

Example 13 Obviously, every finite-dimensional Banach algebra satisfies the condition $(\mathcal{P})$.

The following proposition provides another example of Banach algebra satisfying the condition $(\mathcal{P})$.

Proposition 1.5.1 *If X is a Banach algebra satisfying the condition $(\mathcal{P})$, then $\mathcal{C}(K, X)$ is also a Banach algebra satisfying the condition $(\mathcal{P})$, where K is a compact Hausdorff space.*

Proof. Let $\{x_n\}$ and $\{y_n\}$ be any sequences in $\mathcal{C}(K, X)$, such that $x_n \rightharpoonup x$ and $y_n \rightharpoonup y$. So, for each $t \in K$, we have $x_n(t) \rightharpoonup x(t)$ and $y_n(t) \rightharpoonup y(t)$ (see Theorem 1.4.1). Since X satisfies the condition $(\mathcal{P})$, then

$$x_n(t).y_n(t) \rightharpoonup x(t).y(t),$$

because $(x_n.y_n)_n$ is a bounded sequence. Moreover, this implies with Theorem 1.4.1, that

$$x_n.y_n \rightharpoonup x.y,$$

which shows that the space $\mathcal{C}(K, X)$ verifies condition $(\mathcal{P})$. Q.E.D.

In a Banach algebra X satisfying the condition $(\mathcal{P})$, we have the following fact.

Lemma 1.5.1 *If K and K' are two weakly compact subsets of a Banach algebra X satisfying the condition $(\mathcal{P})$, then*

$$K.K' = \Big\{x.y \ ; \ x \in K \text{ and } x' \in K'\Big\}$$

is a weakly compact subset of X.

Proof. We will show that $K.K'$ is weakly sequentially compact. For this, let $\{x_n\}_{n=0}^{\infty}$ be any sequence of K and let $\{x'_n\}_{n=0}^{\infty}$ be any sequence of K'. By hypothesis, there is a renamed subsequence $\{x_n\}_{n=0}^{\infty}$ of K such that $x_n \rightharpoonup x \in K$. Again, there is a renamed subsequence $\{x'_n\}_{n=0}^{\infty}$ of K' such that $x'_n \rightharpoonup x' \in K'$. This, together with the condition $(\mathcal{P})$, implies that

$$x_n.x'_n \rightharpoonup x.x'.$$

This, in turn, shows that $K.K'$ is weakly sequentially compact. Hence, an application of the Eberlein–Šmulian theorem shows that $K.K'$ is weakly compact. Q.E.D.

In [16], J. Banas has introduced a class of Banach algebras satisfying a certain condition denoted by (m):

$$(m) \qquad \omega(X.Y) \leq ||X||\omega(Y) + ||Y||\omega(X),$$

where ω is a measure of weak noncompactness, X and Y are bounded subsets, and $\|X\| := \sup \{\|x\| \text{ such that } x \in X\}$. In the following lemma, we will show that Banach algebras satisfying the condition $(\mathcal{P})$ verify, in a special but important case, the condition (m) for the De Blasi's measure of weak noncompactness ω:

Lemma 1.5.2 *For any bounded subset V of a Banach algebra X satisfying the condition $(\mathcal{P})$ and for any weakly compact subset K of X, we have*

$$\omega(V.K) \leq ||K||\omega(V).$$

Proof. We may assume that $||K|| > 0$. Let $\varepsilon > 0$ be given. From the definition of ω, we deduce that there exists a weakly compact subset K' of X, such that

$$V \subset K' + \left(\omega(V) + \frac{\varepsilon}{||K||}\right) B_X.$$

Then, we have

$$V.K \subset K'.K + \left(\omega(V) + \frac{\varepsilon}{||K||}\right) B_X.K.$$

From which, we infer that

$$V.K \subset K'.K + \left(\omega(V) + \frac{\varepsilon}{||K||}\right) ||K||B_X.$$

Now, by using Lemma 1.5.1, we have

$$\omega(V.K) \leq ||K||\omega(V) + \varepsilon,$$

which implies, since ε is arbitrary, that

$$\omega(V.K) \leq ||K||\omega(V).$$

Q.E.D.

1.6 Elementary Fixed Point Theorems

In mathematics, a number of fixed point theorems in infinite-dimensional spaces generalize the Brouwer fixed point theorem. They have applications, for example, in the proof of existence theorems for differential equations. The first result in the field was Schauder's fixed point theorem, proved in 1930 by Juliusz Schauder. This theorem still has an enormous influence on fixed point theory and on the theory of differential equations. Several further results followed. Schauder's fixed point theorem states in one version:

Theorem 1.6.1 *(Schauder) If $\mathcal{M}$ is a nonempty, closed, and convex subset of a Banach space X and A is a continuous map from $\mathcal{M}$ to $\mathcal{M}$ whose image is countably compact, then A has, at least, a fixed point.*

The Tikonov (Tychonoff) fixed point theorem is applied to any locally convex topological space X. It states :

Theorem 1.6.2 *(Tychonoff) Let $\mathcal{M}$ be a convex, and compact subset of a locally convex topological space X. If A is a continuous map on $\mathcal{M}$ into $\mathcal{M}$, then A has, at least, a fixed point.*

Remark 1.6.1 *Tychonoff's theorem contains, as a special case, the earlier result in Schauder asserting the existence of a fixed point for each weakly continuous self mapping of a weakly compact, and convex subset $\mathcal{M}$ of a Banach space.*

Brouwer's fixed point theorem is as follows:

Theorem 1.6.3 *(Brouwer) If $x \longrightarrow \varphi(x)$ is a continuous point-to-point mapping of an r-dimensional closed simplex S into itself, then there exists an $x_0 \in S$, such that $x_0 = \varphi(x_0)$.*

The last theorem can be generalized in the following way: Let $\mathcal{P}_{cl,\ cv}(S)$ be the family of all closed convex subsets of S. A point-to-set mapping $x \longrightarrow \Phi(x) \in \mathcal{P}_{cl,\ cv}(S)$ of S into $\mathcal{P}_{cl,\ cv}(S)$ is called upper semi-continuous (in short u.s.c.) if $x_n \to x_0$, $y_n \in \Phi(x_n)$ and $y_n \to y_0$ implying $y_0 \in \Phi(x_0)$. Obviously, this condition is equivalent to saying that the graph of $\Phi(x)$ is a closed subset of $S \times S$, where $\times$ denotes a cartesian product. As a generalization, S. Kakutani [109] obtained the following fixed point theorem.

Theorem 1.6.4 *(Kakutani) If $x \longrightarrow \Phi(x)$ is an upper semi-continuous point-to-set mapping of an r-dimensional closed simplex S into $\mathcal{P}_{cl,\ cv}(S)$, then there exists an $x_0 \in S$ such that $x_0 \in \Phi(x_0)$.*

Proof. Let $S^{(n)}$ be the n-th barycentric simplicial subdivision of S. For each vertex x^n of $S^{(n)}$ let us take an arbitrary point y^n from $\Phi(x^n)$. Then, the mapping $x^n \to y^n$ defined on all vertices of $S^{(n)}$ will define, if it is extended linearly inside each simplex of $S^{(n)}$, a continuous point-to-point mapping $x \to \varphi_n(x)$ of S into itself. Consequently, by using Brouwer's fixed point theorem, there exists an $x_n \in S$ such that $x_n = \varphi_n(x_n)$. Now, if we take a subsequence $(x_{n_v})_v$ $(v = 1, 2, ...)$ of $(x_n)_n$ $(n = 1, 2, ...)$ which converges to a point $x_0 \in S$, then x_0 is the required point. In order to prove this, let Δ_n be an r-dimensional simplex of $S^{(n)}$ which contains the point x_n: (If x_n lies on the lower-dimensional simplex of $S^{(n)}$, then Δ_n is not uniquely determined. In this case, let Δ_n be any one of these simplexes). Let $x_0^n, x_1^n, ..., x_r^n$ be the vertices of Δ_n. Then, it is clear that the sequence $(x_i^{n_v})_v$ $(v = 1, 2, ...)$ converges to x_0 for $i = 0, 1, ..., r$, and we have

$$x_n = \sum_{i=0}^{r} \lambda_i^n x_i^n$$

for suitable (λ_i^n) $(i = 0, 1, ..., r;\ n = 1, 2, ...)$ with $\lambda_i^n \geq 0$ and

$$\sum_{i=0}^{r} \lambda_i^n = 1.$$

Let us further put $y_i^n = \varphi_n(x_i^n)$ $(i = 0, 1, ..., r;\ n = 1, 2, ...)$. Then, we have $y_i^n \in \Phi(x_i^n)$ and

$$x_n = \varphi_n(x_n) = \sum_{i=0}^{r} \lambda_i^n y_i^n \quad for \quad n = 1, 2, ...$$

Now, let us take a further subsequence (n_v') $(v = 1, 2, ...)$ of (n_v) $(v = 1, 2, ...)$ such that $(y_i^{n_v'})$ and $(\lambda_i^{n_v'})$ $(v = 1, 2, ...)$ converge for $i = 0, 1, ..., r$, and let us put $\lim_{v\to\infty} y_i^{n_v'} = y_i^0$ and $\lim_{v\to\infty} \lambda_i^{n_v'} = \lambda_i^0$ for $i = 0, 1, ..., r$. Then, we clearly have

$$\lambda_i^0 \geq 0, \quad \sum_{i=0}^{n} \lambda_i^0 = 1 \text{ and } x_0 = \sum_{i=0}^{r} \lambda_i^0 y_i^0.$$

Since $x_i^{n_v'} \to x_0$, $y_i^{n_v'} \in \Phi(x_i^{n_v'})$ and $y_i^{n_v'} \to y_i^0$, for $i = 0, 1, ..., r$. We must have, by using the upper semi-continuity of $\Phi(x)$, $y_i^0 \in \Phi(x_0)$ for $i = 0, 1, ..., r$, and this implies, by using the convexity of $\Phi(x_0)$, that

$$x_0 = \sum_{i=0}^{r} \lambda_i^0 y_i^0 \in \Phi(x_0).$$

This completes the proof. Q.E.D.

As a consequence of the Brouwer's fixed point theorem, we recall Schauder's fixed point theorem. For a proof, the reader can see [149].

Theorem 1.6.5 *(Schauder) Let Ω be a nonempty, closed, and convex subset of a Banach space X. If T is a completely continuous mapping from Ω into Ω, then T has, at least, a fixed point in Ω.*

We recall that a set is precompact, if every sequence in the set contains a convergent subsequence. A set is compact, if it is precompact and closed.

Theorem 1.6.6 *(Schauder's fixed point theorem) Let X be a Banach space, let B be a closed and convex subset of X, and let $T : B \longrightarrow B$ be a continuous map. Then, T has, at least, one fixed point in B if the range $T(B) := \{Tx; x \in B\}$ is precompact.*

Notice that, if B is compact, then automatically $T(B) \subset B$ is precompact. In both the contraction and the Schauder's fixed point theorems, the condition that T maps B into itself usually constitutes the main difficulty in applications. The central point of the Schauder fixed point theorem is the compactness of the image TB. Quite often, one takes a bounded B and shows that T is compact. In other cases, one simply takes a compact B (then, one has to be very careful about the continuity of T).

Theorem 1.6.7 *(Leray–Schauder) Let X be a Banach space and let $T : X \longrightarrow X$ be compact and continuous. Suppose the existence of a positive constant M, such that*

$$\|x\| \leq M \text{ whenever } \exists\, \sigma \in [0,1] \text{ such that } x = \sigma Tx.$$

Then, T has, at least, a fixed point in X.

The Leray–Schauder fixed point theorem is also known as Schaefer's fixed point theorem. Its advantage over Schauder's fixed point theorem for applications is that we don't have to identify a convex and closed subset where T maps into itself. Several problems arising from the most diverse areas of natural sciences (when modeled under the mathematical point of view) involve the study of solutions for nonlinear equations of the form

$$Au + Bu = u, \qquad u \in \mathcal{M},$$

where $\mathcal{M}$ is a closed and convex subset of a Banach space X; see for example [48, 51, 67, 68, 76]. Krasnosel'skii's fixed point theorem appeared as a prototype for solving equations of the above type. Motivated by the observation that the inversion of a perturbed differential operator could yield the sum of a contraction and a compact operator, M. A. Krasnosel'skii proved the next theorem. A proof can be found in [149].

Theorem 1.6.8 *(Krasnosel'skii) Let $\mathcal{M}$ be a closed, convex, and nonempty subset of a Banach space X. Suppose that A and B map $\mathcal{M}$ into X, and that:*

(i) $Ax + By \in \mathcal{M}$, for all $x, y \in \mathcal{M}$,

(ii) A is continuous on $\mathcal{M}$, and $A(\mathcal{M})$ is contained in a compact subset of X, and

(iii) B is a λ-contraction in X, with $\lambda \in [0,1[$.

Then, there exists y in $\mathcal{M}$, such that

$$Ay + By = y.$$

While it is not always possible to show that a given mapping between functional Banach spaces is weakly continuous, quite often its weak sequential continuity does not create any problem. This is deduced from the fact that the Lebesgue's dominated convergence theorem is valid for sequences but not for nets. A very interesting discussion (including illustrative examples) about the different types of continuity can be found in [13]. The following version of Schauder–Tychonoff's theorem holds [10, Theorem 1] :

Theorem 1.6.9 *(Arino, Gautier, and Penot)(1984) Let X be a metrizable, locally convex topological vector space, and let $\mathcal{M}$ be a weakly compact, and convex subset of X. Then, any weakly sequentially continuous map $A : \mathcal{M} \longrightarrow \mathcal{M}$ has, at least, a fixed point.*

Proof. It is sufficient to prove that A is weakly continuous, so that the Schauder–Tychonoff's fixed point theorem can be applied. Now, for each weakly closed subset E of X, $A^{-1}(E)$ is sequentially closed in M, hence weakly compact by using Eberlein–Šmulian's theorem (Theorem 1.3.3), and $A^{-1}(E)$ is weakly closed. Hence, A is weakly continuous. Q.E.D.

An important fixed point theorem that has been commonly used in the theory of nonlinear differential and integral equations is the following result proved by D. W. Boyd and J. S. W. Wong in [39]. This theorem extends the contractions to nonlinear contractions, and also generalizes the Banach fixed point principle, dating from 1922 (see for example [154]).

Theorem 1.6.10 *(Boyd and Wong's fixed point theorem) Let $A : X \longrightarrow X$ be a nonlinear contraction. Then, A has a unique fixed point x^*, and the sequence $(A^n x)_n$ of successive iterations of A converges to x^* for each $x \in X$.*

1.7 Positivity and Cones

Let X be a real Banach space with a norm $\|\cdot\|$.

Definition 1.7.1 *A closed subset X_+ is called a positive cone if the following holds:*

(i) $X_+ + X_+ \subset X_+$,

(ii) $\lambda X_+ \subset X_+$ *for* $\lambda \geq 0$,

(iii) $X_+ \cap (-X_+) = \{0\}$, *and*

(iv) $X_+ \neq \{0\}$.

Remark 1.7.1 *For all λ, $\mu \geq 0$, and all u, $v \in X_+$, we have $\lambda u + \mu v \in X_+$.*

The positive cone X_+ defines an ordering on a linear space X by the relation $f \leq g$ if, and only if, $g - f \in X_+$, and we write $f < g$ if $g - f \in X_+ \backslash \{0\}$.

Definition 1.7.2 *Let X be a linear space with a positive cone X_+. X is called a vector lattice, if for any f, $g \in X$ there is* $\sup(f,g)$ *and* $\inf(f,g)$ *in X.*

Especially, in a vector lattice X, there is $f^+ := \sup(f,0)$, $f^- := \inf(f,0)$, and $|f| := f^+ + f^-$ for all $f \in X$.

Definition 1.7.3 *Let X be a vector lattice and a Banach space. X is called a Banach lattice, if $|f| \leq |g|$ implies that $\|f\| \leq \|g\|$.*

An example of a Banach lattice is $X = L_p(\chi, d\mu)$, $1 \leq p < \infty$, μ is a positive σ-finite measure on a locally compact space χ. In this case, $f \in X_+$ means $f(x) \geq 0$ μ-a.e.

Definition 1.7.4 *An ideal in a vector lattice X is a linear subspace I for which $g \in I$ whenever $g \in X$ and $|g| \leq |f|$ for some $f \in I$.*

Definition 1.7.5 *An ideal I of X is called a band if $J \subseteq I$ and* $\sup J \in X$, *together imply that* $\sup J \in I$.

Let $X = L_p(\chi, d\mu)$, $1 \leq p < \infty$. Then, ideals are of the form

$$I = \Big\{ f \in L_p(\chi, d\mu) \ : \ f = 0 \text{ a.e. on } Y \Big\},$$

where Y is a measurable subset of χ.

Definition 1.7.6 *A topological vector space X, which is also a vector lattice, is said to have a quasi-interior point, if there is $f \in X_+$ such that the ideal generated by f is dense in X.*

It is known that, if μ is σ-finite, then each of the Banach lattices $L_p(\chi, (\mu))$, $1 \leq p \leq \infty$ possesses quasi-interior points. In this case, $f \in X_+$ is a quasi-interior point means that $f > 0$ μ-a.e. For more details on cones and positive cones and their properties, the reader can refer to the work of D. Guo and V. Lakshmikantham [92] and of S. Heikkilä and V. Lakshmikantham [94].

Theorem 1.7.1 *(Hahn–Banach) Let X be a topological vector space over $\mathbb{R}$, let O be an open, convex, and nonempty subset of X, and let F be a subspace of X such that $F \cap O = \emptyset$. Then, there exists a closed subspace H of codimension 1 such that $H \supseteq F$ and $H \cap O = \emptyset$.*

Corollary 1.7.1 *Let X be a topological vector space over $\mathbb{K}$, let $\mathcal{C} \neq \emptyset$ be open, convex, and let $O \neq \emptyset$ be a convex subset of X. If $\mathcal{C} \cap O = \emptyset$, then there exists a continuous functional φ on X and an $\alpha \in \mathbb{R}$ such that, for the real part, $\Re(\varphi(.))$, of $\varphi(.)$*

$$\forall c \in \mathcal{C} \ \forall d \in O : \Re(\varphi(c)) < \alpha \leq \Re(\varphi(d)).$$

Corollary 1.7.2 *Let X be a topological vector space, let $\mathcal{C}$ be a compact, convex, nonempty set, and let $\mathcal{D}$ be a closed, nonempty, and convex set. If $\mathcal{C} \cap \mathcal{D} = \emptyset$, then there exists $\varphi \in X'$, $\alpha, \beta \in \mathbb{R}$ such that*

$$\forall c \in \mathcal{C}\ \forall d \in \mathcal{D} : \Re(\varphi(c)) \leq \alpha < \beta \leq \Re(\varphi(d)).$$

Chapter 2

Fixed Point Theory under Weak Topology

In the first part of this chapter, we give some variants of the Schauder and Krasnosel'skii fixed point theorems in Dunford–Pettis spaces for weakly compact operators. Precisely, if an operator A acting on a Banach space X having the property of Dunford–Pettis, leaves a subset $\mathcal{M}$ of X invariant, then A has, at least, a fixed point in $\mathcal{M}$. In addition, if B is a contraction map of $\mathcal{M}$ into X, $Ax + By \in \mathcal{M}$ for x, y in $\mathcal{M}$ and if $(I - B)^{-1}A$ is a weakly compact operator, then $A + B$ has, at least, a fixed point in $\mathcal{M}$. Both of these two theorems can be used to resolve some open problems (see Chapter 5). In the second part of this chapter, we establish new variants of fixed point theorems in general Banach spaces. Furthermore, nonlinear Leray–Schauder alternatives for the sum of two weakly sequentially continuous mappings are presented. This notion of weakly sequential continuity seems to be the most convenient in use. Moreover, it is not always possible to show that a given operator between Banach spaces is weakly continuous. Quite often, its weakly sequential continuity presents no problem. Finally, we establish fixed point theorems for multi-valued maps with weakly sequentially closed graphs.

2.1 Fixed Point Theorems in DP Spaces and Weak Compactness

The aim of this section is to give the Schauder and Krasnosel'skii fixed point theorems in the case of Dunford–Pettis spaces to the class of weakly compact operators. We first give an extension form of Schauder's theorem. Then, we present other results which follow by quite simple arguments.

2.1.1 Schauder's fixed point theorem in DP spaces

Let us first recall some nice results due to A. Ben Amar, A. Jeribi, and M. Mnif proved in [32].

Theorem 2.1.1 *Let X be a Dunford–Pettis space, $\mathcal{M}$ be a nonempty, closed, bounded, and convex subset of X and let A be a weakly compact linear operator on X. If A leaves $\mathcal{M}$ invariant, then A has, at least, a fixed point in $\mathcal{M}$.*

Proof. Let $\mathcal{N} = \overline{co}(A(\mathcal{M}))$ be the closed convex hull of $A(\mathcal{M})$. Since $\mathcal{M}$ is a closed convex subset and $A(\mathcal{M}) \subset \mathcal{M}$, then $\mathcal{N} \subset \mathcal{M}$. Therefore,

$$A(\mathcal{N}) \subset A(\mathcal{M}) \subset \overline{co}(A(\mathcal{M})),$$

i.e., A maps $\mathcal{N}$ into itself. Since A is a weakly compact operator on X, it follows that $\overline{A(\mathcal{M})}^w$ is weakly compact. Moreover, by using $\overline{co}(A(\mathcal{M})) \subset \overline{co}(\overline{A(\mathcal{M})}^w)$ and the Krein–Šmulian's theorem (see Theorem 1.3.5), it follows that $\overline{co}(\overline{A(\mathcal{M})}^w)$ is a weakly compact set. Hence, according to Mazur's theorem (see Theorem 1.3.4) one sees that $\mathcal{N}$ is a weakly compact set. We claim that $A(\mathcal{N})$ is compact in X. To see this, let $(\psi_n)_n$ be a sequence in $A(\mathcal{N})$. There is a sequence $(\varphi_n)_n$ in $\mathcal{N}$ such that $\psi_n = A(\varphi_n)$ for each n. Since $\mathcal{N}$ is a weakly compact set, it follows that $(\varphi_n)_n$ has a weak converging subsequence $(\varphi_{n_p})_p$. From the weak compactness of A, it follows that $(A(\varphi_{n_p}))_p$ converges strongly in X. Then, $(\psi_{n_p})_p$ converges strongly in X. As a result, $A(\mathcal{N})$ is a compact set in X. Finally, the use of Schauder's fixed point theorem (see Theorem 1.6.1) shows that A has, at least, one fixed point in $\mathcal{N}$. Q.E.D.

Remark 2.1.1 *Since $\mathcal{K}(X) \subset \mathcal{W}(X)$, Theorem 2.1.1 is a new variant of the Schauder's fixed point theorem in the Dunford–Pettis space under the weak topology.*

Corollary 2.1.1 *Let X be a Dunford–Pettis space and $\mathcal{M}$ be a nonempty, closed, bounded, and convex subset of X. Let $B : X \longrightarrow X$ be a continuous map and C be a weakly compact linear operator on X. If $A = BC$ is a weakly compact operator on X with A leaving $\mathcal{M}$ invariant, then A has, at least, a fixed point in $\mathcal{M}$.*

Proof. Let $\mathcal{N} = \overline{co}(A(\mathcal{M}))$. Since A is a weakly compact operator on X, we find that $\mathcal{N}$ is weakly compact. Now, arguing as in the proof of Theorem 2.1.1, we can see that $C(\mathcal{N})$ is a compact set in X. Since B is a continuous map, then $BC(\mathcal{N})$ is a compact set in X. Finally, the Schauder's fixed point theorem shows that A has, at least, one fixed point in $\mathcal{N}$. Q.E.D.

Corollary 2.1.2 *Let X be a Dunford–Pettis space and $\mathcal{M}$ be a nonempty, closed, bounded, and convex subset of X. Let B be a weakly compact linear operator on X and let $C : X \longrightarrow X$ be continuous and maps bounded sets into bounded ones, and weakly compact sets into weakly compact ones. If $A = BC$ leaves $\mathcal{M}$ invariant, then A has, at least, a fixed point in $\mathcal{M}$.*

Proof. Let $\mathcal{N} = \overline{co}(A(\mathcal{M}))$. Since $C(\mathcal{M})$ is a bounded set and B is a weakly compact operator, we find that $\mathcal{N}$ is a weakly compact set. We claim that $A(\mathcal{N})$ is a compact set in X. To see this, let $(\psi_n)_n$ be a sequence in $A(\mathcal{N})$. There is a sequence $(\varphi_n)_n$ in $\mathcal{N}$ such that $\psi_n = A(\varphi_n)$ for each n. Since $\mathcal{N}$ is weakly compact, it follows that $(\varphi_n)_n$ has a weak converging subsequence $(\varphi_{n_p})_p$. Hence, $(C(\varphi_{n_p}))_p$ is weakly convergent. According to the weak compactness of B, it follows that $(BC(\varphi_{n_p}))_p$ converges strongly in X. Then, $(\psi_{n_p})_p$ converges strongly in X. Therefore, $A(\mathcal{N})$ is a compact set in X. Finally, the use of Schauder's fixed point theorem shows that A has, at least, one fixed point in $\mathcal{N}$. Q.E.D.

2.1.2 Krasnosel'skii's fixed point theorem in DP spaces

Let us consider the sum of a weakly compact operator and a contraction mapping.

Theorem 2.1.2 *Let X be a Dunford–Pettis space, and let $\mathcal{M}$ be a nonempty, closed, bounded, and convex subset of X. Suppose that A and B map $\mathcal{M}$ into X such that :*

(i) $A \in \mathcal{L}(X)$ and is weakly compact,

(ii) B is a contraction mapping,

(iii) $Ax + By \in \mathcal{M}$ for all x, y in $\mathcal{M}$, and

(iv) $(I - B)^{-1}A(\mathcal{M})$ is a weakly compact set.

Then, there exists y in $\mathcal{M}$ such that $Ay + By = y$.

Proof. For each $y \in \mathcal{M}$, the equation

$$z = Bz + Ay$$

has a unique solution $z \in \mathcal{M}$, since $z \longrightarrow Bz + Ay$ defines a contraction mapping of $\mathcal{M}$ into $\mathcal{M}$. Thus, $z = (I - B)^{-1}Ay$ is in $\mathcal{M}$. Now, the use of

Corollary 2.1.1 for the operators A and $(I - B)^{-1}$, shows that $(I - B)^{-1}A$ has, at least, a fixed point in $\mathcal{M}$. This point y is the one required. Q.E.D.

Remark 2.1.2 *(i) The Theorem 2.1.2 is a new variant of the Krasnosel'skii's fixed point theorem in the Dunford–Pettis spaces under the weak topology.*
(ii) Note that Theorem 2.1.2 remains true if we replace the assumption (iii) by a weaker one; (if $u = Bu + Av$ with $v \in M$, then $u \in M$).

In the remaining part of this section, we will briefly discuss the existence of positive solutions. Let X_1 and X_2 be two Banach lattice spaces, with positive cônes X_1^+ and X_2^+, respectively. An operator T from X_1 into X_2 is said to be positive if it carries the positive cône X_1^+ into X_2^+ (i.e., $T(X_1^+) \subset X_2^+$).

Theorem 2.1.3 *Let X be a Dunford–Pettis space, $\mathcal{M}$ be a nonempty, closed, bounded, and convex subset of X such that $\mathcal{M} \cap X^+ \neq \emptyset$ and A is a positive weakly compact linear operator on X. If A leaves $\mathcal{M}$ invariant, then A has, at least, a positive fixed point in $\mathcal{M}$.*

Proof. Let $\mathcal{M}^+ = \mathcal{M} \cap X^+$, then $\mathcal{M}^+$ is a closed, bounded, and convex subset of X^+ and $A(\mathcal{M}^+) \subset \mathcal{M}^+$. Let $\mathcal{N}^+ = \overline{co}(A(\mathcal{M}^+))$. Since $\mathcal{M}^+$ is a closed convex subset and $A(\mathcal{M}^+) \subset \mathcal{M}^+$, we get $\mathcal{N}^+ \subset \mathcal{M}^+$ and therefore,

$$A(\mathcal{N}^+) \subset A(\mathcal{M}^+) \subset \overline{co}(A(\mathcal{M}^+)) = \mathcal{N}^+,$$

i.e., A maps $\mathcal{N}^+$ into itself. Now, the rest of the proof is similar to that of Theorem 2.1.1; it is sufficient to replace the set $\mathcal{N}$ by $\mathcal{N}^+$. Q.E.D.

2.2 Banach Spaces and Weak Compactness

The purpose of this section is to extend the results of the previous section to general Banach spaces under weak topology conditions.

2.2.1 Schauder's fixed point theorem

Theorem 2.2.1 *Let X be a Banach space, $\mathcal{M}$ be a nonempty, closed, and convex subset of X and let $A : \mathcal{M} \longrightarrow \mathcal{M}$ be a weakly sequentially continuous map. If $A(\mathcal{M})$ is relatively weakly compact, then A has, at least, a fixed point in $\mathcal{M}$.*

Proof. Let $\mathcal{N} = \overline{co}(A(\mathcal{M}))$ be the closed convex hull of $A(\mathcal{M})$. Since $A(\mathcal{M})$ is relatively weakly compact, then $\mathcal{N}$ is a weakly compact convex subset of X. Moreover,

$$A(\mathcal{N}) \subset A(\mathcal{M}) \subset \overline{co}(A(\mathcal{M})) = \mathcal{N},$$

i.e., A maps $\mathcal{N}$ into itself. Since A is weakly sequentially continuous, and by using the O. Arino, S. Gautier, and J. P. Penot theorem (see Theorem 1.6.9), it follows that A has, at least, one fixed point in $\mathcal{N}$. Q.E.D.

Corollary 2.2.1 *Let X be a Banach space and $\mathcal{M}$ be a nonempty, closed, bounded, and convex subset of X. Let us assume that $A = BC$, where B is a linear weakly compact operator on X and C is a nonlinear weakly sequentially continuous operator, which maps bounded sets into bounded sets. If $A(\mathcal{M}) \subset \mathcal{M}$, then A has, at least, a fixed point in $\mathcal{M}$.*

Proof. Thanks to Theorem 2.2.1, it is sufficient to show that the operator A is weakly sequentially continuous and that $A(\mathcal{M})$ is relatively weakly compact. Indeed, the fact that B is linear and weakly compact implies that B is weakly sequentially continuous. Hence, A is weakly sequentially continuous. From the boundedness of $\mathcal{M}$ and the properties of C, we have $C(\mathcal{M})$ is bounded. Since B is a weakly compact operator, it follows that $A(\mathcal{M})$ is relatively weakly compact. Q.E.D.

Corollary 2.2.2 *Let X be a Banach space and $\mathcal{M}$ be a nonempty, closed, bounded, and convex subset of X. Let us assume that $A = CB$, where B is a linear weakly compact operator on X and C is a weakly sequentially continuous operator. If $A(\mathcal{M}) \subset \mathcal{M}$, then A has, at least, a fixed point in $\mathcal{M}$.*

Proof. Arguing as in the proof of Corollary 2.2.1, we deduce that A is weakly sequentially continuous. We claim that $A(\mathcal{M})$ is relatively weakly compact. To see this, let $(y_n)_n$ be a sequence in $A(\mathcal{M})$. There is a sequence $(x_n)_n$ in $\mathcal{M}$ such that $y_n = Ax_n$ for each n. Since $\mathcal{M}$ is bounded and B is a weakly compact operator, $(Bx_n)_n$ has a weak converging subsequence $(Bx_{\varphi(n)})_n$. Since C is weakly sequentially continuous, we have $(CBx_{\varphi(n)})_n$ is weakly convergent. So, $(y_{\varphi(n)})_n$ is weakly convergent. Therefore, $A(\mathcal{M})$ is relatively sequentially weakly compact. Using the Eberlein–Šmulian theorem (see Theorem 1.3.3), we deduce the relatively weak compactness of $A(\mathcal{M})$. This proves the claim. Now, the result follows immediately from Theorem 2.2.1. Q.E.D.

Now, we may briefly discuss the existence of positive solutions.

Theorem 2.2.2 *Let X be a Banach space and let $\mathcal{M}$ be a nonempty, closed, and convex subset of X such that $\mathcal{M} \cap X^+ \neq \emptyset$ and A is a positive weakly sequentially continuous map. If A leaves $\mathcal{M}$ invariant and if $A(\mathcal{M})$ is relatively weakly compact, then A has, at least, a positive fixed point in $\mathcal{M}$.*

Proof. Let $\mathcal{M}^+ = \mathcal{M} \cap X^+$, then $\mathcal{M}^+$ is a closed convex subset of X^+. Let $\mathcal{N}^+ = \overline{co}(A(\mathcal{M}^+))$. Since $\mathcal{M}^+$ is a closed convex subset and $A(\mathcal{M}^+) \subset \mathcal{M}^+$, $\mathcal{N}^+ \subset \mathcal{M}^+$ and therefore,

$$A(\mathcal{N}^+) \subset A(\mathcal{M}^+) \subset \overline{co}(A(\mathcal{M}^+)) = \mathcal{N}^+,$$

i.e., A maps $\mathcal{N}^+$ into itself. Moreover, using the fact that $A(\mathcal{M}^+)$ is relatively weakly compact, it follows, by using the Krein–Šmulian's theorem (see Theorem 1.3.5), that $\mathcal{N}^+$ is a weakly compact set. Now, the result follows from O. Arino, S. Gautier, and J. P. Penot's theorem (see Theorem 1.6.9). Q.E.D.

2.2.2 Krasnosel'skii's fixed point theorem

Now, we may establish a fixed point theorem which combines both the Banach contraction mapping principle and Theorem 2.2.1.

Theorem 2.2.3 *Let X be a Banach space and let $\mathcal{M}$ be a nonempty, closed, and convex subset of X. Let us suppose that $A : \mathcal{M} \longrightarrow X$ and $B : X \longrightarrow X$ such that :*

(i) B is a contraction mapping,

(ii) $(I - B)^{-1}A$ is weakly sequentially continuous,

(iii) $(I - B)^{-1}A(\mathcal{M})$ is relatively weakly compact, and

(iv) $Ax + By \in \mathcal{M}$ for all $x, y \in \mathcal{M}$.

Then, there is $x \in \mathcal{M}$ such that $x = Ax + Bx$.

Proof. Thanks to Theorem 2.2.1, it is sufficient to show that $(I - B)^{-1}A$ maps $\mathcal{M}$ into itself. In fact, for each $y \in \mathcal{M}$, the equation $z = Ay + Bz$ has a unique solution $z \in \mathcal{M}$, since $z \longrightarrow Bz + Ay$ defines a contraction mapping of $\mathcal{M}$ into $\mathcal{M}$. Thus, $z = (I - B)^{-1}Ay$ is in $\mathcal{M}$. Q.E.D.

Corollary 2.2.3 *Let X be a Banach space and let $\mathcal{M}$ be a nonempty, closed, and convex subset of X. Suppose that $A : \mathcal{M} \longrightarrow X$ and $B : X \longrightarrow X$ such that :*

(i) A *is weakly sequentially continuous,*

(ii) B *is a contraction and weakly sequentially continuous,*

(iii) $(I-B)^{-1}A(\mathcal{M})$ *is relatively weakly compact, and*

(iv) $Ax + By \in \mathcal{M}$ *for all* $x,\ y \in \mathcal{M}$.

Then, there is $x \in \mathcal{M}$ *such that* $x = Ax + Bx$.

Proof. Arguing as in the proof of Theorem 2.2.3, we have $(I-B)^{-1}A(\mathcal{M}) \subset \mathcal{M}$. Thanks to Theorem 2.2.3, it is sufficient to prove that $(I-B)^{-1}A$ is weakly sequentially continuous. Indeed, let $(u_n)_n$ be a sequence in $\mathcal{M}$ such that $u_n \rightharpoonup u$ in $\mathcal{M}$. Since $\{(I-B)^{-1}Au_n,\ n \in \mathbb{N}\} \subset (I-B)^{-1}A(\mathcal{M})$, by assumption (iii) we get a subsequence $(u_{\rho(n)})_n$ such that $(I-B)^{-1}Au_{\rho(n)} \rightharpoonup v$ in $\mathcal{M}$. The sequential weak continuity of B leads to $B(I-B)^{-1}Au_{\rho(n)} \rightharpoonup Bv$. Also, from the equality

$$B(I-B)^{-1}A = -A + (I-B)^{-1}A,$$

it follows that $-Au_{\rho(n)} + (I-B)^{-1}Au_{\rho(n)} \rightharpoonup -Au + v$. So, $v = (I-B)^{-1}Au$. We claim that $(I-B)^{-1}Au_n \rightharpoonup (I-B)^{-1}Au$. Suppose that this is not the case, then there exists a subsequence $(u_{\rho_1(n)})_n$ and a weak neighborhood V^w of $(I-B)^{-1}Au$ such that $(I-B)^{-1}Au_{\rho_1(n)} \notin V^w$ for all $n \in \mathbb{N}$. Moreover, we have $u_{\rho_1(n)} \rightharpoonup u$, then arguing as before, we find a subsequence $(u_{\rho_1(\rho_2(n))})_n$ such that $(I-B)^{-1}Au_{\rho_1(\rho_2(n))} \rightharpoonup (I-B)^{-1}Au$, which is a contradiction. This proves the claim and achieves the proof. Q.E.D.

2.3 Fixed Point Theorems and MNWC

Throughout this section, X denotes a Banach space. For any $r > 0$, B_r denotes the closed ball in X centered at 0_X with radius r. Ω_X is the collection of all nonempty bounded subsets of X and $\mathcal{K}^w$ is the subset of Ω_X consisting of all weakly compact subsets of X. Recall that the notion of the measure of weak noncompactness (in short MNWC) was introduced by De Blasi [64]; it is the map $\omega : \Omega_X \longrightarrow [0, +\infty)$ defined in the following way:

$$\omega(M) = \inf\{r > 0 : \text{ there exits } K \in \mathcal{K}^w \text{ such that } M \subset K + B_r\}, \quad (2.1)$$

for all $M \in \Omega_X$. For more convenience, we recall the basic properties of $\omega(.)$ presented in Lemma 1.4.1 and needed below.

Lemma 2.3.1 *Let M_1, M_2 be two elements of Ω_X. Then, the following conditions are satisfied:*

(1) *$M_1 \subset M_2$ implies $\omega(M_1) \leq \omega(M_2)$.*

(2) *$\omega(M_1) = 0$ if, and only if, $\overline{M_1}^w \in \mathcal{K}^w$.*

(3) *$\omega(\overline{M_1}^w) = \omega(M_1)$.*

(4) *$\omega(M_1 \cup M_2) = \max\{\omega(M_1), \omega(M_2)\}$.*

(5) *$\omega(\lambda M_1) = |\lambda|\omega(M_1)$ for all $\lambda \in \mathbb{R}$.*

(6) *$\omega(co(M_1)) = \omega(M_1)$, i.e., $co(M_1)$ is the convex hull of M_1.*

(7) *$\omega(M_1 + M_2) \leq \omega(M_1) + \omega(M_2)$.*

(8) *if $(M_n)_{n\geq 1}$ is a decreasing sequence of nonempty, bounded, and weakly closed subsets of X with $\lim\limits_{n\to\infty} \omega(M_n) = 0$, then $M_\infty := \cap_{n=1}^{\infty} M_n$ is nonempty and $\omega(M_\infty) = 0$, i.e., M_∞ is relatively weakly compact.*

2.3.1 Sum of two weakly sequentially continuous mappings

At the beginning of this section, we will state some new fixed point theorems of the Krasnosel'skii type for the sum of two weakly sequentially continuous mappings. First, let us introduce the following definition:

Definition 2.3.1 *A map $A : M \subset X \longrightarrow X$ is said to be ω-contractive (or ω-contraction) if it maps bounded sets into bounded sets, and there exist some $\alpha \in [0, 1)$ such that $\omega(A(N)) \leq \alpha\omega(N)$ for all bounded subsets $N \subset M$.*

Remark 2.3.1 *Notice that every weakly sequentially continuous nonlinear contraction is ω-condensing.*

The first result is formulated as:

Theorem 2.3.1 *Let M be a nonempty, bounded, closed, and convex subset of a Banach space X. Suppose that $A : M \longrightarrow X$ and $B : X \longrightarrow X$ are such that:*

(*i*) *A is weakly sequentially continuous,*

(*ii*) *there exists $\alpha \in [0, 1)$ such that $\omega(A(N) + B(N)) \leq \alpha\omega(N)$ for all $N \subset M$,*

(iii) B is a strict contraction and weakly sequentially continuous, and

(iv) $(x = Bx + Ay,\ y \in M) \Longrightarrow x \in M$.

Then, $A + B$ has, at least, a fixed point in M.

Proof. Since B is a contraction with a constant $k \in (0, 1)$, it follows by Lemma 1.2.2 that the mapping $I - B$ is a homeomorphism from X into X. Let y be fixed in M, the map which assigns to each $x \in X$ the value $Bx + Ay$ defines a contraction from X into X. So, by the Banach fixed point theorem, the equation $x = Ax + By$ has a unique solution $x \in X$. By hypothesis (iv) we have $x \in M$. Hence, $x = (I - B)^{-1}Ay \in M$ which, accordingly, implies the inclusion

$$(I - B)^{-1}A(M) \subset M. \tag{2.2}$$

Now, let us define the sequence $(M_n)_{n\geq 1}$ of subsets of M by:

$$M_1 = M \quad and \quad M_{n+1} = \overline{co}\left((I - B)^{-1}A(M_n)\right). \tag{2.3}$$

We claim that the sequence $(M_n)_{n\geq 1}$ satisfies the condition of property (8) of $\omega(.)$(see Lemma 2.3.1). Indeed, it is clear that the sequence $(M_n)_{n\geq 1}$ consists of nonempty closed convex subsets of M. Using Eq. (2.3), we notice that it is also decreasing. Now, using Eq. (2.2) and the following equality:

$$(I - B)^{-1}A = A + B(I - B)^{-1}A, \tag{2.4}$$

we obtain the inclusions

$$(I - B)^{-1}A(M_n) \subset A(M_n) + B\left(\overline{co}\left((I - B)^{-1}A(M_n)\right)\right) \subset A(M_n) + B(M_n). \tag{2.5}$$

Combining Eq. (2.5) with the properties (1) and (6) of Lemma 1.4.1, we get

$$\omega(M_{n+1}) = \omega(\overline{co}(I-B)^{-1}A(M_n)) = \omega((I-B)^{-1}A(M_n)) \leq \omega(A(M_n)+B(M_n)).$$

Moreover, the use of the assumption (ii) leads to

$$\omega(M_{n+1}) \leq \alpha\omega(M_n).$$

Proceeding by induction, we get

$$\omega(M_n) \leq \alpha^{n-1}\omega(M),$$

and therefore $\lim\limits_{n\to\infty} \omega(M_n) = 0$, because $\alpha \in [0, 1)$. Now, applying property (8) of Lemma 1.4.1, we infer that $M_\infty := \cap_{n=1}^{\infty} M_n$ is a nonempty, closed,

convex, weakly compact subset of M. Moreover, we can easily verify that $(I-B)^{-1}A(M_n) \subset M_n$, for all $n \geq 1$, thus we get $(I-B)^{-1}A(M_\infty) \subset M_\infty$. Consequently, $(I-B)^{-1}A(M_\infty)$ is relatively weakly compact. Now, let us show that $(I-B)^{-1}A : M_\infty \longrightarrow M_\infty$ is weakly sequentially continuous.
Let $(x_n)_{n\geq 1}$ be a sequence in M_∞ such that $x_n \rightharpoonup x$ in M_∞. Since

$$((I-B)^{-1}Ax_n)_{n\geq 1} \subset (I-B)^{-1}A(M_\infty),$$

we get a subsequence $(x_{\rho(n)})_{n\geq 1}$ of $(x_n)_{n\geq 1}$ such that $(I-B)^{-1}Ax_{\rho(n)} \rightharpoonup y$ in M_∞. Going back to Eq. (2.4), the weak sequential continuity of the maps A and B yields to $y = By + Ax$ and thus $y = (I-B)^{-1}Ax$.
We claim that:

$$(I-B)^{-1}Ax_n \rightharpoonup (I-B)^{-1}Ax.$$

Suppose the contrary, then there exists a subsequence $(x_{\rho_1(n)})_{n\geq 1}$ and a weak neighborhood V^w of $(I-B)^{-1}Ax$ such that $(I-B)^{-1}Ax_{\rho_1(n)} \notin V^w$ for all $n \geq 1$. Moreover, $x_{\rho_1(n)} \rightharpoonup x$, then arguing as before, we find a subsequence $(x_{\rho_1(\rho_2(n))})_{n\geq 1}$ of $(x_{\rho_1(n)})_{n\geq 1}$ such that

$$(I-B)^{-1}Ax_{\rho_1(\rho_2(n))} \rightharpoonup (I-B)^{-1}Ax$$

which is absurd, since $(I-B)^{-1}Ax_{\rho_1(\rho_2(n))} \notin V^w$ for all $n \geq 1$. Finally, $(I-B)^{-1}A$ is weakly sequentially continuous. Now, applying the Arino–Gautier–Penot fixed point theorem (see Theorem 1.6.9), we conclude that $(I-B)^{-1}A$ has, at least, one fixed point $x \in M_\infty$, that is, $Ax + Bx = x$. This completes the proof. Q.E.D.

Remark 2.3.2 *The result in Theorem 2.3.1 remains valid for any arbitrary measure of weak noncompactness on X.*

It should be noticed that the case $B = 0$ in Theorem 2.3.1 corresponds to the following well-known result of D. O'Regan.

Corollary 2.3.1 *Let M be a nonempty, bounded, closed, and convex subset of a Banach space X. Assume that $A : M \longrightarrow M$ is weakly sequentially continuous and ω-contractive, then A has, at least, a fixed point in M.*

Corollary 2.3.2 *Let M be a nonempty, bounded, closed, and convex subset of a Banach space X. Suppose that $A : M \longrightarrow X$ and $B : X \longrightarrow X$ are two mappings such that:*

(i) A is weakly sequentially continuous and $A(M)$ is relatively weakly compact,

(ii) B is a strict contraction with a constant k and is weakly sequentially continuous, and

(iii) $(x = Bx + Ay,\ y \in M) \Longrightarrow x \in M$.

Then, $A + B$ has, at least, a fixed point in M.

Proof. Let $N \subset M$. In view of Theorem 2.3.1, it is sufficient to establish that

$$\omega(A(N) + B(N)) \leq k\omega(N).$$

Taking into account that $A(M)$ is relatively weakly compact and using the subadditivity of the De Blasi measure of weak noncompactness, we get

$$\omega(A(N) + B(N)) \leq \omega(A(N)) + \omega(B(N)) \leq \omega(B(N)). \tag{2.6}$$

Now, let $r > 0$ and let K be a weakly compact subset of M such that $N \subset K + B_r$. We show that

$$B(N) \subset B(K) + B_{kr}.$$

To see this, let $x \in N$. Then, there is a $y \in K$ such that $\|x - y\| \leq r$. Since B is a strict contraction with a constant k then, $\|Bx - By\| \leq k\|x - y\| \leq kr$. As a result, $Bx - By \in B_{kr}$. Hence, $Bx \in B(K) + B_{kr}$. Accordingly,

$$B(N) \subset B(K) + B_{kr} = \overline{B(K)}^{w} + B_{kr},$$

where $\overline{B(K)}^{w}$ is the weak closure of $B(K)$. By hypothesis, B is weakly sequentially continuous and K is weakly compact. Hence, $B(K)$ is weakly compact. By using Eq. (2.1), we show that:

$$\omega(B(N)) \leq k\omega(N). \tag{2.7}$$

Now, combining Eqs. (2.6) and (2.7), we get $\omega(A(N) + B(N)) \leq k\omega(N)$. Q.E.D.

Corollary 2.3.3 *Let M be a nonempty, bounded, closed, and convex subset of a Banach space X. Assume that $A : M \longrightarrow M$ is weakly sequentially continuous. If $A(M)$ is relatively weakly compact, then A has, at least, a fixed point in M.*

Theorem 2.3.2 *Let Ω be a nonempty, closed, bounded, and convex subset of a Banach space X. In addition, let $A : \Omega \longrightarrow X$ be a weakly sequentially continuous mapping and $B : X \longrightarrow X$ satisfying:*

(i) $A(\Omega)$ is relatively weakly compact,

(ii) B is linear, bounded and there exists $p \in \mathbb{N}^*$ such that B^p is a separate contraction, and

(iii) $(x = Bx + Ay,\ y \in \Omega) \Longrightarrow x \in \Omega$.

Then, there exists $x \in \Omega$ such that $x = Ax + Bx$.

Proof. Since B is linear, bounded and B^p is a separate contraction, it follows by Lemma 1.2.2 that $(I - B^p)^{-1}$ exists on X. Hence,

$$(I - B)^{-1} = (I - B^p)^{-1} \sum_{k=0}^{p-1} B^k. \tag{2.8}$$

By Eq. (2.8), we have $(I - B)^{-1} \in \mathcal{L}(X)$, so $(I - B)^{-1}$ is weakly continuous. Define the mapping $F := (I - B)^{-1}A$. Since $A : \Omega \longrightarrow X$, then from assumption *(iii)* it follows that $F : \Omega \longrightarrow \Omega$. Since $(I - B)^{-1}$ is weakly continuous and A is weakly sequentially continuous, so F is weakly sequentially continuous. Moreover, we have A maps bounded sets into relatively weakly compact sets and $(I - B)^{-1}$ is weakly continuous, then F maps bounded sets into relatively weakly compact sets. Hence, F fulfills the conditions of Corollary 2.3.3. Q.E.D.

We should notice that Theorem 2.3.2 remains true if we suppose that there exists $p \in \mathbb{N}^*$ such that B^p is a nonlinear contraction.

Theorem 2.3.3 *Let Ω be a closed, bounded, and convex subset of a Banach space X. In addition, let $A : \Omega \longrightarrow X$ be a weakly sequentially continuous mapping and $B : X \longrightarrow X$ satisfying:*

(i) $A(\Omega)$ is relatively weakly compact,

(ii) B is linear, bounded and there exists $p \in \mathbb{N}^*$ such that B^p is a nonlinear contraction, and

(iii) $(x = Bx + Ay,\ y \in \Omega) \Longrightarrow x \in \Omega$.

Then, there exists $x \in \Omega$ such that $x = Ax + Bx$.

Proof. Since B is linear, bounded and B^p is a nonlinear contraction, then $(I - B^p)^{-1}$ exists on X. Now, reasoning as in the proof of Theorem 2.3.2, we get the desired result. Q.E.D.

Remark 2.3.3 *Since nonlinear contraction mappings do not generate separate contraction mappings, so Theorems 2.3.2 and 2.3.3 are two different new generalizations of Krasnoselskii's fixed point theorem.*

Theorem 2.3.4 *Let Ω be a nonempty, convex, and closed set in a Banach space X. Assume that $F : \Omega \longrightarrow \Omega$ is a weakly sequentially continuous map and condensing with respect to ω. In addition, suppose that $F(\Omega)$ is bounded. Then, F has, at least, a fixed point.*

Proof. Let $x_0 \in \Omega$. We consider the family $\mathcal{F}$ of all closed bounded convex subsets D of Ω such that $x_0 \in D$ and $F(D) \subset D$. Obviously, $\mathcal{F}$ is nonempty, since $\overline{conv}(F(\Omega) \cup \{x_0\}) \in \mathcal{F}$. We denote

$$K = \bigcap_{D \in \mathcal{F}} D.$$

Let us notice that K is closed and convex and $x_0 \in K$. If $x \in K$ then, $Fx \in D$ for all $D \in \mathcal{F}$ and hence, $F(K) \subset K$. Therefore, $K \in \mathcal{F}$. We will prove that K is weakly compact. Denoting by:

$$K_* = \overline{conv}(F(K) \cup \{x_0\}),$$

we have $K_* \subset K$, which implies that $F(K_*) \subset F(K) \subset K_*$. Therefore, $K_* \in \mathcal{F}$, $K \subset K_*$. Hence, $K = K_*$. Since K is weakly closed, it is sufficient to show that K is relatively weakly compact. If $\omega(K) > 0$, we get

$$\omega(K) = \omega(\overline{conv}(F(K) \cup \{x_0\})) \leq \omega(F(K)) < \omega(K)$$

which is a contradiction. Hence, $\omega(K) = 0$ and so, K is relatively weakly compact. Now, F is weakly sequentially continuous of K into itself. Applying Theorem 1.6.9, we conclude that F has, at least, a fixed point in $K \subset \Omega$. Q.E.D.

Theorem 2.3.5 *Let Ω be a nonempty, bounded, closed, and convex subset of a Banach space $(X, \|.\|)$. Suppose that $A : \Omega \longrightarrow X$ and $B : X \longrightarrow X$ are two weakly sequentially continuous mappings such that:*

(i) A is weakly compact,

(ii) B is a nonlinear contraction, and

(iii) $(A + B)(\Omega) \subset \Omega$.

Then, there exists $x \in \Omega$ such that $x = Ax + Bx$.

Proof. First, we claim that B is ω-condensing. Indeed, let D be a bounded subset of X such that $\omega(D) = d > 0$. Let $\varepsilon > 0$, then there exists a weakly compact set K of X satisfying $D \subseteq K + B_{d+\varepsilon}$. So, for $x \in D$ there exists $y \in K$ and $z \in B_{d+\varepsilon}$ such that $x = y + z$ and so

$$\|Bx - By\| \leq \varphi(\|x - y\|) \leq \varphi(\varepsilon + d).$$

It follows immediately, that

$$B(D) \subseteq B(K) + B_{\varphi(\varepsilon+d)}.$$

Moreover, since B is a weakly sequentially continuous mapping and K is weakly compact, then $B(K)$ is weakly compact. Therefore,

$$\omega(B(D)) \leq \varphi(d + \varepsilon).$$

Since $\varepsilon > 0$ is arbitrary, then

$$\omega(B(D)) \leq \varphi(d) < d = \omega(D).$$

Hence, B is ω-condensing, which ends the proof of the claim. On the other hand, it is easy to see that $A + B$ is weakly sequentially continuous. Thanks to Theorem 2.3.4, it suffices to show that $A + B$ is ω-condensing. To see this, let D be a bounded subset of Ω. Taking into account the fact that $A(D)$ is relatively weakly compact and using the subadditivity of the De Blasi measure of weak noncompactness we get

$$\omega((A + B)(D)) \leq \omega(A(D) + B(D)) \leq \omega(A(D)) + \omega(B(D)) \leq \omega(B(D)).$$

So, if $\omega(B(D) \neq 0$ then

$$\omega((A + B)(D)) < \omega(D),$$

and hence $A + B$ is ω-condensing, which ends the proof of the theorem. Q.E.D.

We point out that Theorem 2.3.5 remains valid if we replace the assumption

$$(A + B)(\Omega) \subset \Omega$$

with the following one

$$(x = Bx + Ay,\ y \in \Omega) \Longrightarrow x \in \Omega.$$

Theorem 2.3.6 *Let Ω be a nonempty, bounded, closed, and convex subset of a Banach space X. Suppose that $A : \Omega \longrightarrow X$ and $B : X \longrightarrow X$ are two weakly sequentially continuous mappings such that:*

(i) A is weakly compact,

(ii) B is a nonlinear contraction, and

(iii) $(x = Bx + Ay,\ y \in \Omega) \Longrightarrow x \in \Omega$.

Then, there exists $x \in \Omega$ such that $x = Ax + Bx$.

Proof. Let y be fixed in Ω. The map which assigns to each $x \in \Omega$ the value $Bx + Ay$ defines a nonlinear contraction from Ω into Ω. So, using Theorem 1.6.10 together with assumption *(iii)*, the equation $x = Bx + Ay$ has a unique solution $x = (I - B)^{-1}Ay \in \Omega$. Therefore,

$$(I - B)^{-1}A(\Omega) \subset \Omega.$$

Now, define the mapping $F : \Omega \longrightarrow \Omega$ by:

$$F(x) := (I - B)^{-1}Ax.$$

Let $K = \overline{conv}(F(\Omega))$ be the closed convex hull of $F(\Omega)$. Clearly, K is closed, convex, bounded, and $F(K) \subset K \subset \Omega$. We claim that K is weakly compact. If it is not the case, then $\omega(K) > 0$. Since $F(\Omega) \subseteq A(\Omega) + BF(\Omega)$, we obtain

$$\omega(K) = \omega(F(\Omega)) \leq \omega(A(\Omega) + BF(\Omega)) \leq \omega(A(\Omega)) + \omega(BF(\Omega)).$$

Taking into account the fact that A is weakly compact and B is ω-condensing, we obtain

$$\omega(K) = \omega(F(\Omega)) \leq \omega(B(F(\Omega))) < \omega(F(\Omega)),$$

which is absurd. Hence, K is weakly compact. In view of Corollary 2.3.3, it remains to show that $F : K \longrightarrow K$ is weakly sequentially continuous. In fact, let $(x_n)_n \subset K$ such that $x_n \rightharpoonup x$. Because $F(K)$ is relatively weakly compact, it follows by the Eberlein–Šmulian theorem that there exists a subsequence $(x_{n_k})_k$ of $(x_n)_n$ such that $F(x_{n_k}) \rightharpoonup y$. The weakly sequentially continuity of B leads to $BF(x_{n_k}) \rightharpoonup By$. Also, from the equality $BF = -A + F$, it results that

$$-A(x_{n_k}) + F(x_{n_k}) \rightharpoonup -A(x) + y.$$

So, $y = F(x)$. We claim that $F(x_n) \rightharpoonup F(x)$. Suppose that this is not the case, then there exists a subsequence $(x_{\varphi_1(n)})_n$ and a weak neighborhood V^w

of $(I-B)^{-1}Ax$ such that $(I-B)^{-1}Ax_{\varphi_1(n)} \notin V^w$, for all $n \in \mathbb{N}$. On the other hand, we have $x_{\varphi_1(n)} \rightharpoonup x$, then arguing as before, we find a subsequence $(x_{\varphi_1(\varphi_2(n))})_n$ such that $(I-B)^{-1}Ax_{\varphi_1(\varphi_2(n))}$ converges weakly to $(I-B)^{-1}Ax$, which is a contradiction and hence F is weakly sequentially continuous. Q.E.D.

2.3.2 Leray–Schauder's alternatives for weakly sequentially continuous mappings

At the beginning of this section, we will state some new variants of Krasnosel'skii–Leray–Schauder for different classes of weakly sequentially continuous mappings. We will need the following nonlinear alternatives of the Leray–Schauder type for single valued mappings.

Theorem 2.3.7 *Let Ω be a nonempty, closed, and convex subset of a Banach space X. In addition, let U be a weakly open subset of Ω with $\theta \in U$ such that $\overline{U^w}$ is a weakly compact subset of Ω and $F : \overline{U^w} \longrightarrow \Omega$ is a weakly sequentially continuous mapping. Then, either*

(i) F has, at least, a fixed point; or

(ii) there is a point $x \in \partial_\Omega^w U$ (the weak boundary of U in Ω) and a $\lambda \in (0,1)$ with $x = \lambda F(x)$.

Proof. Suppose that *(ii)* does not hold. We notice that this supposition is also satisfied for $\lambda = 0$ (since $\theta \in U$). If *(ii)* is satisfied for $\lambda = 1$ then, in this case, we have a fixed point in $u \in \partial_\Omega^w U$ and there is nothing to prove. In conclusion, we can consider that the supposition is satisfied for any $x \in \partial_\Omega^w U$ and any $\lambda \in [0,1]$. Let D be the set defined by

$$D = \Big\{x \in \overline{U^w} \text{ such that } x = \lambda F(x), \text{ for some } \lambda \in [0,1]\Big\}.$$

The set D is nonempty because $\theta \in U$. We will show that D is weakly compact. The weak sequential continuity of F implies that D is weakly sequentially closed. For that, let $(x_n)_n$ be a sequence of D such that $x_n \rightharpoonup x$, $x \in \overline{U^w}$. For all $n \in \mathbb{N}$, there exists a $\lambda_n \in [0,1]$ such that $x_n = \lambda_n F(x_n)$. Since $\lambda_n \in [0,1]$, we can extract a subsequence $(\lambda_{n_j})_j$ such that $\lambda_{n_j} \to \lambda \in [0,1]$. So, $\lambda_{n_j}F(x_{n_j}) \rightharpoonup \lambda F(x)$. Hence, $x = \lambda F(x)$ and $x \in D$. Let $x \in \overline{U^w}$ be weakly adherent to D. Since $\overline{D^w}$ is weakly compact by the Eberlein–Šmulian theorem, there exists a sequence $(x_n)_n \subset D$ such that $x_n \rightharpoonup x$, so $x \in D$. Hence, $\overline{D^w} = D$ and D is weakly closed. Therefore, D is weakly compact.

Because X endowed with its weak topology is a Hausdorff locally convex space, then X is completely regular [146, p.16]. Since $D \cap (\Omega \setminus U) = \emptyset$, then by Lemma 1.2.1, there is a weakly continuous function $\varphi : \Omega \longrightarrow [0,1]$, such that $\varphi(x) = 1$ for $x \in D$ and $\varphi(x) = 0$ for $x \in \Omega \setminus U$. Let $F^* : \Omega \longrightarrow \Omega$ be the mapping defined by:

$$F^*(x) = \varphi(x)F(x).$$

Because $\partial_\Omega^w U = \partial_\Omega^w \overline{U^w}$, φ is weakly continuous and F is weakly sequentially continuous, we deduce that F^* is weakly sequentially continuous. In addition,

$$F^*(\Omega) \subset \overline{conv}(F(\overline{U^w}) \cup \{\theta\}).$$

Let

$$D_* = \overline{conv}(F(\overline{U^w}) \cup \{\theta\}).$$

Using the Krein–Šmulian theorem (see Theorem 1.3.5) and the weakly sequential continuity of F, it follows that D_* is a weakly compact and convex set. Moreover, $F^*(D_*) \subset D_*$. Since F^* is weakly sequentially continuous, and by using Theorem 1.6.9, it follows that F^* has, at least, a fixed point $x_0 \in \Omega$. If $x_0 \notin U$, $\varphi(x_0) = 0$ and $x_0 = 0$, which contradicts the hypothesis $\theta \in U$. Then, $x_0 \in U$ and $x_0 = \varphi(x_0)F(x_0)$, which implies that $x_0 \in D$. Hence, $\varphi(x_0) = 1$ and the proof is complete. Q.E.D.

Remark 2.3.4 *The condition "$\overline{U^w}$ is weakly compact" in the statement of Theorem 2.3.7 can be removed if we suppose that $F(\overline{U^w})$ is relatively weakly compact.*

Theorem 2.3.8 *Let Ω be a nonempty, closed, and convex subset of a Banach space X. In addition, let U be a weakly open subset of Ω with $\theta \in U$ and let $F : \overline{U^w} \longrightarrow \Omega$ be weakly sequentially continuous and ω-condensing mapping such that $F(\overline{U^w})$ is bounded. Then, either*

(i) F has, at least, a fixed point; or

(ii) there is a point $x \in \partial_\Omega^w U$ (the weak boundary of U in Ω) and a scalar $\lambda \in (0,1)$ with $x = \lambda F(x)$.

Proof. Suppose that (ii) does not hold and F does not have a fixed point in $\partial_\Omega^w U$ (otherwise, we have finished, i.e., (i) occurs). Let D be the set defined by:

$$D = \Big\{x \in \overline{U^w} \text{ such that } x = \lambda F(x), \text{ for some } \lambda \in [0,1]\Big\}.$$

D is nonempty and bounded, because $\theta \in D$ and $F(\overline{U^w})$ is bounded. We have

$$D \subset conv(\{\theta\} \cup F(D)).$$

So, $\omega(D) \neq 0$ which implies

$$\omega(D) \leq \omega(conv(\{\theta\} \cup F(D))) \leq \omega(F(D)) < \omega(D)$$

which is a contradiction. Hence, $\omega(D) = 0$ and D is relatively weakly compact. Now, we prove that D is weakly closed. Arguing as in the proof of Theorem 2.3.7, we prove that D is weakly sequentially closed. Let $x \in \overline{U^w}$, and weakly adherent to D. Since $\overline{D^w}$ is weakly compact, according to the Eberlein–Šmulian theorem, there exists a sequence $(x_n)_n \subset D$ such that $x_n \rightharpoonup x$, so $x \in D$. Hence, $\overline{D^w} = D$ and D is weakly closed. Therefore, D is weakly compact. Because X endowed with its weak topology is a Hausdorff locally convex space, then X is completely regular [146, p.16]. Since $D \cap (\Omega \setminus U) = \emptyset$, then by Lemma 1.2.1, there is a weakly continuous function $\varphi : \Omega \longrightarrow [0, 1]$, such that $\varphi(x) = 1$ for $x \in D$ and $\varphi(x) = 0$ for $x \in \Omega \setminus U$. Let $F^* : \Omega \longrightarrow \Omega$ be the mapping defined by:

$$F^*(x) = \varphi(x)F(x).$$

Clearly, $F^*(\Omega)$ is bounded. Because φ is weakly continuous and F is weakly sequentially continuous, we show that F^* is weakly sequentially continuous. Let $X \subset \Omega$ be bounded. Since

$$F^*(X) \subset conv(\{\theta\} \cup F(X \cap \Omega)),$$

then, we have $\omega(F^*(X)) \leq \omega(F(X \cap \Omega)) \leq \omega(F(X))$ and $\omega(F^*(X)) < \omega(X)$ if $\omega(X) \neq 0$. So, F^* is condensing with respect to ω. Therefore, Theorem 2.3.4 implies that F^* has, at least, a fixed point $x_0 \in \Omega$. If $x_0 \notin U$, $\varphi(x_0) = 0$ and $x_0 = 0$, which contradicts the hypothesis $\theta \in U$. Then, $x_0 \in U$ and $x_0 = \varphi(x_0)F(x_0)$, which implies that $x_0 \in D$, and so $\varphi(x_0) = 1$ and the proof is complete. Q.E.D.

2.4 Fixed Point Theorems for Multi-Valued Mappings

We define a subset Ω of a locally convex space X to be almost convex if, for any neighborhood V of θ and for any finite set $\{w_1, ..., w_n\}$ of points of Ω, there exist $z_1, ..., z_n \in \Omega$ such that $z_i - w_i \in V$ for all i, and

$$conv\{z_1, ..., z_n\} \subset \Omega.$$

The following theorem can be found in [95].

Theorem 2.4.1 *Let K be a nonempty compact subset of a separated locally convex space X, and $G : K \longrightarrow K$ be an u.s.c. multi-function such that $G(x)$ is closed for all z in K and convex for all x in some dense almost convex subset Ω of K. Then, G has, at least, a fixed point.*

Proof. Let $\mathcal{V}$ be a local base of neighborhoods of θ consisting of closed, convex, and symmetric sets. For each $V \in \mathcal{V}$, let

$$F_V = \{x \in K \text{ such that } x \in G(x) + V\}.$$

To find a fixed point of G, it is clearly sufficient (and necessary) to show that

$$\bigcap \{F_V \text{ such that } V \in \mathcal{V}\} \neq \emptyset.$$

Since $F_U \cap F_V \supset F_{U \cap V}$ for all U, $V \in \mathcal{V}$, it is sufficient, by the compactness of K, to show that each F_V is closed and nonempty. So, let $V \in \mathcal{V}$. Let us define the multi-functions $G_V : K \longrightarrow K$ and $R_V : K \longrightarrow K$ by:

$$G_V(x) = (G(x) + V) \cap K, \quad R_V(x) = (x + V) \cap K, \quad x \in K.$$

Then, $G_V = R_V \circ G$. Moreover, R_V is a closed subset of $K \times K$ since $R_V = \{(x, y) \in K \times K \text{ such that } y - x \in V\}$ and V is closed. Since K is compact, it follows that both R_V and G are u.s.c. Hence, G_V is u.s.c. and, in particular, is a closed subset of $K \times K$. Let Δ be the diagonal in $K \times K$. Then, F_V is obtained by projecting the compact set $\Delta \cap G_V$ onto the domain of G_V. It follows that F_V is closed. Now, let us choose $z_1, ..., z_m \in \Omega$ such that $K \subset \cup\{z_i + V \text{ such that } 1 \leq i \leq m\}$, and $C = conv\{z_1, ..., z_m\} \subset \Omega$. Let us define $H_V \subset C \times C$ by:

$$H_V = G_V \cap (C \times C).$$

For each $x \in C$, $H_V(x)$ is closed, convex (since $C \subset \Omega$), and nonempty (since $G(x) + V$ contains some z_i). Moreover, H_V is a closed subset of $C \times C$ since G_V is closed. Thus, H_V has, at least, a fixed point by Kakutani's fixed point theorem (see Theorem 1.6.4). It belongs to F_V which is then nonempty. Q.E.D.

As a consequence of Theorem 2.4.1, we have the following:

Corollary 2.4.1 *Let Ω be a nonempty, convex, and closed subset of a locally convex space X. Let $F : \Omega \longrightarrow \mathcal{P}_{cl,\ cv}(\Omega)$ be an upper semicontinuous multi-valued mapping such that $F(\Omega)$ is relatively compact. Then, F has, at least, a fixed point.*

2.4.1 Multi-valued maps with a weakly sequentially closed graph

Theorem 2.4.2 *Let Ω be a nonempty, weakly compact subset of a Banach space X. Suppose that $F : \Omega \longrightarrow \mathcal{P}(X)$ has a weakly sequentially closed graph and $F(\Omega)$ is relatively weakly compact. Then, F has a weakly closed graph.*

Proof. Since $(X \times X)_w = X_w \times X_w$ (X_w the space X endowed with its weak topology), it follows that $\Omega \times \overline{F(\Omega)}^w$ is a weakly compact subset of $X \times X$. Also, $Gr(F) \subset \Omega \times \overline{F(\Omega)}^w$. So, $Gr(F)$ is relatively weakly compact. Let $(x, y) \in \Omega \times \overline{F(\Omega)}^w$ be weakly adherent to $Gr(F)$, then by the Eberlein–Šmulian theorem, we can find $((x_n), (y_n))_n \subset Gr(F)$ such that $y_n \in F(x_n)$, $x_n \rightharpoonup x$ and $y_n \rightharpoonup y$ in X. Because F has a weakly sequentially closed graph, $y \in F(x)$ and so, $(x, y) \in Gr(F)$. Therefore, $Gr(F)$ is weakly closed. Q.E.D.

The next theorem needs the following lemma.

Lemma 2.4.1 *Let us assume that $\varphi : X \longrightarrow Y$ is a multi-valued map such that $\varphi(X) \subset K$ and the graph $Gr(\varphi)$ of φ is closed, where K is a compact set. Then, φ is u.s.c.*

Proof. Let us assume the contrary, i.e., φ is not u.s.c. Then, there exists an open neighborhood $V_{\varphi(x)}$ of $\varphi(x)$ in Y such that, for every open neighborhood U_x of x in X, we have $\varphi(U_x)$ is not contained in $V_{\varphi(x)}$. We take $U_x = B(x, \frac{1}{n})$, $n = 1, 2, ...$ Then, for every n, we get a point $x_n \in B(x, \frac{1}{n})$ such that $\varphi(x_n)$ is not contained in $V_{\varphi(x)}$. Let y_n be a point in Y such that $y_n \in \varphi(x_n)$ and $y_n \notin V_{\varphi(x)}$. Then, we have $\lim_{n\to\infty} x_n = x$ and $(y_n)_n \subset K$. Since K is compact, we can assumes without loss of generality, that $\lim_{n\to\infty} y_n = y \in K$. We see that $y \notin V_{\varphi(x)}$. Then, for every n, we have $(x_n, y_n) \in Gr(\varphi)$ and $(x_n, y_n) \to (x, y)$. So $(x, y) \in Gr(\varphi)$ because $Gr(\varphi)$ is a closed subset of $X \times Y$ but it contradicts $y \notin V_{\varphi(x)}$ and the proof is completed. Q.E.D.

Theorem 2.4.3 *Let Ω be a nonempty, closed, and convex subset of a Banach space X. Suppose $F : \Omega \longrightarrow \mathcal{P}_{cv}(\Omega)$ has a weakly sequentially closed graph and $F(\Omega)$ is weakly relatively compact. Then, F has, at least, a fixed point.*

Proof. Set $K = \overline{conv}(F(\Omega))$. It follows, using the Krein–Šmulian theorem (see Theorem 1.3.5), that K is a weakly compact and convex set. We have $F(\Omega) \subset K \subset \Omega$. Notice also that $F : K \longrightarrow \mathcal{P}_{cv}(K)$. By Theorem 2.4.2, F has a weakly closed graph, and so $F(x)$ is weakly closed for every $x \in K$. Thus, by Lemma 2.4.1, F is weakly upper semicontinuous. Because X endowed with its weak topology is a Hausdorff locally convex space, we apply Corollary 2.4.1 to ensure that F has, at least, a fixed point $x \in K \subset \Omega$. Q.E.D.

Recall that in Theorems 2.3.2, 2.3.3, and 2.3.6 our arguments were based on the invertibility of the mapping $I - B$ and our strategy consists in proving the fixed point property of the mapping $(I-B)^{-1}A$. Hence, it would be interesting to investigate the case when $I - B$ may not be injective.

In that line, the following result presents a critical type of Krasnoselskii's fixed point theorem. Having obtained these results, we shall now study the fixed point property for a larger class of weakly sequentially continuous mappings under weaker assumptions. Besides, we will focus on the case invertible and we investigate this kind of generalization by looking for the multi-valued mapping $(I - B)^{-1}A$.

Theorem 2.4.4 *Let Ω be a nonempty, closed, and convex subset of a Banach space X. Suppose that A and B are weakly sequentially continuous mappings from Ω into X such that:*

(i) $A(\Omega) \subset (I - B)(X)$ and $(x = Bx + Ay,\ y \in \Omega) \Longrightarrow x \in \Omega$ (or $A(\Omega) \subset (I - B)(\Omega)$),

(ii) $A(\Omega)$ is a relatively weakly compact subset of X,

(iii) if $(I - B)x_n \rightharpoonup y$, then there exists a weakly convergent subsequence of $(x_n)_n$, and

(iv) for every y in the range of $I - B$, $D_y = \{x \in \Omega$ such that $(I - B)x = y\}$ is a convex set.

Then, there exists $x \in \Omega$ such that $x = Ax + Bx$.

Proof. First, we assume that $I - B$ is invertible. For any given $y \in \Omega$, define $F : \Omega \longrightarrow \Omega$ by:

$$Fy := (I - B)^{-1}Ay.$$

F is well defined by assumption (i).

Step 1: $F(\Omega)$ is relatively weakly compact. For any $(y_n)_n \subset F(\Omega)$, we choose

$(x_n)_n \subset \Omega$ such that $y_n = F(x_n)$. Taking into account assumption (ii), together with the Eberlein–Šmulian theorem, we get a subsequence $(y_{\varphi_1(n)})_n$ of $(y_n)_n$ such that $(I-B)y_{\varphi_1(n)} \rightharpoonup z$, for some $z \in \Omega$. Thus, by assumption (iii), there exists a subsequence $y_{\varphi_1(\varphi_2(n))}$ converging weakly to $y_0 \in \Omega$.

Step 2: F is weakly sequentially continuous. The result can be checked in the same way as in Theorem 2.3.6. Consequently, using Corollary 2.3.3, we get the desired result.

Second, if $I - B$ is not invertible, $(I-B)^{-1}$ could be seen as a multi-valued mapping. For any given $y \in \Omega$, define $H : \Omega \longrightarrow P(\Omega)$ by:

$$Hy := (I-B)^{-1}Ay.$$

H is well defined by assumption (i). We should prove that H fulfills the hypotheses of Theorem 2.4.3.

Step 1: $H(x)$ is a convex set for each $x \in \Omega$. This is an immediate consequence of assumption (iv).

Step 2: H has a weakly sequentially closed graph. Let $x \in \Omega$ and $(x_n)_n \subset \Omega$ such that $x_n \rightharpoonup x$ and $y_n \in H(x_n)$ such that $y_n \rightharpoonup y$. By the definition of H, we have $(I-B)y_n = Ax_n$. Since A and $I-B$ are weakly sequentially continuous, we obtain $(I-B)y = Ax$. Thus $y \in (I-B)^{-1}Ax$.

Step 3: $H(x)$ is closed for each $x \in \Omega$. This assertion follows from Steps 1 and 2 by setting $(x_n)_n \equiv x$.

Step 4: $H(\Omega)$ is relatively weakly compact. This assertion is proved by using the same reasoning as the one in Step 1 of the first part of the proof.
In view of Theorem 2.4.3, we get $x \in H(x)$, for some $x \in \Omega$. Thus, there exists $x \in \Omega$ such that $x = Ax + Bx$. Q.E.D.

Remark 2.4.1 *We shall emphasize the fact that if $B : X \longrightarrow X$ is a ω-condensing weakly sequentially continuous mapping so that $B(X)$ is a bounded subset of X and $I - B$ is invertible, then the assumptions (iii) and (iv) of Theorem 2.4.4 are satisfied. Indeed, suppose that $(I-B)x_n \rightharpoonup y$, for some $(x_n)_n \subset \Omega$ and $y \in \Omega$. Writing x_n as $x_n = (I-B)x_n + Bx_n$ and using the subadditivity of the De Blasi measure of weak noncompactness, we get*

$$\omega(\{x_n\}) \leq \omega(\{(I-B)x_n\}) + \omega(\{Bx_n\}).$$

Since $\overline{\{(I-B)x_n\}}^w$ is weakly compact, we obtain $\omega(\{x_n\}) \leq \omega(\{Bx_n\})$. Now, *we show that* $\omega(\{x_n\}) = 0$. *If we suppose the contrary, then using the fact that* B *is* ω*-condensing, we obtain*

$$\omega(\{x_n\}) \leq \omega(\{Bx_n\}) < \omega(\{x_n\}),$$

which is absurd. So, $\omega(\{x_n\}) = 0$. *Consequently,* $\overline{\{x_n\}}^w$ *is weakly compact and then by the Eberlein–Šmulian theorem, there exists a weakly convergent subsequence of* $(x_n)_n$. *Hence, the assumption* (*iii*) *is satisfied. On the other hand, since* $I - B$ *is invertible, we have for every* y *in the range of* $I - B$, *the set* D_y *is reduced to* $\{(I - B)^{-1}y\}$, *which is convex.*

In particular, if $B : X \longrightarrow X$ is a weakly sequentially continuous nonlinear contraction so that $B(X)$ is bounded, we get the following corollary:

Corollary 2.4.2 *Let* Ω *be a nonempty, closed, and convex subset of a Banach space* X. *Suppose that* $A : \Omega \longrightarrow X$ *and* $B : X \longrightarrow X$ *are two weakly sequentially continuous mappings such that:*

(*i*) $A(\Omega)$ *is relatively weakly compact,*

(*ii*) B *is a nonlinear contraction such that* $B(X)$ *is bounded, and*

(*iii*) $(x = Bx + Ay,\ y \in \Omega) \Longrightarrow x \in \Omega$.

Then, there exists $x \in X$ *such that* $x = Ax + Bx$.

Proof. Since B is a nonlinear contraction, so $I - B$ is invertible and $(I - B)(X) = X$ (see Theorem 1.6.10), hence the first part of assumption (*i*) of Theorem 2.4.4 is fulfilled. Moreover, we have already proved that every weakly sequentially continuous nonlinear contraction is ω-condensing (see proof of Theorem 2.3.3). Hence, in view of Remark 2.4.1 (*ii*), we deduce that B satisfies the assumptions (*iii*) and (*iv*) of Theorem 2.4.4. Q.E.D.

Using the technique used in the proof of Theorem 2.4.4, we have the following result.

Theorem 2.4.5 *Let* Ω *be a nonempty, bounded, closed, and convex subset of a Banach space* X. *Suppose that* A *and* B *are weakly sequentially continuous and map* Ω *into* X *such that:*

(*i*) $(I - A)(\Omega) \subset B(\Omega)$,

(*ii*) $(I - A)(\Omega)$ *is contained in a weakly compact subset of* X,

(*iii*) *if* $Bx_n \rightharpoonup x$, *then there exists a weakly convergent subsequence of* $(x_n)_n$, *and*

(iv) for every y in the range of B, $D_y = \{x \in \Omega$ such that $Bx = y\}$ is a convex set.

Then, there exists $x \in \Omega$ such that $x = Ax + Bx$.

In the remaining part of this section, we prove some fixed point theorems for a Browder class of multi-valued mappings with a weakly sequentially closed graph, in which the operators have the property that the image of any set is, in a certain sense, more weakly compact than the original set itself.

Theorem 2.4.6 *Let Ω be a nonempty, closed, and convex subset of a Banach space X. Assume Φ is a measure of weak noncompactness on X and $F : \Omega \longrightarrow \mathcal{P}_{cv}(\Omega)$ has a weakly sequentially closed graph. In addition, suppose that F is Φ-condensing and $F(\Omega)$ is bounded. Then, F has, at least, a fixed point.*

Proof. Let $x_0 \in \Omega$. We consider the family $\mathcal{F}$ of all closed, bounded, and convex subsets D of Ω such that $x_0 \in D$ and $F(x) \subset D$ for all $x \in D$. Obviously, $\mathcal{F}$ is nonempty since $\overline{conv}(F(\Omega) \cup \{x_0\}) \in \mathcal{F}$. We denote

$$K = \bigcap_{D \in \mathcal{F}} D.$$

We observe that K is closed and convex and $x_0 \in K$. If $x \in K$, then $F(x) \subset D$ for all $D \in \mathcal{F}$ and hence, $F(x) \subset K$. Consequently, we have that $K \in \mathcal{F}$. We will prove that K is weakly compact. Denoting by:

$$K_* = \overline{conv}(F(K) \cup \{x_0\}),$$

we have $K_* \subset K$, which implies that $F(x) \subset F(K) \subset K_*$. Therefore, $K_* \in \mathcal{F}$, and $K \subset K_*$. Hence $K = K_*$. Since K is weakly closed, it is sufficient to show that K is relatively weakly compact. If $\Phi(K) > 0$, we obtain

$$\Phi(K) = \Phi(\overline{conv}(F(K) \cup \{x_0\})) \leq \Phi(F(K)) < \Phi(K)$$

which is a contradiction. Hence, $\Phi(K) = 0$ and so, K is relatively weakly compact. Now, $F : K \longrightarrow \mathcal{P}_{cv}(K)$ has a weakly sequentially closed graph. From Theorem 2.4.3, F has, at least, a fixed point in $K \subset \Omega$. Q.E.D.

2.4.2 Leray–Schauder's and Furi–Pera's types of fixed point theorems

In applications, the construction of the set Ω such that $F(\Omega) \subset \Omega$ is very difficult and sometimes impossible. That is why we investigate maps $F : \Omega \longrightarrow \mathcal{P}(X)$ with a weakly sequentially closed graph.

Lemma 2.4.2 *Let Ω be a weakly closed subset of a Banach space X with $\theta \in \Omega$. Assume that $F : \Omega \longrightarrow \mathcal{P}(X)$ has a weakly sequentially closed graph with $F(\Omega)$ being bounded. Let $(x_n)_n \subset \Omega$ and $(\lambda_n)_n$ be a real sequence. If $x_n \rightharpoonup x$ and $\lambda_n \to \lambda \in \mathbb{R}$, then the condition $x_n \in \lambda_n F(x_n)$ for all n implies that $x \in \lambda F(x)$.*

Proof. For all n, there exists $y_n \in F(x_n)$ such that $x_n = \lambda_n y_n$. If $\lambda = 0$ then, $x_n \rightharpoonup \theta$ ($F(\Omega)$ is bounded) and $x \in \{\theta\} \subset \Omega$. If $\lambda \neq 0$ then, without loss of generality, we can suppose that $\lambda_n \neq 0$ for all n. So, $\lambda_n^{-1} x_n = y_n$ for all n implies that $y_n \rightharpoonup \lambda^{-1} x$. Since F has a weakly sequentially closed graph, we have $y \in F(x)$, which means that $x \in \lambda F(x)$. Q.E.D.

Theorem 2.4.7 *Let X be a Banach space, Ω be a nonempty, closed, and convex subset of X and U be a weakly open subset of Ω with $\theta \in U$. Assume that Φ is a measure of weak noncompactness on X and $F : \overline{U^w} \longrightarrow \mathcal{P}_{cv}(\Omega)$ has a weakly sequentially closed graph. In addition, suppose that F is Φ-condensing and $F(\overline{U^w})$ is bounded. Then, either*
(i) F has, at least, a fixed point, or

(ii) there is a point $x \in \partial_\Omega^w U$ (the weak boundary of U in Ω) and $\lambda \in (0,1)$ with $x \in \lambda F(x)$.

Proof. Suppose that (ii) does not hold and F does not have a fixed point in $\partial_\Omega^w U$ (otherwise, we have finished, i.e., (i) occurs). Let D be the set defined by:

$$D = \Big\{x \in \overline{U^w} \text{ such that } x \in \lambda F(x), \text{ for some } \lambda \in [0,1]\Big\}.$$

D is nonempty and bounded, because $\theta \in D$ and $F(\overline{U^w})$ is bounded. We have $D \subset conv(\{\theta\} \cup F(D))$. So, $\Phi(D) \neq 0$ which implies the following:

$$D \subset conv(\{\theta\} \cup F(D)) = \Big\{x \in \overline{U^w} \text{ such that } x \in \lambda F(x), \text{ for some } \lambda \in [0,1]\Big\}$$

which is a contradiction. Hence, $\Phi(D) = 0$ and D is weakly relatively compact. Now, we prove that D is weakly sequentially closed. For this, let $(x_n)_n$ be a sequence of D such that $x_n \rightharpoonup x$, $x \in \overline{U^w}$. For all $n \in \mathbb{N}$, there exists a $\lambda_n \in [0,1]$ such that $x_n = \lambda_n F(x_n)$. Since $\lambda_n \in [0,1]$, we can extract a subsequence $(\lambda_{n_j})_j$ such that $\lambda_{n_j} \to \lambda \in [0,1]$. We put $x_{n_j} = \lambda_{n_j} y_{n_j}$, where $y_{n_j} \in F(x_{n_j})$. Applying Lemma 2.4.2, we deduce that $x \in D$. Let $x \in \overline{U^w}$ be weakly adherent to D. Since $\overline{D^w}$ is weakly compact, by the Eberlein–Šmulian theorem (Theorem 1.3.3), there exists a sequence $(x_n)_n \subset D$ such that $x_n \rightharpoonup x$, so $x \in D$. Hence, $\overline{D^w} = D$ and D is a weakly closed subset of the weakly compact set U^w. Therefore, D is weakly compact. Because X endowed with its weak topology is a Hausdorff locally convex space, we deduce that X is completely regular [146, p. 16]. Since $D \cap (\Omega \setminus U) = \emptyset$ then, by Lemma 1.2.1, there is a weakly continuous function $\varphi : \Omega \longrightarrow [0,1]$, such that $\varphi(x) = 1$ for $x \in D$ and $\varphi(x) = 0$ for $x \in \Omega \setminus U$. Since Ω is convex, $\theta \in \Omega$, and F with nonempty convex values, we can define the multi-valued map $F^* : \Omega \longrightarrow P_{cv}(\Omega)$ by:

$$F^*(x) = \begin{cases} \varphi(x)F(x), & \text{if } x \in \overline{U^w}, \\ \{\theta\}, & \text{if } x \in \Omega \setminus \overline{U^w}. \end{cases}$$

Clearly, $F^*(\Omega)$ is bounded. Because $\partial^w_\Omega U = \partial^w_\Omega \overline{U^w}$, $[0,1]$ is compact, φ is weakly continuous and F has a weakly sequentially closed graph. Using Lemma 2.4.2, we notice that F^* has a weakly sequentially closed graph. Let $X \in \Omega$ be bounded. Then, since

$$F^*(X) \subset conv\left(\{\theta\} \cup F(X \cap \overline{U^w})\right),$$

we have

$$\Phi(F^*(X)) \leq \Phi(X \cap \overline{U^w}) \leq \Phi(F(X))$$

and $\Phi(F^*(X)) < \Phi(X)$ if $\Phi(X) \neq 0$. So, F^* is Φ-condensing. Therefore, all the assumptions of Theorem 2.4.6 are satisfied for F^*. Consequently, there exists $x_0 \in \Omega$ with $x_0 \in F^*(x_0)$. If $x_0 \notin U$, $\Phi(x_0) = 0$ and $x_0 = \theta$, which contradicts the hypothesis $\theta \in U$. Then, $x_0 \in U$ and $x_0 \in \Phi(x_0)F(x_0)$, which implies that $x_0 \in D$. Hence, $\varphi(x_0) = 1$ and the proof is complete. Q.E.D.

Corollary 2.4.3 *Let X be a Banach space, Ω be a nonempty, closed, and convex subset of X and U be a weakly open subset of Ω with $\theta \in U$. Assume that $H : \overline{U^w} \longrightarrow \mathcal{P}_{cv}(\Omega)$ is a weakly completely continuous map with $H(\overline{U^w})$ being bounded. In addition, suppose that H satisfies the Leray–Schauder boundary condition*

$$x \neq \lambda H(x) \text{ for every } x \in \partial^w_\Omega U \text{ and } \lambda \in (0,1).$$

Then, H has, at least, a fixed point in $\overline{U}^w$.

Proof. Since H is weakly completely continuous, it follows that H is Φ-condensing on Ω for any measure of weak noncompactness on X. Now, it is sufficient to apply Theorem 2.4.7. Q.E.D.

The following theorem is a Furi–Pera alternative (see [85]) for a multi-valued mapping having a weakly sequentially closed graph.

Theorem 2.4.8 *Let Ω be a nonempty, closed, and convex subset of a separable Banach space $(X, \|.\|)$, M be a closed convex subset of Ω with $\theta \in M$ and let $H : M \longrightarrow \mathcal{P}(\Omega)$ be a multi-valued mapping, such that:*

(i) H has a weakly sequentially closed graph,

(ii) H is a weakly compact map, and

(iii) the set $H(x)$ is nonempty and convex for all $x \in M$.

In addition, assume that:

(iv) There exists a weakly sequentially continuous retraction $r : X \longrightarrow M$,

(v) there exists $\delta > 0$ and a weakly compact set M_δ with $K_\delta = \{x \in X \text{ such that } d(x, M) \leq \delta\} \subset M_\delta$ here, $d(x, y) = \|x - y\|$, and

(vi) for any $K_\varepsilon = \{x \in X \text{ such that } d(x, M) \leq \varepsilon, 0 < \varepsilon \leq \delta\}$, if $\{(x_j, \lambda_j)\}_{j=1}^{\infty}$ is a sequence in $M \times [0, 1]$ with $x_j \rightharpoonup x \in \partial_{K_\varepsilon} M$, $\lambda_j \to \lambda$ and $x \in \lambda H(x)$, $0 \leq \lambda < 1$, then $\lambda_j H(x_j) \subset M$ for j sufficiently large, here $\partial_{K_\varepsilon} M$ denotes the weak boundary of M in K_ε.

Then, H has, at least, a fixed point in M.

Proof. Let us consider

$$N = \Big\{x \in \Omega \text{ such that } x \in Fr(x)\Big\}.$$

First, we show that $N \neq \emptyset$. Notice that $Fr : \Omega \longrightarrow \mathcal{P}_{cv}(\Omega)$ is weakly compact. Since r is weakly sequentially continuous, M is weakly closed, and H has a weakly sequentially closed graph, it follows that Fr has a weakly sequentially closed graph. Theorem 2.4.3 implies that Fr has, at least, a fixed point, so $N \neq \emptyset$. Next, we show that N is weakly compact. Indeed, $N \subset Fr(\Omega) \subset H(M)$, so N is relatively weakly compact. Now, let $(x_n)_n$ be a sequence of N such that $x_n \rightharpoonup x$, $x \in X$. For all $n \in \mathbb{N}$, we have $x_n \in Fr(x_n)$ and $r(x_n) \rightharpoonup r(x)$ in M. Because H has a weakly sequentially closed graph, $x \in Fr(x)$. Hence, $x \in N$ and N is weakly sequentially closed. Applying again the Eberlein–Šmulian theorem (see Theorem 1.3.3), we deduce that N is weakly compact. Now, we show that $M \cap N = \emptyset$. To do this, we argue by contradiction. Suppose that $M \cap N \neq \emptyset$. Then, since N is compact and M is closed, we get $d(N, M) = \inf\{\|x - y\|$ such that $x \in N,\ y \in M\} > 0$. Thus, there exists ε, $0 < \varepsilon \leq \delta$ with $K_\varepsilon \cap N = \emptyset$. Here,

$$K_\varepsilon = \Big\{x \in X \text{ such that } d(x, M) \leq \varepsilon\Big\}.$$

We have K_ε is closed and convex, hence weakly closed and $K_\varepsilon \subset M_\delta$. Using (v), we deduce that K_ε is weakly compact. Because X is separable, the weak topology on K_ε is metrizable (see [154]), and let ρ denote this metric. For $i \in \{1, 2, ...\}$, let

$$U_i = \Big\{x \in K_\varepsilon \text{ such that } \rho(x, M) < \frac{\varepsilon}{i}\Big\}.$$

We fix $i \in \{1, 2, ...\}$. Now, U_i is open in K_ε, with respect to the topology generated by ρ and so, U_i is weakly open in K_ε. Also, we have

$$\overline{U_i}^w = \overline{U_i}^\rho = \Big\{x \in K_\varepsilon \text{ such that } \rho(x, M) \leq \frac{\varepsilon}{i}\Big\}$$

and

$$\partial_{K_\varepsilon} U_i = \Big\{x \in K_\varepsilon \text{ such that } \rho(x, M) = \frac{\varepsilon}{i}\Big\}.$$

Since $d(N, M) > \varepsilon$, we get $N \cap \overline{U_i}^w = \emptyset$. Applying Corollary 2.4.3, we deduce that there exists $\lambda \in (0, 1)$ and $y_i \in \partial_{K_\varepsilon} U_i$ such that $y_i \in \lambda_i Fr(y_i)$. In particular, since $y_i \in \partial_{K_\varepsilon} U_i$, it follows that

$$\lambda_i Fr(y_i) \not\subset M \text{ for each } i \in \{1, 2, ...\}. \tag{2.9}$$

Now, we investigate

$$R = \Big\{x \in X \text{ such that } x \in \lambda Fr(x), \text{ for some } \lambda \in [0, 1]\Big\}.$$

R is nonempty, because $\theta \in R$. Also, $R \subset \overline{conv}(H(M) \cup \{\theta\})$. Hence, using the Krein–Šmulian theorem (see Theorem 1.3.5), we deduce that R is relatively weakly compact. Since Fr has a weakly sequentially closed graph and $[0, 1]$ is compact, we deduce by Lemma 2.4.2 that R is weakly sequentially closed. Combining this result with the following:

$$\rho(y_j, M) = \frac{\varepsilon}{j},\ \lambda_j \in [0, 1] \text{ for } j \in \{1, 2, ...\}$$

implies that we may assume, without loss of generality, that $\lambda_j \to \lambda_0$ and $y_j \rightharpoonup y_0 \in \overline{M}^w \cap \overline{K_\varepsilon \backslash M}^w = \partial_{K_\varepsilon} M$. Moreover, since $y_j \in \lambda_j Fr(y_j)$ we have

$y_0 \in \lambda_0 Fr(y_0)$. If $\lambda_0 = 1$ then, $y_0 \in Fr(y_0)$ which contradicts $M \cap N \neq \emptyset$. Thus, $\lambda_0 \in [0, 1)$. But, the assumption (vi) with $x_j = r(y_j) \in M$, $x = y_0 = r(y_0) \in \partial_{K_\varepsilon} M$ implies that $\lambda_j Fr(y_j) \subset M$ for j sufficiently large. This contradicts Eq. (2.9). Hence, $M \cap N = \emptyset$. As a result, there exists $x \in M$ such that $x \in Fr(x) = H(x)$. Q.E.D.

Since every weakly sequentially continuous single valued mapping can be identified as a multi-valued mapping having weakly sequentially closed graph, we obtain the following corollary:

Corollary 2.4.4 *Let Ω be a nonempty, closed, and convex subset of a separable Banach space $(X, \|.\|)$, M be a closed and convex subset of Ω with $\theta \in M$ and $F : M \longrightarrow \Omega$ be a weakly sequentially continuous mapping such that $F(M)$ is relatively weakly compact. In addition, suppose that:*

(i) There exists a weakly sequentially continuous retraction $r : X \longrightarrow M$,

(ii) there exists $\delta > 0$ and a weakly compact set M_δ with:

$K_\delta = \{x \in X$ such that $d(x, M) \leq \delta\} \subset M_\delta$, here $d(x, y) = \|x - y\|$, and

(iii) for any $K_\varepsilon = \{x \in X$ such that $d(x, M) \leq \varepsilon, 0 < \varepsilon \leq \delta\}$, if $\{(x_j, \lambda_j)\}_{j=1}^{\infty}$ is a sequence in $M \times [0, 1]$ with $x_j \rightharpoonup x \in \partial_{K_\varepsilon} M$, $\lambda_j \to \lambda$ and $x = \lambda F(x)$, $0 \leq \lambda < 1$, then, $\lambda_j F(x_j) \subset M$ for j sufficiently large, here, $\partial_{K_\varepsilon} M$ denotes the weak boundary of M in K_ε.

Then, F has, at least, a fixed point in M.

2.5 Some Leray–Schauder's Alternatives

The first result is formulated as:

Theorem 2.5.1 *Let Ω be a nonempty, closed, and convex subset of a Banach space X. In addition, let U be a weakly open subset of Ω with $\theta \in U$, $A : \overline{U^w} \longrightarrow X$ weakly sequentially continuous and $B : X \longrightarrow X$ satisfying:*

(i) $A(\overline{U^w})$ is relatively weakly compact,

(ii) B is linear, bounded and there exists $p \in \mathbb{N}^*$ such that B^p is a separate contraction, and

(iii) $(x = Bx + Ay,\ y \in \overline{U^w}) \Longrightarrow x \in \Omega$.

Then, either $A+B$ has, at least, a fixed point or there is a point $x \in \partial_{\Omega}^{w} U$ (the weak boundary of U in Ω) and a scalar $\lambda \in (0,1)$ with $x = Bx + \lambda Ax$.

Proof. Since B is linear, bounded and B^p is a separate contraction, by using Lemma 1.2.2, $(I-B^p)^{-1}$ exists on X. Hence,

$$(I-B)^{-1} = (I-B^p)^{-1}\sum_{k=0}^{p-1} B^k. \tag{2.10}$$

Using Eq. (2.10), we have $(I-B)^{-1} \in \mathcal{L}(X)$. So, $(I-B)^{-1}$ is weakly continuous. Let us set $F := (I-B)^{-1}A$. Since A acts from $\overline{U^w}$ into X then, from assumption (iii), the mapping F acts from $\overline{U^w}$ into Ω. Since $(I-B)^{-1}$ is weakly continuous and A is weakly sequentially continuous, we deduce that F is weakly sequentially continuous. Moreover, since $A(\overline{U^w})$ is relatively weakly compact and $(I-B)^{-1}$ is weakly continuous, we get $F(\overline{U^w})$ is relatively weakly compact. Consequently, using Theorem 2.3.7 and Remark 2.3.4, we get either F has, at least, a fixed point or there exists a point $x \in \partial_{\Omega}^{w} U$ and a $\lambda \in (0,1)$ such that $x = \lambda F(x)$. This yields, either $A+B$ has, at least, a fixed point or there is a point $x \in \partial_{\Omega}^{w} U$ and a $\lambda \in (0,1)$ such that

$$\frac{x}{\lambda} = (I-B)^{-1}A(x). \tag{2.11}$$

Eq. (2.11) implies that $(I-B)(\frac{x}{\lambda}) = A(x)$. Since B is linear, we get $x = \lambda Ax + Bx$. Q.E.D.

2.5.1 Leray–Schauder's alternatives involving nonlinear contraction mappings

We start this subsection by showing that Theorem 2.5.1 remains true if we suppose that there exists $p \in \mathbb{N}^*$ such that B^p is a nonlinear contraction.

Theorem 2.5.2 *Let Ω be a nonempty, closed, and convex subset of a Banach space X. In addition, let U be a weakly open subset of Ω with $\theta \in U$, $A : \overline{U^w} \longrightarrow X$ weakly sequentially continuous and $B : X \longrightarrow X$ satisfying:*

(i) $A(\overline{U^w})$ is relatively weakly compact,

(ii) B is linear, bounded and there exists $p \in \mathbb{N}^*$ such that B^p is a nonlinear contraction, and

(iii) $(x = Bx + Ay,\ y \in \overline{U^w}) \Longrightarrow x \in \Omega$.

Then, either $A + B$ has, at least, a fixed point or there is a point $x \in \partial_{\Omega}^{w} U$ (the weak boundary of U in Ω) and a scalar $\lambda \in (0, 1)$ with $x = Bx + \lambda Ax$.

Proof. Since B is linear, bounded and B^p is a nonlinear contraction, then $(I - B^p)^{-1}$ exists on X. Now, reasoning as in the proof of Theorem 2.5.1, we get the desired result. Q.E.D.

The next result asserts:

Theorem 2.5.3 *Let Ω be a nonempty, closed, and convex subset of a Banach space X. In addition, let U be a weakly open subset of Ω with $\theta \in U$, $A : \overline{U^w} \longrightarrow X$ and $B : X \longrightarrow X$ two weakly sequentially continuous mappings satisfying:*

(i) A is weakly compact,

(ii) B is a nonlinear contraction mapping, and

(iii) $(A + B)(\overline{U^w})$ is a bounded subset of Ω.

Then, either $A + B$ has, at least, a fixed point or there is a point $x \in \partial_{\Omega}^{w} U$ (the weak boundary of U in Ω) and a scalar $\lambda \in (0, 1)$ with $x = \lambda Ax + \lambda Bx$.

Proof. Let D be a bounded subset of $\overline{U^w}$ such that $\omega(D) > 0$. Taking into account the fact that $A(D)$ is relatively weakly compact, and using the sub-additivity of the De Blasi measure of weak noncompactness, we get

$$\omega((A + B)(D)) \leq \omega(A(D) + B(D)) \leq \omega(A(D)) + \omega(B(D)) \leq \omega(B(D)).$$

Since B is ω-condensing (see Remark 2.3.1), we obtain

$$\omega((A + B)(D)) < \omega(D).$$

Then, $A+B$ is ω-condensing. Moreover, it is easy to show that $A+B$ is weakly sequentially continuous. Hence, the result follows immediately from Theorem 2.3.8. Q.E.D.

Now, let us replace the assumption $(A+B)(\overline{U^w}) \subset \Omega$ by the following weaker one $(x = Bx + Ay,\ y \in \overline{U^w}) \Longrightarrow x \in \Omega$.

Theorem 2.5.4 *Let Ω be a nonempty, closed, and convex subset of a Banach space X. In addition, let U be a weakly open subset of Ω with $\theta \in U$, $A : \overline{U^w} \longrightarrow X$ and $B : X \longrightarrow X$ two weakly sequentially continuous mappings* *satisfying:*

(i) $A(\overline{U^w})$ *is relatively weakly compact,*

(ii) B *is a nonlinear contraction,*

(iii) $(x = Bx + Ay,\ y \in \overline{U^w}) \Longrightarrow x \in \Omega$, *and*

(iv) $(I - B)^{-1}A(\overline{U^w})$ *is a bounded subset of* E.

Then, either $A + B$ *has, at least, a fixed point or there is a point* $x \in \partial_\Omega^w U$ *(the weak boundary of* U *in* Ω*) and a scalar* $\lambda \in (0, 1)$ *with* $x = \lambda Ax + \lambda B(\frac{x}{\lambda})$.

Proof. Let y be fixed in $\overline{U^w}$. The map which assigns to each $x \in \Omega$ the value $Bx + Ay$ defines a nonlinear contraction from Ω into Ω. So, taking into account the fact that $I - B$ is a homeomorphism together with the assumption (iii), the equation $x = Bx + Ay$ has a unique solution $x = (I - B)^{-1}Ay \in \Omega$. Therefore,

$$(I - B)^{-1}A(\overline{U^w}) \subset \Omega.$$

Now, let us define the mapping $F : \overline{U^w} \longrightarrow \Omega$ by:

$$F(x) := (I - B)^{-1}Ax.$$

We claim that set $F(\overline{U^w})$ is relatively weakly compact. Indeed, if it is not the case, then $\omega(F(\overline{U^w})) > 0$. Since

$$F(\overline{U^w}) \subset A(\overline{U^w}) + BF(\overline{U^w}),$$

we get

$$\omega(F(\overline{U^w})) \leq \omega(A(\overline{U^w})) + \omega(BF(\overline{U^w})).$$

Taking into account the fact that $A(\overline{U^w})$ is relatively weakly compact and B is ω-condensing (see Remark 2.3.1), we obtain

$$\omega(F(\overline{U^w})) \leq \omega(BF(\overline{U^w})) < \omega(F(\overline{U^w})),$$

which is absurd. Hence, $F(\overline{U^w})$ is relatively weakly compact. In view of both Theorem 2.3.7 and Remark 2.3.4, we still have to show that $F : \overline{U^w} \longrightarrow \Omega$ is weakly sequentially continuous. In fact, let $(x_n)_n \subset \overline{U^w}$ such that $x_n \rightharpoonup x$. Since $F(\overline{U^w})$ is relatively weakly compact, and using the Eberlein–Šmulian theorem, it follows that there exists a subsequence $(x_{n_k})_k$ of $(x_n)_n$ such that $F(x_{n_k}) \rightharpoonup y$. The weakly sequential continuity of B leads to $BF(x_{n_k}) \rightharpoonup By$. Also, from the equality $BF = -A + F$, it results that

$$-A(x_{n_k}) + F(x_{n_k}) \rightharpoonup -A(x) + y.$$

Hence, $y = F(x)$. We claim that $F(x_n) \rightharpoonup F(x)$. Suppose that this is not the case, then there exists a subsequence $(x_{\varphi_1(n)})_n$ and a weak neighborhood V^w of $(I-B)^{-1}Ax$ such that $(I-B)^{-1}Ax_{\varphi_1(n)} \notin V^w$, for all $n \in \mathbb{N}$. Moreover, we have $x_{\varphi_1(n)} \rightharpoonup x$. Then arguing as before, we find a subsequence $(x_{\varphi_1(\varphi_2(n))})_n$ such that $(I-B)^{-1}Ax_{\varphi_1(\varphi_2(n))}$ converges weakly to $(I-B)^{-1}Ax$, which is a contradiction. Hence, F is weakly sequentially continuous. Consequently, either F has, at least, a fixed point or there is a point $x \in \partial_\Omega^w U$ (the weak boundary of U in Ω) and a $\lambda \in (0,1)$ with $x = \lambda F(x)$. Q.E.D.

2.5.2 Leray–Schauder's alternatives for the sum of two weakly sequentially continuous mappings

Now, we state some new variants of the Leray–Schauder type fixed point theorem for the sum of two weakly sequentially continuous mappings A and B. In that line, we will investigate the case when $I-B$ may not be invertible by looking for the multi-valued mapping $(I-B)^{-1}A$.

Theorem 2.5.5 *Let Ω be a nonempty, closed, and convex subset of a Banach space X. In addition, let U be a weakly open subset of Ω with $\theta \in U$, $A : \overline{U^w} \longrightarrow X$ and $B : \Omega \longrightarrow X$ two weakly sequentially continuous mappings satisfying:*

(i) $A(\overline{U^w})$ is relatively weakly compact,

(ii) $A(\overline{U^w}) \subset (I-B)(\Omega)$,

(iii) if $(I-B)x_n \rightharpoonup y$, then there exists a weakly convergent subsequence of $(x_n)_n$, and

(iv) for every y in the range of $I-B$, $D_y = \{x \in \Omega$ such that $(I-B)x = y\}$ is a convex set.

Then, either $A+B$ has, at least, a fixed point or there is a point $x \in \partial_\Omega^w U$ (the weak boundary of U in Ω) and a $\lambda \in (0,1)$ with $x = \lambda A(x) + \lambda B(\frac{x}{\lambda})$.

Proof. *Case when $I-B$ is invertible*: For any given $y \in \overline{U^w}$, let us define $F : \overline{U^w} \longrightarrow \Omega$ by:

$$Fy := (I-B)^{-1}Ay.$$

F is well defined by assumption (ii).

Step 1: $F(\overline{U^w})$ is relatively weakly compact. For any $(y_n)_n \subset F(\overline{U^w})$, we choose $(x_n)_n \subset \overline{U^w}$ such that $y_n = F(x_n)$. Taking into account assumption

(i), together with Eberlein–Šmulian's theorem (see Theorem 1.3.3), we get a subsequence $(y_{\varphi_1(n)})_n$ of $(y_n)_n$ such that $(I-B)y_{\varphi_1(n)} \rightharpoonup z$, for some $z \in \Omega$. Thus, by assumption (iii), there exists a subsequence $y_{\varphi_1(\varphi_2(n))}$ converging weakly to $y_0 \in \Omega$. Hence, $F(\overline{U^w})$ is relatively weakly compact.

Step 2: F is weakly sequentially continuous. Let $(x_n)_n \subset \overline{U^w}$ such that $x_n \rightharpoonup x$. Because $F(\overline{U^w})$ is relatively weakly compact, and using Eberlein–Šmulian's theorem, it follows that there exists a subsequence $(x_{n_k})_k$ of $(x_n)_n$ such that $F(x_{n_k}) \rightharpoonup y$, for some $y \in \Omega$. The weakly sequentially continuity of B leads to $BF(x_{n_k}) \rightharpoonup By$. Also, from the equality $BF = -A + F$, it results that

$$-A(x_{n_k}) + F(x_{n_k}) \rightharpoonup -A(x) + y.$$

So, $y = F(x)$. We claim that $F(x_n) \rightharpoonup F(x)$. Suppose that this is not the case, then there exists a subsequence $(x_{\varphi_1(n)})_n$ and a weak neighborhood V^w of $(I-B)^{-1}A(x)$ such that $(I-B)^{-1}A\left(x_{\varphi_1(n)}\right) \notin V^w$, for all $n \in \mathbb{N}$. Moreover, we have $x_{\varphi_1(n)} \rightharpoonup x$. Then, arguing as before, we find a subsequence $(x_{\varphi_1(\varphi_2(n))})_n$ such that $(I-B)^{-1}A(x_{\varphi_1(\varphi_2(n))})$ converges weakly to $(I-B)^{-1}A(x)$, which is a contradiction. Hence, F is weakly sequentially continuous. Consequently, combining Theorem 2.3.7 and Remark 2.3.4, we get either F has, at least, a fixed point or there exists a point $x \in \partial_\Omega^w U$ and a $\lambda \in (0,1)$ such that $x = \lambda F(x)$. This yields, either $A+B$ has, at least, a fixed point or there is a point $x \in \partial_\Omega^w U$ and a scalar $\lambda \in (0,1)$ such that

$$\frac{x}{\lambda} = (I-B)^{-1}A(x). \tag{2.12}$$

Eq. (2.12) implies $(I-B)(\frac{x}{\lambda}) = A(x)$. So, $x = \lambda A(x) + \lambda B(\frac{x}{\lambda})$.

Case when $I-B$ is not invertible: The map $(I-B)^{-1}$ could be seen as a multi-valued mapping. For any given $y \in \overline{U^w}$, define $H : \overline{U^w} \longrightarrow P(\Omega)$ by:

$$H(y) := (I-B)^{-1}A(y).$$

H is well defined by assumption (ii). We should prove that H fulfills the hypotheses of Theorem 2.4.7.

Step 1: $H(x)$ is a convex set for each $x \in \overline{U^w}$. This is an immediate consequence of assumption (iv).

Step 2: H has a weakly sequentially closed graph. Let $(x_n)_n \subset \overline{U^w}$ such that $x_n \rightharpoonup x$ and $y_n \in H(x_n)$ such that $y_n \rightharpoonup y$. By the definition of H, we have $(I-B)(y_n) = A(x_n)$. Since A and $I-B$ are weakly sequentially continuous, we obtain $(I-B)(y) = A(x)$. Thus, $y \in (I-B)^{-1}A(x)$.

Step 3: $H(\overline{U^w})$ is relatively weakly compact. This assertion is proved by using the same reasoning as in Step 1 of the first part of the proof. Hence, H is ω-condensing. In view of Theorem 2.4.7, either H has, at least, a fixed point; or there is a point $x \in \partial_{\Omega}^{w} U$ and a scalar $\lambda \in (0,1)$ with $x \in \lambda H(x)$. By the definition of H, the last assertion implies that either there is a point $x \in \partial_{\Omega}^{w} U$ such that $(I - B)(x) = A(x)$; or there is a point $x \in \partial_{\Omega}^{w} U$ and a scalar $\lambda \in (0,1)$ such that $(I - B)(\frac{x}{\lambda}) = A(x)$. This leads to either $A + B$ has, at least, a fixed point or there is a point $x \in \partial_{\Omega}^{w} U$ and a scalar $\lambda \in (0,1)$ with $x = \lambda A(x) + \lambda B(\frac{x}{\lambda})$. Q.E.D.

Corollary 2.5.1 *Let Ω be a nonempty, closed, and convex subset of a Banach space X. In addition, let U be a weakly open subset of Ω with $\theta \in U$, $A : \overline{U^w} \longrightarrow X$ and $B : \Omega \longrightarrow X$ are two weakly sequentially continuous mappings satisfying:*

(i) $A(\overline{U^w})$ is relatively weakly compact,

(ii) B is a contraction mapping such that $B(\Omega)$ is bounded, and

(iii) $A(\overline{U^w}) + B(\Omega) \subset \Omega$.

Then, either $A+B$ has, at least, a fixed point or there is a point $x \in \partial_{\Omega}^{w} U$ (the weak boundary of U in Ω) and a scalar $\lambda \in (0,1)$ with $x = \lambda A(x) + \lambda B(\frac{x}{\lambda})$.

Proof. The result follows immediately from Theorem 2.5.5. Indeed, since B is a nonlinear contraction, then taking into account Remarks 2.3.1 and 2.4.1, we get that B satisfies the assumption *(iii)* of Theorem 2.5.5. Moreover, we have $I - B$ is a homeomorphism. So, for every y in the range of $I - B$, the set $D_y = \{x \in \Omega \text{ such that } (I - B)x = y\}$ is reduced to $\{(I - B)^{-1} y\}$, which is convex. Q.E.D.

The next theorem extends a result of H. Schaefer [145] to the case of multi-valued mappings in the context of weak topology, dealing with the method of a priori estimate in the Leray–Schauder theory.

Theorem 2.5.6 *Let X be a Banach space and $H : X \longrightarrow \mathcal{P}(X)$ a multi-valued mapping. Suppose that*

(i) H has a weakly sequentially closed graph,

(ii) there exists a closed convex, balanced, and absorbing weak neighborhood U of θ such that the set $H(mU)$ is relatively weakly compact for all $m \in \mathbb{N}$, and

(iii) the set $H(x)$ is closed, convex and nonempty for all $x \in X$.

Then, either for any $\lambda \in [0,1]$ there exists an x such that

$$x \in \lambda H(x) \tag{2.13}$$

or the set $\{x \in X : \exists\ \lambda \in]0,1[\ ,\ x \in \lambda H(x)\}$ is unbounded.

Proof. Denote by p the Minkowski functional of the set U. Since X endowed with its weak topology is locally convex, we get p is a weakly continuous seminorm and $U = \{x \in X$ such that $p(x) \leq 1\}$. Clearly, θ is the unique solution of Eq. (2.13) for $\lambda = 0$. If for $\lambda_0 \in (0,1]$, there is no solution of Eq. (2.13) for $\lambda = \lambda_0$, we consider the weakly sequentially closed multi-valued mapping G defined by:

$$G(x) = \lambda_0 H(x),$$

for all $x \in X$ and we shall show that for any natural m there exists $y_m \in \eta_m H(y_m)$ with $0 < \eta_m < 1$ and $p(y_m) = n$. To do this, let m be a natural number and define a weakly continuous retraction $r_m : X \longrightarrow mU$ by $r_m(x) = x$ for all $x \in mU$ and $r_m(x) = \frac{mx}{p(x)}$ for all x such that $p(x) > m$. Consider the composition $H_m = G \circ r_m$. In the following, we will prove that H_m satisfies the conditions of Theorem 2.4.3.

Step 1: H_m is weakly sequentially closed. Let $x \in X$, $(x_n)_n \in X$ such that $x_n \rightharpoonup x$ and $y_n \in H_m(x_n)$ such that $y_n \rightharpoonup y$. Since the retraction r_m is weakly continuous, we get $r_m(x_n) \rightharpoonup r_m(x)$. On the other hand, we have $y_n \in G(r_m(x_n))$, $y_n \rightharpoonup y$ and G is weakly sequentially closed. So, $y \in G(r_m(x))$, i.e., $y \in H_m(x)$. Consequently, H_m has a weakly sequentially closed graph.

Step 2: $H_m(X)$ is relatively weakly compact. The assumption follows from the fact that $H_m(X) = G(mU)$ and the hypothesis (ii).

Step 3: $H_m(x)$ is closed, convex, and nonempty for all $x \in X$. This is an immediate consequence of (iii).

Consequently, by Theorem 2.4.3, H_m has, at least, a fixed point x_m in X, i.e., there exists an x_m such that $x_m \in \lambda_0 H(r_m(x_m))$. Notice that the case $p(x_m) \leq m$ cannot occur, otherwise we get $x_m = \lambda_0 H(x_m)$ which contradicts our assumption. Hence, $p(x_m) > m$ and thus

$$\frac{r_m(x_m)p(x_m)}{n} = \lambda_0 H(r_m(x_m)).$$

This gives that $y_m = \eta_m H(y_m)$ with $y_m = r_m(x_m)$, $\eta_m = \frac{n\lambda_0}{p(x_m)} < 1$ and $p(y_m) = n$. Q.E.D.

Since every weakly sequentially continuous single-valued mapping can be identified with a multi-valued mapping having a weakly sequentially closed graph, we obtain the following corollary.

Corollary 2.5.2 *Let X be a Banach space and $F : X \longrightarrow X$ be a weakly sequentially continuous mapping. Assume that there exists a closed, convex, balanced, and absorbing weak neighborhood U of θ such that the set $F(mU)$ is relatively weakly compact for all $m \in \mathbb{N}$. Then, either for any $\lambda \in [0,1]$ there exists an x such that*

$$x = \lambda F(x)$$

or the set $\{x \in X : \exists\ \lambda \in]0,1[\ ,\ x = \lambda F(x)\}$ is unbounded.

Having obtained these results, we are ready to state the following Krasnosel'skii–Schaefer type fixed point theorem in the setting of weak topology.

Theorem 2.5.7 *Let X be a Banach space and A, $B : X \longrightarrow X$ two weakly sequentially continuous mappings satisfying:*

(i) $A(X) \subset (I - B)(X)$,

(ii) there exists a closed convex, balanced and absorbing weak neighborhood U of θ such that the set $A(nU)$ is relatively weakly compact for all $n \in \mathbb{N}$,

(iii) if $(I - B)x_n \rightharpoonup y$, then there exists a weakly convergent subsequence $(x_{n_k})_k$ of $(x_n)_n$, and

(iv) for every y in the range of $I - B$, $D_y = \{x \in X : (I - B)x = y\}$ is convex.

Then, either for any $\lambda \in [0,1]$ there exists an $x \in X$ such that $x = \lambda B(\frac{x}{\lambda}) + \lambda Ax$ or the set $\{x \in X : \exists\ \lambda \in]0,1[,\ x = \lambda B(\frac{x}{\lambda}) + \lambda Ax\}$ is unbounded.

Proof. First, we assume that $I - B$ is invertible. For any given $y \in \Omega$, define $F : X \longrightarrow X$ by:

$$Fy := (I - B)^{-1}Ay.$$

F is well defined by assumption (i).

Step 1: F is weakly sequentially continuous. Let $(y_n)_n = ((I - B)x_n)_n$ be a sequence in $(I - B)(X)$ such that $y_n \rightharpoonup y$. By item (iii), there exists a subsequence $(x_{\varphi(n)})_n$ converging weakly to $x' \in X$. The weakly sequentially continuity of $I - B$ leads to $(I - B)x_{\varphi(n)} \rightharpoonup (I - B)x'$. So, $y = (I - B)x'$ and then $x' = (I - B)^{-1}y$. Using the same reasoning as in the proof of Theorem 2.3.6, we get $x_n \rightharpoonup (I-B)^{-1}y$. Then, $(I-B)^{-1}$ is weakly sequentially continuous. Since A is weakly sequentially continuous, then it is so for F.

Step 2: $F(nU)$ is relatively weakly compact. The result can be seen in the same way as in Step 1 of the first part of the proof of Theorem 2.4.4. Consequently, using Corollary 2.5.2, we get the desired result.

Second, if $I-B$ is not invertible, $(I-B)^{-1}$ could be seen as a multi-valued mapping. For any given $y \in \Omega$, define $H : X \longrightarrow P(X)$ by:

$$Hy := (I-B)^{-1}Ay.$$

H is well defined by assumption (i). Now, arguing as in the proof of the second part of Theorem 2.4.4, we prove that H satisfies the hypotheses of Theorem 2.5.6. So, using this theorem, we get the desired result. Q.E.D.

Corollary 2.5.3 *Let E be a Banach space and A, $B : X \longrightarrow X$ two weakly sequentially continuous mappings satisfying:*

(i) $A(X) \subset (I-B)(X)$,

(ii) there exists a closed convex, balanced, and absorbing weak neighborhood U of θ such that the set $A(nU)$ is relatively weakly compact for all $n \in \mathbb{N}$, and

(iii) B is a nonlinear contraction so that $B(X)$ is bounded.

Then, either for any $\lambda \in [0,1]$ there exists an $x \in X$ such that $x = \lambda B(\frac{x}{\lambda}) + \lambda Ax$ or the set $\{x \in X : \exists\ \lambda \in]0,1[,\ x = \lambda B(\frac{x}{\lambda}) + \lambda Ax\}$ is unbounded.

Proof. The result follows immediately from Theorem 2.5.7 and Remark 2.4.1 (i). Q.E.D.

2.5.3 Furi–Pera's fixed point theorem for the sum of two weakly sequentially continuous mappings

We end this section with a Furi–Pera fixed point theorem for the sum of two weakly sequentially continuous mappings.

Theorem 2.5.8 *Let Ω be a nonempty, closed, and convex subset of a Banach space $(X, \|.\|)$. Assume that M is a closed convex subset of Ω with $\theta \in M$, $A : M \longrightarrow X$ and $B : \Omega \longrightarrow X$ two weakly sequentially continuous mappings such that:*

(i) $A(M)$ is relatively weakly compact and $A(M) \subset (I-B)(\Omega)$,

(ii) if $(I-B)x_n \rightharpoonup y$*, then there exists a weakly convergent subsequence of* $(x_n)_n$*,*

(iii) for every y *in the range of* $I-B$*,* $D_y = \{x \in \Omega \text{ such that } (I-B)x = y\}$ *is a convex set.*

In addition, suppose that:

(iv) There exists a weakly sequentially continuous retraction $r : X \longrightarrow M$*,*

(v) there exist $\delta > 0$ *and a weakly compact set* M_δ *with* $K_\delta = \{x \in X \text{ such that } d(x,M) \leq \delta\} \subset M_\delta$*, here,* $d(x,y) = \|x-y\|$*, and*

(vi) for any $K_\varepsilon = \{x \in X \text{ such that } d(x,M) \leq \varepsilon, 0 < \varepsilon \leq \delta\}$*, if* $\{(x_j,\lambda_j)\}_{j=1}^{\infty}$ *is a sequence in* $M \times [0,1]$ *with* $x_j \rightharpoonup x \in \partial_{K_\varepsilon} M$*,* $\lambda_j \to \lambda$ *and* $x = \lambda A(x) + \lambda B(\frac{x}{\lambda})$*,* $0 < \lambda < 1$*, then* $\lambda_j (I-B)^{-1} A(x_j) \subset M$ *for* j *sufficiently large, here* $\partial_{K_\varepsilon} M$ *denotes the weak boundary of* M *in* K_ε*.*

Then, $A+B$ *has, at least, a fixed point in* M*.*

Proof. *Case when* $I-B$ *is invertible*: For any given $y \in M$, define $F : M \longrightarrow \Omega$ by:

$$Fy := (I-B)^{-1} Ay.$$

Step 1: Arguing as in Step 1 of the first part of the proof of Theorem 2.5.5, we get $F(M)$ is relatively weakly compact.

Step 2: F is weakly sequentially continuous. Using Step 1 and making the same reasoning as in Step 2 of the first part of the proof of Theorem 2.5.5, we get F is weakly sequentially continuous.

Step 3: For any $K_\varepsilon = \{x \in X : d(x,M) \leq \varepsilon, 0 < \varepsilon \leq \delta\}$, if $\{(x_j,\lambda_j)\}_{j=1}^{\infty}$ is a sequence in $M \times [0,1]$ with $x_j \rightharpoonup x \in \partial_{K_\varepsilon} M$, $\lambda_j \to \lambda$ and $x = \lambda F(x)$, $0 < \lambda < 1$, then $\lambda_j F(x_j) \subset M$ for j sufficiently large. This is an immediate consequence of assumption (vi). Consequently, taking into account the assumptions (iv) and (v), and using Corollary 2.4.4, we deduce that there exists $x \in M$ such that $x = F(x)$. This implies that $A+B$ has, at least, a fixed point in M.

Case when $I-B$ *is not invertible*: The map $(I-B)^{-1}$ could be seen as a multi-valued mapping. For any given $y \in M$, define $H : M \longrightarrow P(\Omega)$ by:

$$Hy := (I-B)^{-1} Ay.$$

We should prove that H fulfills the hypotheses of Theorem 2.4.8.

Step 1: $H(x)$ is a convex set for each $x \in M$. This is an immediate consequence of assumption (iii).

Step 2: H has a weakly sequentially closed graph. The assumption may be seen in the same way as the one in Step 2 of the second part of the proof of Theorem 2.5.5.

Step 3: $H(M)$ is relatively weakly compact.

Step 4: For any $K_\varepsilon = \{x \in X : d(x, M) \leq \varepsilon, 0 < \varepsilon \leq \delta\}$, if $\{(x_j, \lambda_j)\}_{j=1}^{\infty}$ is a sequence in $M \times [0, 1]$ with $x_j \rightharpoonup x \in \partial_{K_\varepsilon} M$, $\lambda_j \to \lambda$ and $x \in \lambda H(x)$, $0 < \lambda < 1$, then $\lambda_j H(x_j) \subset M$ for j sufficiently large. The assumption follows from assumption (vi). In view of Theorem 2.4.8, we deduce that H has, at least, a fixed point in M. Q.E.D.

Chapter 3

Fixed Point Theory in Banach Algebras

In the present chapter, we investigate a class of Banach algebras satisfying the condition $(\mathcal{P})$ (see Definition 1.5.2). The central goal is to prove some new fixed point theorems under a weak topology setting for maps acting on a nonempty, closed, and convex subset of a Banach algebra satisfying or not the condition $(\mathcal{P})$. Our main conditions are formulated in terms of weak sequential continuity, dealing with three nonlinear operators. Moreover, no weak continuity conditions are required for this work. In addition, some fixed results using the notion of WC–Banach algebras are discussed.

3.1 Fixed Point Theorems Involving Three Operators

In 1988, B. C. Dhage in [66] proved a fixed point theorem involving three operators in a Banach algebra by combining the Banach's fixed point theorem with Schauder's fixed point principle.

Theorem 3.1.1 *Let S be a closed, convex, and bounded subset of a Banach algebra X and let A, B, $C : S \longrightarrow S$ be three operators such that:*

(i) A and C are Lipschitzian with Lipschitz constants α and β, respectively,

(ii) $\left(\frac{I-C}{A}\right)^{-1}$ exists on $B(S)$, I being the identity operator on X,

(iii) B is completely continuous, and

(iv) $Ax.By + Cx \in S \quad \forall\ x,\, y \in S$.

Then, the operator equation

$$Ax.Bx + Cx = x \tag{3.1}$$

has a solution whenever $\alpha M + \beta < 1$, *where*

$$M = \|B(S)\| = \sup\{\|Bx\| \ : \ x \in S\}.$$

Remark 3.1.1 *Notice that the symbol* $\left(\frac{I-C}{A}\right)$ *means the mapping defined by:*

$$\left(\frac{I-C}{A}\right)(x) = (x - Cx).(Ax)^{-1},$$

where $(Ax)^{-1}$ *denotes the inverse of* Ax *in the Banach algebra* X.

Theorem 3.1.2 *Let* X *be a Banach algebra and let* S *be a nonempty, closed, and convex subset of* X. *Let* $A, C : X \longrightarrow X$ *and* $B : S \longrightarrow X$ *be three operators such that:*

(i) $\left(\frac{I-C}{A}\right)^{-1}$ *exists on* $B(S)$,

(ii) $\left(\frac{I-C}{A}\right)^{-1} B$ *is weakly sequentially continuous,*

(iii) $\left(\frac{I-C}{A}\right)^{-1} B(S)$ *is relatively weakly compact, and*

(iv) $x = Ax.By + Cx \Longrightarrow x \in S$, *for all* $y \in S$.

Then, Eq. (3.1) has, at least, one solution in S.

Proof. From assumption (i), it follows that, for each y in S, there exists a unique $x_y \in X$ such that

$$\left(\frac{I-C}{A}\right)x_y = By. \tag{3.2}$$

or, equivalently

$$Ax_y.By + Cx_y = x_y. \tag{3.3}$$

Since the hypothesis (iv) holds, then $x_y \in S$. Therefore, we can define

$$\begin{cases} \mathcal{N} : S \longrightarrow S \\ \qquad y \longrightarrow \mathcal{N}y = \left(\dfrac{I-C}{A}\right)^{-1} By. \end{cases}$$

By using the hypotheses (ii), (iii), combined with Theorem 2.2.1, we conclude that $\mathcal{N}$ has, at least, a fixed point y in S. Hence, y verifies Eq. (3.1). Q.E.D.

3.1.1 Fixed point theorems for $\mathcal{D}$-Lipschitzian mappings

Theorem 3.1.3 *Let* S *be a nonempty, closed, convex, and bounded subset of a Banach algebra* X, *and let* $A : X \longrightarrow X$, $B : S \longrightarrow X$ *be two operators such that:*

(i) A is $\mathcal{D}$-Lipschitzian with a $\mathcal{D}$-function Φ,

(ii) $\left(\frac{I}{A}\right)^{-1}$ exists on $B(S)$, I being the identity operator on X,

(iii) B is completely continuous, and

(iv) $x = Ax.By \Longrightarrow x \in S$, for all $y \in S$.

Then, the operator equation $Ax.Bx = x$ has a solution whenever $M\Phi(r) < r$, $r > 0$ where $M = \|B(S)\|$.

Proof. Let us define an operator T on S by:

$$T = \left(\frac{I}{A}\right)^{-1} B,$$

where I represents the identity operator on X. The conclusion of the theorem follows if we show that T is well defined and also maps S into itself. Since $\left(\frac{I}{A}\right)^{-1}$ exists on $B(S)$, the composition $\left(\frac{I}{A}\right)^{-1} B$ is a well-defined map from S into X. We claim that

$$\left(\frac{I}{A}\right)^{-1} B : S \longrightarrow S. \tag{3.4}$$

In order to prove this, it is sufficient to show that

$$B(S) \subset \left(\frac{I}{A}\right)(S). \tag{3.5}$$

For this purpose, let $y \in S$ be fixed and let us define the operator

$$\begin{cases} A_y : X \longrightarrow X \\ \qquad x \longrightarrow Ax.By. \end{cases}$$

For any $x_1,\ x_2 \in X$, by using hypothesis (i), we have

$$\|A_y(x_1) - A_y(x_2)\| \leq \|Ax_1 - Ax_2\|\|By\| \leq M\Phi(\|x_1 - x_2\|).$$

Since $M\Phi(r) < r$, $r > 0$, this shows that A_y is a nonlinear contraction on X. Hence, by applying a fixed point theorem of Boyd and Wong (see Theorem 1.6.10), there exists a unique point x^* in X, such that

$$A_y(x^*) = x^* = Ax^*.By. \tag{3.6}$$

By using hypothesis (iv), $x^* \in S$. Moreover, Eq. (3.6) yields

$$\left(\frac{I}{A}\right) x^* = By.$$

This proves the claim (3.5) and consequently, (3.4). It is easy to check that $\left(\frac{I}{A}\right)^{-1}$ is well defined on $B(S)$. Since T is a composition of a continuous and a completely continuous operator, then it is a completely continuous operator on S. Now, we may apply Schauder's fixed point theorem in order to deduce the desired result. Q.E.D.

Corollary 3.1.1 *Let S be a nonempty, closed, convex, and bounded subset of a Banach algebra X, and let $A : X \longrightarrow X$, $B : S \longrightarrow X$ be two operators, such that:*

(i) A is Lipschitzian with a Lipschitz constant α,

(ii) $\left(\frac{I}{A}\right)$ is well defined and one to one,

(iii) B is completely continuous, and

(iv) $x = Ax.By \Longrightarrow x \in S$, for all $y \in S$.

Then, the operator equation $Ax.Bx = x$ has a solution whenever $\alpha M < 1$, where $M = \|B(S)\|$.

Below, we give a sufficient condition that guarantees hypothesis (iv) of Theorem 3.1.3. Let us consider the equation

$$x = Ax.By,$$

which implies that

$$\left(\frac{I}{A}\right) x = By.$$

Moreover, this also implies that

$$\left\|\left(\frac{I}{A}\right) x\right\| = \|By\|.$$

Proposition 3.1.1 *Let S be a nonempty, closed, convex, and bounded subset of a Banach algebra X such that $S = \{y \in X, \ \|y\| \leq r\}$ for some real number $r > 0$. Let $A : X \longrightarrow X$, $B : S \longrightarrow X$ be two operators satisfying the hypotheses (i)–(iii) of Theorem 3.1.3. Further, if*

$$\|x\| \leq \left\|\left(\frac{I}{A}\right) x\right\|, \tag{3.7}$$

for all $x \in X$, then $x \in S$.

Proof. The proof follows immediately from Eq. (3.7). Q.E.D.

Proposition 3.1.2 *Let X be a Banach algebra, and let S be a nonempty, closed, and convex subset of X. Let A, $C : X \longrightarrow X$ and $B : S \longrightarrow X$ be three operators such that:*

(i) A and C are $\mathcal{D}$-Lipschitzian with the $\mathcal{D}$-functions ϕ_A and ϕ_C, respectively,

(ii) A is regular on X, i.e., A maps X into the set of all invertible elements of X, and

(iii) B is a bounded function with a bound M.

Then, $\left(\frac{I-C}{A}\right)^{-1}$ exists on $B(S)$ whenever $M\phi_A(r) + \phi_C(r) < r$, for $r > 0$.

Proof. Let y be fixed in S, and let us define the mapping

$$\begin{cases} \varphi_y : X \longrightarrow X \\ \qquad x \longrightarrow \varphi_y(x) = Ax.By + Cx. \end{cases}$$

Let x_1, $x_2 \in X$. The use of the assumption (i) leads to

$$\begin{aligned} \|\varphi_y(x_1) - \varphi_y(x_2)\| &\leq \|Ax_1.By - Ax_2.By\| + \|Cx_1 - Cx_2\| \\ &\leq \|Ax_1 - Ax_2\|\|By\| + \|Cx_1 - Cx_2\| \\ &\leq M\phi_A(\|x_1 - x_2\|) + \phi_C(\|x_1 - x_2\|). \end{aligned}$$

Now, by applying a fixed point theorem of Boyd and Wong (see Theorem 1.6.10), we deduce that there exists a unique element $x_y \in X$ such that

$$\varphi_y(x_y) = x_y.$$

Hence, x_y verifies Eq. (3.3) and so, by virtue of the hypothesis (ii), x_y verifies Eq. (3.2). Therefore, the mapping $\left(\frac{I-C}{A}\right)^{-1}$ is well defined on $B(S)$, and

$$\left(\frac{I-C}{A}\right)^{-1} By = x_y,$$

which gives the desired result. Q.E.D.

In what follows, we will combine Theorem 3.1.2 and Proposition 3.1.2 in order to obtain the following fixed point theorems in Banach algebras.

Theorem 3.1.4 *Let X be a Banach algebra and let S be a nonempty, closed,* *and convex subset of X. Let A, $C : X \longrightarrow X$ and $B : S \longrightarrow X$ be three operators such that:*

(i) A and C are $\mathcal{D}$-Lipschitzian with the $\mathcal{D}$-functions ϕ_A and ϕ_C respectively,

(ii) A is regular on X,

(iii) B is strongly continuous,

(iv) $B(S)$ is bounded with a bound M,

(v) $\left(\frac{I-C}{A}\right)^{-1}$ is weakly compact on $B(S)$, and

(vi) $x = Ax.By + Cx \Longrightarrow x \in S$, for all $y \in S$.

Then, Eq. (3.1) has, at least, one solution in S whenever $M\phi_A(r)+\phi_C(r) < r$, for all $r > 0$.

Proof. From Proposition 3.1.2, it follows that $\left(\frac{I-C}{A}\right)^{-1}$ exists on $B(S)$. By virtue of assumption (vi), we obtain

$$\left(\frac{I-C}{A}\right)^{-1} B(S) \subset S.$$

Moreover, the use of the hypotheses (iv) and (v) implies that

$$\left(\frac{I-C}{A}\right)^{-1} B(S)$$

is relatively weakly compact. Now, we may show that

$$\left(\frac{I-C}{A}\right)^{-1} B$$

is weakly sequentially continuous. To do so, let $\{u_n\}$ be any sequence in S such that $u_n \rightharpoonup u$ in S. By virtue of the assumption (iii), we have

$$Bu_n \to Bu.$$

Since $\left(\frac{I-C}{A}\right)^{-1}$ is a continuous mapping on $B(S)$, we deduce that

$$\left(\frac{I-C}{A}\right)^{-1} Bu_n \to \left(\frac{I-C}{A}\right)^{-1} Bu.$$

This shows that $\left(\frac{I-C}{A}\right)^{-1} B$ is weakly sequentially continuous. Finally, an application of Theorem 2.2.1 shows that Eq. (3.1) has a solution in S. Q.E.D.

Theorem 3.1.5 *Let S be a nonempty, closed, and convex subset of a Banach algebra X. Let A, $C : X \longrightarrow X$ and $B : S \longrightarrow X$ be three operators such that:*

(i) A and C are $\mathcal{D}$-Lipschitzian with the $\mathcal{D}$-functions ϕ_A and ϕ_C, respectively,

(ii) B is weakly sequentially continuous and $B(S)$ is relatively weakly compact,

(iii) A is regular on X,

(iv) $\left(\frac{I-C}{A}\right)^{-1}$ is weakly sequentially continuous on $B(S)$, and

(v) $x = Ax.By + Cx \Longrightarrow x \in S$, for all $y \in S$.

Then, Eq. (3.1) has, at least, one solution in S whenever $M\phi_A(r)+\phi_C(r) < r$, for all $r > 0$.

Proof. Similarly to the proof of the preceding Theorem 3.1.4, we show that $\left(\frac{I-C}{A}\right)^{-1}$ exists on $B(S)$, and

$$\left(\frac{I-C}{A}\right)^{-1} B(S) \subset S.$$

Since $\left(\frac{I-C}{A}\right)^{-1}$ and B are weakly sequentially continuous, then by composition, we show that

$$\left(\frac{I-C}{A}\right)^{-1} B$$

is weakly sequentially continuous. Finally, we claim that

$$\left(\frac{I-C}{A}\right)^{-1} B(S)$$

is relatively weakly compact. To see this, let $\{u_n\}$ be any sequence in S, and let

$$v_n = \left(\frac{I-C}{A}\right)^{-1} Bu_n.$$

Since $B(S)$ is relatively weakly compact, we deduce that there is a renamed subsequence $\{Bu_n\}$ weakly converging to an element w. This fact, together with hypothesis (iv), implies that

$$v_n = \left(\frac{I-C}{A}\right)^{-1} Bu_n \rightharpoonup \left(\frac{I-C}{A}\right)^{-1} w.$$

We infer that $\left(\frac{I-C}{A}\right)^{-1} B(S)$ is sequentially relatively weakly compact. An application of Eberlein–Šmulian's theorem (see Theorem 1.3.3) implies that

$$\left(\frac{I-C}{A}\right)^{-1} B(S)$$

is relatively weakly compact, which proves our claim. The result is deduced immediately from Theorem 3.1.2. Q.E.D.

In [70], B. C. Dhage gave a proof of the next theorem in the case of a Lipschitzian mapping. Here, we give a proof for the case of $\mathcal{D}$-Lipschitzian maps.

Theorem 3.1.6 *Let S be a closed, convex, and bounded subset of a Banach algebra X, and let A, $C : X \longrightarrow X$ and $B : S \longrightarrow X$ be three operators such that:*

(i) A and C are $\mathcal{D}$-Lipschitzian with the $\mathcal{D}$-functions ϕ_A and ϕ_C, respectively,

(ii) B is completely continuous, and

(iii) $x = Ax.By + Cx \Longrightarrow x \in S$, for all $y \in S$.

Then, the operator $A.B+C$ has, at least, a fixed point in S whenever $M\phi_A(r)+\phi_C(r) < r$, for $r > 0$ and $Id_{\mathbb{R}} - \big(M\phi_A + \phi_C\big)$ is strictly increasing, where $M = \|B(S)\|$.

Proof. Let $y \in S$ and let us define a mapping

$$\begin{cases} A_y : X \longrightarrow X \\ \qquad x \longrightarrow Ax.By + Cx. \end{cases}$$

Notice that this operator is a nonlinear contraction on X, with a $\mathcal{D}$-function ψ given by:

$$\psi(r) = M\phi_A(r) + \phi_C(r) < r, \text{ for } r \in \mathbb{R}^+.$$

To see this, let us observe that

$$\begin{aligned} \|A_y(x_1) - A_y(x_2)\| &\leq \|Ax_1 - Ax_2\|\|By\| + \|Cx_1 - Cx_2\| \\ &\leq (M\phi_A + \phi_C)(\|x_1 - x_2\|) \end{aligned}$$

for any x_1, $x_2 \in X$. Now, by applying a fixed point theorem of Boyd and Wong (see Theorem 1.6.10), we deduce that there exists a unique point $z \in X$ such that

$$A_y(z) = z$$

or, equivalently

$$Az.By + Cz = z.$$

Since the hypothesis (iii) holds for all $y \in S$, then we have $z \in S$. Let us define a mapping

$$\begin{cases} N : S \longrightarrow X \\ \qquad y \longrightarrow z \end{cases}$$

where z is the unique solution of the equation

$$Az.By + Cz = z, \ \ y \in S.$$

Now, let us show that N is continuous. To do this, let $\{y_n\}$ be any sequence in S converging to a point y and set $z_n = Ny_n$. Since S is closed, then $y \in S$. Moreover, let us notice that

$$\begin{aligned}\|Ny_n - Ny\| &= \|Az_n.By_n + Cz_n - Az.By - Cz\| \\ &\leq \|Az_n.By_n - Az.By\| + \|Cz_n - Cz\| \\ &\leq (M\phi_A + \phi_C)(\|z_n - z\|) + \|ANy\|\|By_n - By\|.\end{aligned}$$

Hence,

$$\overline{L} \leq (M\phi_A + \phi_C)\,\overline{L} + \|ANy\| \limsup_n \|By_n - By\|,$$

where

$$\overline{L} := \limsup_n \|Ny_n - Ny\|.$$

This shows that,

$$\lim_{n \to +\infty} \|Ny_n - Ny\| = 0$$

and consequently, N is continuous on S. Next, we may show that N is a compact operator on S. In fact, for any point $z \in S$, we have

$$\|Az\| \leq \|Aa\| + \phi_A(\|z - a\|) < \|Aa\| + \frac{\|z - a\|}{M} \leq c,$$

where

$$c = \|Aa\| + \frac{\operatorname{diam}(S)}{M}$$

for some point a in S. Let $\varepsilon > 0$ be given. Since $B(S)$ is a totally bounded subset, there exists a subset $Y = \{y_1, \ldots, y_n\}$ of points in S such that

$$B(S) \subset \bigcup_{i=1}^{n} B_\delta(w_i),$$

where $w_i = By_i$ and $\delta = \frac{1}{c}(\varepsilon - (M\phi_A(\varepsilon) + \phi_C(\varepsilon)))$, and $B_\delta(w_i)$ is an open ball in X centered at w_i of radius δ. Therefore, for any y in S, we have a y_k in Y such that

$$c\|By_k - By\| < \varepsilon - \big(M\phi_A(\varepsilon) + \phi_C(\varepsilon)\big). \tag{3.8}$$

We also have

$$\begin{aligned}\|Ny_k - Ny\| &\leq \|Az_k.By_k - Az.By\| + \|Cz_k - Cz\| \\ &\leq \|Az_k - Az\|\|By_k\| + \|Az\|\|By_k - By\| + \|Cz_k - Cz\| \\ &\leq (M\phi_A + \phi_C)(\|z_k - z\|) + \|Az\|\|By_k - By\| \\ &< (M\phi_A + \phi_C)(\|z_k - z\|) + c\|By_k - By\|.\end{aligned}$$

Then,

$$\Big(Id_{\mathbb{R}} - (M\phi_A + \phi_C)\Big)(\|Ny_k - Ny\|) < c\|By_k - By\|.$$

So, by using Inequality (3.8), we have

$$\|Ny_k - Ny\| < \varepsilon.$$

This is true for every $y \in S$, and so

$$N(S) \subset \bigcup_{i=1}^{n} B_\varepsilon(z_i),$$

where $z_i = Ny_i$. As a result, $N(S)$ is totally bounded. Up to now, N is a continuous operator on S and $N(S)$ is totally bounded. Summing up, N is completely continuous on S. Hence, an application of Schauder's fixed point theorem shows that N has, at least, a fixed point in S.
Then, by using the definition of N, we obtain

$$x = Nx = Ax.Bx + Cx$$

and so, the operator equation $x = Ax.Bx + Cx$ has a solution in S. Q.E.D.

Now, if A, B, and C are maps on a bounded, closed, and convex nonempty subset S of a Banach algebra X into itself, and if the following assumption holds

$$(\aleph) \qquad \left(\frac{I - C}{A}\right)^{-1} = \left(\frac{I}{A}\right)^{-1} (I - C)^{-1},$$

then we will have the following result.

Theorem 3.1.7 *Let S be a nonempty, closed, convex, and bounded subset of a Banach algebra X, and let A, B, and $C : S \longrightarrow S$ be three operators such that:*

(i) A and C are $\mathcal{D}$-Lipschitzian with the $\mathcal{D}$-functions ϕ_A and ϕ_C, respectively,

(ii) $\left(\frac{I}{A}\right)^{-1}$ exists on $B(S)$ satisfying $(\aleph)$, I being the identity operator on X,

(iii) B is completely continuous, and

(iv) $Ax.By + Cx \in S$, for all x, $y \in S$.

Then, the operator $A.B+C$ has, at least, a fixed point in S whenever $M\phi_A(r)+\phi_C(r) < r$, where $M = \|B(S)\|$.

Proof. Let us define a mapping

$$\begin{cases} N : S \longrightarrow S \\ \qquad x \longrightarrow \left(\dfrac{I-C}{A}\right)^{-1} Bx. \end{cases}$$

Since $\phi_C(r) < r$ for all $r > 0$, then $(I - C)^{-1}$ exists on S. Again, the operator $\left(\frac{I}{A}\right)^{-1}$ exists in view of the hypothesis (ii). By using the assumption $(\aleph)$, we deduce that

$$\left(\frac{I-C}{A}\right)^{-1} \text{ exists on } B(S).$$

Let us show that N is well defined. It is sufficient to prove that

$$B(S) \subset \left(\frac{I-C}{A}\right)(S).$$

Let $y \in S$ be a fixed point. We define a mapping $A_y : S \longrightarrow S$ by:

$$A_y(x) = Ax.By + Cx.$$

Let x_1, $x_2 \in S$. Then, we have

$$\|A_y(x_1) - A_y(x_2)\| \leq (M\phi_A + \phi_C)(\|x_1 - x_2\|),$$

where $M\phi_A(r) + \phi_C(r) < r$ for all $r > 0$. Hence, by applying a fixed point theorem of Boyd and Wong (see Theorem 1.6.10), we deduce that there exists a unique point $x^* \in S$, such that

$$x^* = Ax^*.By + Cx^*$$

or, equivalently

$$By = \left(\frac{I-C}{A}\right) x^*.$$

Hence, $\left(\frac{I-C}{A}\right)^{-1} B$ defines a mapping

$$\left(\frac{I-C}{A}\right)^{-1} B : S \longrightarrow S.$$

Next, let us show that the operator $\left(\frac{I-C}{A}\right)^{-1}$ is continuous. For this purpose, let $\{y_n\}$ be any sequence in $B(S)$ converging to a point y, and let

$$\begin{cases} x_n = \left(\frac{I-C}{A}\right)^{-1} (y_n) \\ x = \left(\frac{I-C}{A}\right)^{-1} (y). \end{cases}$$

So,

$$\begin{cases} x_n = Ax_n.y_n + Cx_n \\ x = Ax.y + Cx. \end{cases}$$

Now, we have

$$\begin{aligned} \|x_n - x\| &= \|Ax_n.y_n + Cx_n - Ax.y - Cx\| \\ &\leq \|Ax_n.y_n - Ax.y\| + \|Cx_n - Cx\| \\ &\leq \|Ax_n.y_n - Axy_n\| + \|Ax_n.y_n - Ax.y\| + \|Cx_n - Cx\| \\ &\leq (M\phi_A + \phi_C)(\|x_n - x\|) + \|Ax\|\|y_n - y\|. \end{aligned}$$

This shows that $\lim_{n\to+\infty} \|x_n - x\| = 0$ and consequently, $\left(\frac{I-C}{A}\right)^{-1}$ is continuous on $B(S)$. From the hypothesis (iii), it follows that N is completely continuous on S into itself. Hence, an application of Schauder's fixed point theorem implies that N has, at least, a fixed point in S. Q.E.D.

Remark 3.1.2 *Note that $\left(\frac{I}{A}\right)^{-1}$ exists if $\left(\frac{I}{A}\right)$ is well defined and one-to-one on X. Further, $\left(\frac{I}{A}\right)$ is well defined, if A is regular, i.e., A maps S into the set of all invertible elements of X.*

3.1.2 Fixed point theorems in Banach algebras satisfying the condition $(\mathcal{P})$

Theorem 3.1.8 *Let X be a Banach algebra satisfying condition $(\mathcal{P})$. Let S be a nonempty, closed, and convex subset of X. Let A, $C : X \longrightarrow X$ and $B : S \longrightarrow X$ be three operators such that:*

(i) A and C are $\mathcal{D}$-Lipschitzian with the $\mathcal{D}$-functions ϕ_A and ϕ_C, respectively,

(ii) A is regular on X,

(iii) A, B, and C are weakly sequentially continuous on S,

(iv) B(S) is bounded with a bound M,

(v) $\left(\frac{I-C}{A}\right)^{-1}$ *is weakly compact on B(S), and*

(vi) $x = Ax.By + Cx \Longrightarrow x \in S$, *for all* $y \in S$.

Then, Eq. (3.1) has, at least, one solution in S whenever $M\phi_A(r)+\phi_C(r) < r$, *for all* $r > 0$.

Proof. Similarly to the proof of Theorem 3.1.4, we may deduce that $\left(\frac{I-C}{A}\right)^{-1}$ exists on $B(S)$,

$$\left(\frac{I-C}{A}\right)^{-1} B(S) \subset S,$$

and

$$\left(\frac{I-C}{A}\right)^{-1} B(S)$$

is relatively weakly compact. In view of Theorem 2.2.1, it is sufficient to establish that

$$\left(\frac{I-C}{A}\right)^{-1} B$$

is weakly sequentially continuous. For this purpose, let $\{u_n\}$ be a weakly convergent sequence of S to a point u in S. Now, we define the sequence $\{v_n\}$ of the subset S by:

$$v_n = \left(\frac{I-C}{A}\right)^{-1} Bu_n.$$

Since $\left(\frac{I-C}{A}\right)^{-1} B(S)$ is relatively weakly compact, then there is a renamed subsequence such that

$$v_n = \left(\frac{I-C}{A}\right)^{-1} Bu_n \rightharpoonup v.$$

However, the subsequence $\{v_n\}$ verifies

$$v_n - Cv_n = Av_n.Bu_n.$$

Therefore, from the assumption (iii), and in view of condition $(\mathcal{P})$, we deduce that v verifies the following equation

$$v - Cv = Av.Bu,$$

or, equivalently

$$v = \left(\frac{I-C}{A}\right)^{-1} Bu.$$

Next, we claim that the whole sequence $\{u_n\}$ verifies

$$\left(\frac{I-C}{A}\right)^{-1} Bu_n = v_n \rightharpoonup v.$$

Indeed, let us suppose that this is not the case. Then, there is V^w, a weakly neighborhood of v, satisfying for all $n \in \mathbb{N}$, the existence of an $N \geq n$ such that $v_N \notin V^w$. Hence, there is a renamed subsequence $\{v_n\}$ verifying the property

$$\text{for all } n \in \mathbb{N}, \; v_n \notin V^w. \tag{3.9}$$

However,

$$\text{for all } n \in \mathbb{N}, \; v_n \in \left(\frac{I-C}{A}\right)^{-1} B(S).$$

Again, there is a renamed subsequence such that

$$v_n \rightharpoonup v'.$$

According to the preceding, we have

$$v' = \left(\frac{I-C}{A}\right)^{-1} Bu,$$

and consequently,

$$v = v',$$

which is a contradiction with the property (3.9). This implies that

$$\left(\frac{I-C}{A}\right)^{-1} B$$

is weakly sequentially continuous. Q.E.D.

An interesting corollary of Theorem 3.1.8 is the following.

Corollary 3.1.2 *Let X be a Banach algebra satisfying condition $(\mathcal{P})$, and let S be a nonempty, closed, and convex subset of X. Let A, $C : X \longrightarrow X$ and $B : S \longrightarrow X$ be three operators such that:*

(i) A and C are $\mathcal{D}$-Lipschitzian with the $\mathcal{D}$-functions ϕ_A and ϕ_C, respectively,

(ii) A is regular on X,

(iii) A, B, and C are weakly sequentially continuous on S,

(iv) $A(S)$, $B(S)$, and $C(S)$ are relatively weakly compact, and

(v) $x = Ax.By + Cx \Longrightarrow x \in S$, for all $y \in S$.

Then, Eq. (3.1) has, at least, one solution in S whenever $M\phi_A(r)+\phi_C(r) < r$, for all $r > 0$.

Proof. In view of Theorem 3.1.8, it is sufficient to prove that

$$\left(\frac{I-C}{A}\right)^{-1} B(S)$$

is relatively weakly compact. To do this, let $\{u_n\}$ be any sequence in S, and let

$$v_n = \left(\frac{I-C}{A}\right)^{-1} Bu_n. \tag{3.10}$$

Since $B(S)$ is relatively weakly compact, then there is a renamed subsequence $\{Bu_n\}$ weakly converging to an element w. Moreover, by using Eq. (3.10), we get

$$v_n = Av_n.Bu_n + Cv_n. \tag{3.11}$$

Since $\{v_n\}$ is a sequence in S, so, by using assumption (iv), there is a renamed subsequence such that $Av_n \rightharpoonup x$ and $Cv_n \rightharpoonup y$. Hence, in view of the condition $(\mathcal{P})$ and Eq. (3.11), we obtain

$$v_n \rightharpoonup x.w + y.$$

This shows that

$$\left(\frac{I-C}{A}\right)^{-1} B(S)$$

is sequentially relatively weakly compact. By applying the Eberlein–Šmulian theorem (see Theorem 1.3.3), we deduce that $\left(\frac{I-C}{A}\right)^{-1} B(S)$ is relatively weakly compact. Q.E.D.

3.1.3 Existence of positive solutions

Now, we may briefly discuss the existence of positive solutions. Let X_1 and X_2 be two Banach algebras, with positive closed cones X_1^+ and X_2^+, respectively. An operator $\mathcal{G}$ from X_1 into X_2 is said to be positive if it carries the positive cone X_1^+ into X_2^+ i.e., $\mathcal{G}(X_1^+) \subset X_2^+$.

Theorem 3.1.9 *Let X be a Banach algebra satisfying condition $(\mathcal{P})$ and let S be a nonempty, closed, and convex subset of X such that $S^+ = S \cap X^+ \neq \emptyset$. Let A, $C : X \longrightarrow X$ and $B : S \longrightarrow X$ be three operators such that:*

(i) A and C are $\mathcal{D}$-Lipschitzian with the $\mathcal{D}$-functions ϕ_A and ϕ_C, respectively,

(ii) A is regular on X,

(iii) A, B, and C are weakly sequentially continuous on S^+,

(iv) $A(S^+)$, $B(S^+)$, and $C(S^+)$ are relatively weakly compact, and

(v) $x = Ax.By + Cx \Longrightarrow x \in S^+$, for all $y \in S^+$.

Then, Eq. (3.1) has, at least, one solution in S^+ whenever $M^+\phi_A(r)+\phi_C(r) < r$, for all $r > 0$, where $M^+ = \|B(S^+)\|$.

Proof. Obviously, $S^+ = S \cap X^+$ is a closed and convex subset of X. From Proposition 3.1.2, it follows that

$$\left(\frac{I - C}{A}\right)^{-1}$$

exists on $B(S^+)$. By virtue of assumption (v), we have

$$\left(\frac{I - C}{A}\right)^{-1} B(S^+) \subset S^+.$$

Then, we can define the mapping:

$$\begin{cases} \mathcal{N} : S^+ \longrightarrow S^+ \\ \qquad y \longrightarrow \mathcal{N}y = \left(\dfrac{I - C}{A}\right)^{-1} By. \end{cases}$$

Now, by applying Corollary 3.1.2, we deduce that $\mathcal{N}$ has, at least, a fixed point in S^+. As a result, by using the definition of $\mathcal{N}$, Eq. (3.1) has a solution in S^+. Q.E.D.

3.1.4 Fixed point theorems in Banach algebras and MNWC

In the remainder of this chapter, ω denotes the De Blasi measure of weak noncompactness.

Theorem 3.1.10 *Let X be a Banach space and let Ω be a nonempty and* *weakly closed subset of X. Suppose that there exists a weakly sequentially continuous operator $F:\Omega\longrightarrow\Omega$. Let $x_0\in\Omega$. If the following implication:*

$$(V=F(V)\cup\{x_0\})\Longrightarrow V \text{ is relatively weakly sequentially compact} \quad (3.12)$$

holds for every subset V of Ω, then F has, at least, one fixed point in Ω.

Proof. Let us define a sequence $\{x_n\}_{n=0}^{\infty}$ by the formula

$$x_{n+1}=F(x_n) \quad \text{for} \quad n=0,1,2,....$$

Let $S=\{x_n:\ n=0,1,2,...\}$. Clearly, $S=F(S)\cup\{x_0\}$. Then, from Eq. (3.12), the set S is relatively weakly sequentially compact. Therefore, there exists a renamed subsequence $\{x_n\}_{n=0}^{\infty}$ such that $x_n \rightharpoonup x\in\Omega$. Since F is weakly sequentially continuous, it follows that $F(x_n)=x_{n+1}$ converges weakly to both x and Fx, so that $Fx=x$. Q.E.D.

Theorem 3.1.11 *Let X be a Banach space, and let Ω be a nonempty, convex, and closed subset of X. Suppose there is an operator $F:\Omega\longrightarrow\Omega$ which is weakly sequentially continuous and condensing with respect to ω. In addition, assume that $F(\Omega)$ is bounded. Then, F has, at least, one fixed point in Ω.*

Proof. we will provide a constructive proof based on the superposition method recursively. Let us fix an arbitrary x_0 in Ω. Let

$$Q_0=\{x_0\},\ Q_n=\text{co}\left(\{x_0\}\cup F(Q_{n-1})\right) \text{ for } n=1,2,....$$

Notice that for $n=1,2,....$

$$Q_0\subseteq Q_1\subseteq ...\subseteq Q_n\subseteq ...\subseteq\Omega.$$

Let $Q=\bigcup\limits_{n=0}^{\infty}Q_n$. Since $Q_{n-1}\subseteq Q_n$ for $n=1,2,....$, hence, Q is convex. Now, we claim that $Q=\text{co}\left(\{x_0\}\cup F(Q)\right)$. Let us denote by $Q_*=\text{co}\left(\{x_0\}\cup F(Q)\right)$. Since $F(Q_n)\subseteq Q$ for $n=0,1,2,....$, then,

$$F(Q)=\bigcup_{n=0}^{\infty}F(Q_n)\subseteq Q,$$

which implies that

$$\{x_0\}\cup F(Q)\subseteq Q.$$

Since Q is convex, then we have

$$Q_* \subseteq Q. \tag{3.13}$$

Notice also that $Q_{n-1} \subseteq Q$ for $n = 1, 2, ...$ and consequently,

$$Q = \bigcup_{n=1}^{\infty} \text{co}\,(\{x_0\} \cup F(Q_{n-1})) \subseteq \text{co}\,(\{x_0\} \cup F(Q)) = Q_*. \tag{3.14}$$

Now, by combining Eqs. (3.13) and (3.14), we conclude that

$$Q_* = Q.$$

Therefore, Q is a bounded and convex subset of X. Consequently,

$$\overline{Q^w} = \overline{Q} = \overline{co}\,(\{x_0\} \cup F(Q)).$$

If $\omega(Q) > 0$, then

$$\omega(Q) = \omega(\overline{Q}) = \omega(F(Q)) < \omega(Q),$$

which represents a contradiction. This implies that $\omega(Q) = 0$ and consequently, $\overline{Q^w}$ is weakly compact. Hence, $F : \overline{Q^w} \longrightarrow \Omega$ is weakly continuous (see [10]). Knowing that $F(\overline{Q^w}) \subseteq \overline{F(Q)}^w \subseteq \overline{Q^w}(\subseteq \Omega)$, we may apply the Arino–Gautier–Penot theorem (see Theorem 1.6.9) (consider the locally convex topological vector space $S = (S, w)$ and note that $F : \overline{Q^w} \longrightarrow \overline{Q^w}$ is continuous where $\overline{Q^w}$ is a compact, convex subset of S) to infer that F has, at least, one fixed point in $\overline{Q^w}$. Q.E.D.

Remark 3.1.3 *Without using the Schauder–Tychonoff theorem or the Arino–Gautier–Penot theorem, we can prove that Theorem 3.1.11 is a consequence of Theorem 3.1.10. In fact, since Ω is a nonempty, convex, and closed subset of X then, Ω is a nonempty and weakly closed subset of X. Now, let V be a subset of Ω such that $V = F(V) \cup \{x_0\}$. The use of a property of the measure ω leads to $\omega(V) = \omega(F(V))$. Since F is a condensing map with respect to ω, we must have $\omega(V) = 0$. Therefore, V is relatively weakly compact. By using Eberlein–Šmulian's theorem (see Theorem 1.3.3), it follows that V is weakly sequentially relatively compact. Now, by applying Theorem 3.1.10, we deduce that there exists $x \in \Omega$ such that $x = F(x)$.*

Remark 3.1.4 *Under the assumptions of the above Theorem 3.1.11, the set of fixed points of F belonging to Ω, is relatively weakly compact.*

In the remainder of this section, we assume that the Banach space X has the structure of a Banach algebra satisfying the condition $(\mathcal{P})$.

Theorem 3.1.12 *Let Ω be a nonempty subset of X, and suppose that the operator $F : \Omega \longrightarrow X$ is of the form $Fx = Ax + Lx.Ux$, where:*

(i) $L : \Omega \longrightarrow X$ is a λ-set-contraction with respect to the measure of weak noncompactness ω, and

(ii) A, $U : \Omega \longrightarrow X$ are weakly compact. Suppose that $\gamma := ||U(\Omega)|| < \infty$.

Then, F is a strict set-contraction with respect to ω whenever $\lambda\gamma < 1$.

Proof. Let us arbitrarily take a bounded subset V of Ω. Then,

$$F(V) \subset A(V) + L(V).U(V).$$

The use of a property of ω leads to

$$\begin{aligned} \omega(F(V)) &\leq \omega(A(V)) + \omega(L(V).U(V)) \\ &\leq \omega(\overline{A(V)^w}) + \omega(L(V).\overline{U(V)^w}). \end{aligned}$$

Now, by using the hypothesis (ii) and in view of Lemma 1.5.2, we deduce that

$$\begin{aligned} \omega(F(V)) &\leq \gamma\omega(L(V)) \\ &\leq \lambda\gamma\omega(V). \end{aligned}$$

Since $0 \leq \lambda\gamma < 1$, we infer that F is a strict set-contraction with respect to the measure ω. Q.E.D.

By combining Theorems 3.1.11 and 3.1.12, we get the following fixed point result:

Theorem 3.1.13 *Let Ω be a nonempty, convex, and closed subset of X, and let us suppose that the operator $F : \Omega \longrightarrow X$ is of the form $Fx = Ax + Lx.Ux$, where:*

(i) $L : \Omega \longrightarrow X$ is weakly sequentially continuous on Ω and λ-set-contraction with respect to the measure of weak noncompactness ω,

(ii) A, and $U : \Omega \longrightarrow X$ are weakly sequentially continuous on Ω and weakly compact, and

(iii) $Ax + Lx.Ux \in \Omega$, $x \in \Omega$.

If $F(\Omega) = (A + L.U)(\Omega)$ and $U(\Omega)$ are bounded subsets of X, then F has, at least, one fixed point in Ω whenever $0 \leq \lambda\gamma < 1$, where $\gamma := ||U(\Omega)||$.

Notice that when the scalar λ (used in the previous theorem) vanishes, we have the following result:

Corollary 3.1.3 *Let Ω be a nonempty, convex, and closed subset of X and suppose that the operator $F : \Omega \longrightarrow X$ is of the form $Fx = Ax + Lx.Ux$, where:*

(i) A, L, and $U : \Omega \longrightarrow X$ are weakly sequentially continuous on Ω and also weakly compact, and

(ii) $Ax + Lx.Ux \in \Omega$, $x \in \Omega$.

If $F(\Omega) = (A + L.U)(\Omega)$ and $U(\Omega)$ are bounded subsets of X, then F has, at least, one fixed point in Ω.

Theorem 3.1.14 *Let Ω be a nonempty subset of X and suppose that the operator $F : \Omega \longrightarrow X$ is of the form $Fx = Ax + Lx.Ux$, where:*

(i) $L : \Omega \longrightarrow X$ is a condensing map with respect to ω, and

(ii) A, $U : \Omega \longrightarrow X$ are weakly compact.

If $0 \leq \gamma \leq 1$, where $\gamma := ||U(\Omega)||$, then F is a condensing map with respect to ω.

Proof. Let us take an arbitrary bounded subset V of Ω. Similarly to Theorem 3.1.12, one has

$$\begin{aligned} \omega(F(V)) &\leq \omega(L(V).\overline{U(V)}^{w}) \\ &\leq \gamma\omega(L(V)). \end{aligned}$$

Then,

$$\omega(F(V)) \leq \omega(L(V)).$$

Knowing that L is a condensing map with respect to ω, it follows that F is also a condensing map with respect to ω. Q.E.D.

Combining Theorems 3.1.11 and 3.1.14, we have the following result:

Theorem 3.1.15 *Let Ω be a nonempty, convex, and closed subset of X. Suppose that the operator $F : \Omega \longrightarrow X$ is of the form $Fx = Ax + Lx.Ux$, where:*

(i) $L : \Omega \longrightarrow X$ *is weakly sequentially continuous on* Ω *and is a condensing map with respect to the measure of weak noncompactness* ω,

(ii) A *and* $U : \Omega \longrightarrow X$ *are weakly sequentially continuous on* Ω *and also weakly compact, and*

(iii) $Ax + Lx.Ux \in \Omega$, $x \in \Omega$.

If $F(\Omega) = (A + L.U)(\Omega)$ *and* $U(\Omega)$ *are bounded subsets of* X, *then* F *has, at least, one fixed point in* Ω *whenever* $0 \leq \gamma \leq 1$, *where* $\gamma := ||U(\Omega)||$.

Notice that, there is a relation between α-Lipschitzian and λ-set-contraction maps with respect to ω:

Proposition 3.1.3 *If* $L : X \longrightarrow X$ *is Lipschitzian with a Lipschitz constant* α *and is weakly sequentially continuous on* X, *then* L *is* α*-set-contraction with respect to* ω.

Proof. Let V be a bounded subset of X. We may assume that $\alpha > 0$. Let $\varepsilon > 0$ be given. From the definition of ω, it follows that there exists a weakly compact subset K of X such that $V \subset K + (\omega(V) + \alpha^{-1}\varepsilon)B_X$. Then, $L(V) \subset L(K) + \alpha(\omega(V) + \alpha^{-1}\varepsilon)B_X$, since L is α Lipschitzian. Knowing that K is weakly compact and that L is weakly sequentially continuous on X, then $L : K \longrightarrow X$ is weakly continuous. Hence, $L(K)$ is weakly compact. We infer that

$$\omega(L(V)) \leq \alpha\omega(V) + \varepsilon,$$

and since ε is arbitrary, this implies that

$$\omega(L(V)) \leq \alpha\omega(V).$$

Q.E.D.

Now, by combining Theorem 3.1.12 with Proposition 3.1.3, we deduce the following result:

Theorem 3.1.16 *Let* Ω *be a nonempty subset of* X *and suppose that the operator* $F : \Omega \longrightarrow X$ *is of the form* $Fx = Ax + Lx.Ux$, *where:*

(i) $L : X \longrightarrow X$ *is Lipschitzian with a Lipschitz constant* α *and is weakly sequentially continuous on* X, *and*

(ii) A, $U : \Omega \longrightarrow X$ *are weakly compact.*

Suppose that $\gamma := ||U(\Omega)|| < \infty$. *Then,* F *is a strict set-contraction with respect to* ω *whenever* $\alpha\gamma < 1$.

Proof. In view of the previous Proposition 3.1.3, L is an α-set-contraction with respect to ω. Now, our desired result follows immediately from Theorem 3.1.12. Q.E.D.

Theorem 3.1.17 *Let Ω be a nonempty, convex, closed, and bounded subset of X. Let A, L, and U be three operators such that:*

(i) $L : X \longrightarrow X$ is Lipschitzian with a Lipschitz constant α and weakly sequentially continuous on X,

(ii) A, and $U : \Omega \longrightarrow X$ are weakly sequentially continuous on Ω and are weakly compact, and

(iii) $Ax + Lx.Ux \in \Omega$, $\forall x \in \Omega$.

Then, Eq. (3.1) has, at least, one fixed point in Ω whenever $\alpha\gamma < 1$, where $\gamma := ||U(\Omega)||$.

Proof. We will show that the operator F satisfies all the conditions of Theorem 3.1.11, where F is defined by:

$$\begin{cases} F : \Omega \longrightarrow X \\ \qquad x \longrightarrow Fx = Ax + Lx.Ux. \end{cases}$$

First, since A, L, and U are weakly sequentially continuous on Ω, and according to condition $(\mathcal{P})$, we infer that F is weakly sequentially continuous. Clearly, F is a strict set-contraction with respect to ω, since $\alpha\gamma < 1$. It follows that F is a condensing map with respect to ω. Finally, the use of the hypothesis (iii) implies that $F(\Omega) \subseteq \Omega$ and consequently, $F(\Omega)$ is bounded. Then, Theorem 3.1.11 allow us to reach the desired result. Q.E.D.

Corollary 3.1.4 *Suppose that U is weakly sequentially continuous and is a weakly compact operator on X, and suppose that $x_0 \in X$. If there exists a nonempty, convex, closed, and bounded subset Ω of X such that $\gamma := ||U(\Omega)|| < 1$, and $x_0 + x.Ux \in \Omega$, for each $x \in \Omega$, then the equation*

$$x = x_0 + x.Ux \tag{3.15}$$

has, at least, one solution in Ω.

Proof. It is sufficient to take L as the identity map on X, A is the constant map x_0, then the desired result is deduced immediately from the preceding Theorem 3.1.17. Q.E.D.

In what follows, we will discuss the existence of positive solutions for Eq. (3.1) in an ordered Banach algebra $(X, ||.||, \leq)$ satisfying condition $(\mathcal{P})$, with a positive closed cone X^+. We recall that X^+ verifies (i) $X^+ + X^+ \subseteq X^+$, (ii) $\lambda X^+ \subseteq X^+$ for all $\lambda \in \mathbb{R}^+$, (iii) $\{-X^+\} \cap X^+ = \{0\}$, where 0 is the zero element of X, and (iv) $X^+ \cdot X^+ \subseteq X^+$, where "$\cdot$" is a multiplicative composition in X. We recall the following lemma proved in [72].

Lemma 3.1.1 *Let K be a positive cone in the ordered Banach algebra X. If u_1, u_2, v_1, $v_2 \in K$ are such that $u_1 \leq v_1$ and $u_2 \leq v_2$, then $u_1.u_2 \leq v_1.v_2$.*

Theorem 3.1.18 *Let Ω be a nonempty, convex, and closed subset of X such that $\Omega^+ = \Omega \cap X^+ \neq \emptyset$. Let A, L, $U : \Omega^+ \longrightarrow X$ be three operators such that:*

(i) A, L, and U are weakly sequentially continuous on Ω^+,

(ii) $A(\Omega^+)$, $L(\Omega^+)$, and $U(\Omega^+)$ are relatively weakly compact, and

(iii) $Ax + Lx.Ux \in \Omega^+$, $x \in \Omega^+$.

Then, Eq. (3.1) has, at least, one fixed point in Ω^+.

Proof. Obviously, Ω^+ is a nonempty, convex, and closed subset of X. The use of assumption (ii) implies that $(A + L.U)(\Omega^+)$ and $U(\Omega^+)$ are bounded subsets of X. Now, we may apply Corollary 3.1.3 to infer that Eq. (3.1) has a solution in Ω^+. Q.E.D.

By using Theorem 3.1.15, we have the following result:

Theorem 3.1.19 *Let Ω be a nonempty, convex, and closed subset of X such that $\Omega^+ = \Omega \cap X^+ \neq \emptyset$. Suppose that the operator $F : \Omega \longrightarrow X$ is of the form $Fx = Ax + Lx.Ux$, where:*

(i) $L : \Omega^+ \longrightarrow X$ is weakly sequentially continuous on Ω^+ and is a condensing map with respect to the measure of weak noncompactness ω,

(ii) A, $U : \Omega^+ \longrightarrow X$ are weakly sequentially continuous on Ω^+ and are also weakly compact, and

(iii) $Ax + Lx.Ux \in \Omega^+$, $\forall x \in \Omega^+$.

If $F(\Omega^+) = (A + L.U)(\Omega^+)$ and $U(\Omega^+)$ are bounded subsets of X, then F has, at least, one fixed point in Ω^+ whenever $0 \leq \gamma^+ \leq 1$, where $\gamma^+ := \sup_{x \in \Omega^+} ||Ux|| = ||U(\Omega^+)||$.

By using Theorem 3.1.17, we deduce the following result:

Theorem 3.1.20 *Let Ω be a nonempty, convex, closed, and bounded subset of X such that: $\Omega^+ = \Omega \cap X^+ \neq \emptyset$. Let A, L, and U be three operators such that:*

(i) $L : X \longrightarrow X$ is Lipschitzian with a Lipschitz constant α and weakly sequentially continuous on X,

(ii) A, and $U : \Omega^+ \longrightarrow X$ are weakly sequentially continuous on Ω^+ and are weakly compact, and

(iii) $Ax + Lx.Ux \in \Omega^+$, $x \in \Omega^+$.

Then, Eq. (3.1) has a solution x in Ω^+ whenever $\alpha\gamma^+ < 1$, where $\gamma^+ := ||U(\Omega^+)||$.

To close this section, we will prove the existence of positive solutions for Eq. (3.15) in the Banach algebra $\mathcal{C}(K, X)$, the space of continuous functions from K into X, endowed with the sup-norm $||.||_\infty$ defined by $||f||_\infty := \sup\{||f(t)|| : t \in K\}$, where K is a compact Hausdorff space. In the remainder of this section, we suppose that X^+ verifies the following condition $(\mathcal{H})$:

$(\mathcal{H})$ Let $x,\ y \in X^+$. If $x \leq y$ (*i.e.*, $y - x \in X^+$), then $||x|| \leq ||y||$.

In this case, $\|.\|$ is called monotone increasing or nondecreasing and X^+ is normal. It is known that if the cone X^+ is normal, then every order-bounded subset is bounded in norm. We denote the cone of nonnegative functions in $\mathcal{C}(K, X)$ by $\mathcal{C}_+(K)$ (i.e., $\mathcal{C}_+(K) = \mathcal{C}(K, X^+)$) and for $f_1, f_2 \in \mathcal{C}(K, X)$, we will say that $f_1 \leq f_2$ or $(f_2 \geq f_1)$ provided that $f_2 - f_1 \in \mathcal{C}_+(K)$. A map $\mathcal{F} : \mathcal{C}(K, X) \longrightarrow \mathcal{C}(K, X)$ will be called isotone if $f_1 \leq f_2$, then $\mathcal{F}(f_1) \leq \mathcal{F}(f_2)$.

Theorem 3.1.21 *Suppose that U is weakly sequentially continuous, weakly compact, and is an isotone map of $\mathcal{C}_+(K)$ into itself. For an arbitrary x_0 in $\mathcal{C}_+(K)$, define a sequence $\{x_n\}_{n=0}^\infty$ by:*

$$x_{n+1} = x_0 + x_n.Ux_n,\ n = 0, 1, 2, ...$$

If $\gamma = \sup\limits_{n \in \mathbb{N}} ||Ux_n||_\infty < 1$, then there exists an increasing subsequence $\{x_n\}_{n=0}^\infty$ of $\mathcal{C}_+(K)$ which is weakly convergent to a point x in $\mathcal{C}_+(K)$, and x is a solution of Eq. (3.15) satisfying

$$||x||_\infty \leq (1 - \gamma)^{-1}||x_0||_\infty,$$

and $x_n \leq x$, for all $n \in \mathbb{N}$.

Proof. Let us define F by:

$$\begin{cases} F: \mathcal{C}_+(K) \longrightarrow \mathcal{C}_+(K) \\ \qquad x \longrightarrow Fx = x_0 + x.Ux. \end{cases}$$

Then,

$$x_{n+1} = Fx_n, \ n = 0, 1, 2, ...$$

Hence, F maps the subset $Q = \{x_0, x_1, x_2, ...\}$ into itself. Proceeding by induction and using the fact that U is isotone together with Lemma 3.1.1, we get for each $t \in K$:

$$x_n(t) \leq x_{n+1}(t), \ n = 0, 1, 2, ...$$

and since hypothesis $(\mathcal{H})$ holds, we deduce that

$$\begin{aligned} ||x_n(t)|| &\leq ||x_{n+1}(t)|| \\ &\leq ||x_0(t)|| + ||x_n(t)|| \ ||Ux_n(t)|| \\ &\leq ||x_0||_\infty + \gamma ||x_n||_\infty. \end{aligned}$$

Taking the supremum over t, we obtain:

$$||x_n||_\infty \leq ||x_0||_\infty + \gamma ||x_n||_\infty,$$

and since $0 \leq \gamma < 1$, this implies that

$$||x_n||_\infty \leq (1-\gamma)^{-1} ||x_0||_\infty, \tag{3.16}$$

and consequently, Q is bounded. Moreover, notice that $Q = \{x_0\} \cup F(Q)$, so that $\omega(F(Q))$, the measure of weak noncompactness of $F(Q)$, is just $\omega(Q)$. Now, observe that F is a strict set-contraction with respect to ω, since $\alpha = 1$ and $0 \leq \gamma < 1$. Hence, $\omega(Q) = 0$. This, in turn, shows that Q is relatively weakly compact. By using Eberlein–Šmulian's theorem (Theorem 1.3.3), we deduce that Q is weakly sequentially relatively compact. Consequently, there is a renamed subsequence $\{x_n\}_{n=0}^{\infty}$ which converges weakly to a point x in $\mathcal{C}_+(K)$ (since X^+ and consequently $\mathcal{C}_+(K)$ are weakly closed convex). This fact, together with (3.16) leads to

$$||x||_\infty \leq (1-\gamma)^{-1} ||x_0||_\infty.$$

Since U is weakly sequentially continuous, and using the fact that the Banach algebra $\mathcal{C}(K, X)$ satisfies condition $(\mathcal{P})$, we infer that F is weakly sequentially continuous. Therefore, $Fx_n(=x_{n+1})$ converges weakly to both x and Fx, so that $Fx = x$. Now, it remains to prove that $x_n \leq x$, for $n = 0, 1, 2, ...$ To show it, let us denote by $\{x_{\varphi(n)}\}_{n=0}^{\infty}$ the subsequence in $\mathcal{C}_+(K)$ such that $x_{\varphi(n)} \rightharpoonup x$. Then, $\{x_{\varphi(n)}\}_{n=0}^{\infty}$ is bounded. By using Dobrakov's theorem (see Theorem 1.4.1), we deduce that, for each $t \in K : x_{\varphi(n)}(t) \rightharpoonup x(t)$. Now, let us fix an arbitrary $n \in \mathbb{N}$. Then, for all $p \geq n$, we get

$$x_n(t) \leq x_p(t) \leq x_{\varphi(p)}(t).$$

Hence,

$$x_{\varphi(p)}(t) - x_n(t) \in X^+.$$

Therefore,

$$x(t) - x_n(t) \in X^+, \text{ as } p \to \infty.$$

This implies that

$$x_n(t) \leq x(t),$$

which leads to

$$x_n \leq x.$$

As a result, x satisfies the conclusion of Theorem 3.1.21, which completes the proof. Q.E.D.

Theorem 3.1.22 *Let x_0 be in $\mathcal{C}_+(K)$ and let $\Omega := \{y \in \mathcal{C}_+(K) : y \leq x_0\}$. Let $U : \Omega \longrightarrow \mathcal{C}_+(K)$ be weakly sequentially continuous and a weakly compact operator. If $U(\Omega)$ is bounded, then the equation:*

$$x_0 = x + x.Ux \tag{3.17}$$

has, at least, one fixed point in Ω whenever $\gamma := \sup_{x \in \Omega} ||Ux||_\infty < 1$.

Proof. Clearly, Ω is a nonempty, convex, closed, and bounded subset with a bound $||x_0||$ of the Banach algebra $\mathcal{C}(K, X)$. Eq. (3.17) is equivalent to the equation

$$x = x_0 + x.(-Ux).$$

Let us fix $x \in \Omega$. By definition, we have $x \in \mathcal{C}_+(K)$ and $Ux \in \mathcal{C}_+(K)$. Then,

$$x.Ux \in \mathcal{C}_+(K).$$

For each $t \in K$, this is equivalent to

$$x(t).Ux(t) \in X^+,$$

and then,

$$x_0(t) - (x_0(t) - x(t).Ux(t)) \in X^+,$$

which implies that

$$x_0(t) \geq x_0(t) - x(t).Ux(t),$$

or, equivalently

$$x_0 \geq x_0 + x.(-Ux). \tag{3.18}$$

Moreover, since $x \in \Omega$, then $x \leq x_0$. Keeping in mind that $Ux \in \mathcal{C}_+(K)$, we get

$$x.Ux \leq x_0.Ux,$$

and since $0 \leq \gamma < 1$, we get

$$x.Ux \leq x_0. \tag{3.19}$$

By using both (3.18) and (3.19), we conclude that

$$x_0 + x.(-Ux) \in \Omega,$$

for each $x \in \Omega$. Now, the use of Corollary 3.1.4 allows us to achieve the proof. Q.E.D.

Corollary 3.1.5 *Let x_0 be in $\mathcal{C}_+(K)$ and let $\Omega := \{y \in \mathcal{C}_+(K) : y \leq x_0\}$. Assume that $U : \Omega \longrightarrow \mathcal{C}_+(K)$ is such that:*

(i) U is weakly sequentially continuous on Ω,

(ii) U is Lipschitzian with a Lipschitz constant α, and

(iii) U is weakly compact.

Then, Eq. (3.17) has a unique solution x in Ω whenever $(\alpha||x_0|| + \gamma) < 1$.

Proof. The existence of the solution is proved in the above Theorem 3.1.22. Now, we have to show its uniqueness. For this purpose, let us assume that x_1 and x_2 are two solutions of Eq. (3.17). Hence, it follows that

$$x_1 + x_1.Ux_1 = x_2 + x_2.Ux_2.$$

Thus,

$$x_1 - x_2 = x_2.(Ux_2 - Ux_1) + (x_2 - x_1).Ux_1.$$

Then,

$$\begin{aligned}||x_1 - x_2|| &\leq ||x_2||\ ||Ux_2 - Ux_1|| + \gamma||x_2 - x_1|| \\ &\leq ||x_0||\ ||Ux_2 - Ux_1|| + \gamma||x_2 - x_1|| \\ &\leq (\alpha||x_0|| + \gamma)||x_2 - x_1||.\end{aligned}$$

Since $(\alpha||x_0||+\gamma) < 1$, we must have $x_1 = x_2$, which achieves the proof of the uniqueness of the solution. Q.E.D.

Remark 3.1.5 *The element x_0 is invertible if, and only if, the solution x of Eq. (3.17) is invertible. Indeed, Eq. (3.17) is equivalent to the equation $x_0 = x(I+Ux)$ where I represents the identity operator defined by $Ix = x$, $x \in \mathcal{C}_+(K)$. Since $\gamma := \sup_{y \in \Omega} ||Uy||_\infty < 1$, then $||Ux||_\infty < 1$ and consequently, $(I + Ux)$ is invertible (we recall that $(I + Ux)^{-1} = \sum_{n=0}^{\infty}(-1)^n(Ux)^n$).*

3.2 WC–Banach Algebras

In this section, we will prove some fixed point theorems for the sum and the product of nonlinear weakly sequentially continuous operators acting on a WC–Banach algebra. The main goal is to establish new variants of Theorem 3.1.1 for three operators acting on WC–Banach algebras, without using the sequential condition $(\mathcal{P})$. These new results are due to A. Jeribi, B. Krichen, and B. Mefteh (see [105]).

3.2.1 Fixed point theorems in WC–Banach algebras

First, let us recall the following definition.

Definition 3.2.1 *Let X be a Banach algebra. We say that X is a WC–Banach algebra, if the product $W.W'$ of arbitrary weakly compact subsets W and W' of X is weakly compact.*

Clearly, every finite-dimensional Banach algebra is a WC–Banach algebra. Even if X is a WC–Banach algebra, the set $\mathcal{C}(K, X)$ of all continuous functions from K to X is also a WC–Banach algebra, where K is a compact Hausdorff space. The proof is based on Dobrakov's theorem (see Theorem 1.4.1).

Lemma 3.2.1 *[21] Let M and M' be two bounded subsets of a WC–Banach algebra X. Then, we have the following inequality*

$$\omega(M.M') \leq \|M'\|\omega(M) + \|M\|\omega(M') + \omega(M)\omega(M').$$

Proof. Let us assume that M and M' are arbitrary bounded subsets of a WC–Banach algebra X. Let r and t be fixed numbers with $r > \omega(M)$ and $t > \omega(M')$. Then, we can find two weakly compact subsets W_1 and W_2 of X, such that

$$M \subset W_1 + B_r, \tag{3.20}$$

and

$$M' \subset W_2 + B_t. \tag{3.21}$$

Now, let $z \in M.M'$. Then, z can be represented in the form $z = x.y$ with $x \in M$ and $y \in M'$. In view of (3.20) and (3.21), there exist $w_1 \in W_1$, $w_2 \in W_2$, $u \in B_r$, and $v \in B_t$ such that $x = w_1 + u$ and $y = w_2 + v$. Hence, we get

$$\begin{aligned} z = x.y &= (w_1 + u).(w_2 + v) \\ &= w.w_2 + w_1.v + u.w_2 + u.v \\ &= w_1.w_2 + (x - u).v + u.(y - v) + u.v \\ &= w_1.w_2 + x.v + u.y - u.v. \end{aligned}$$

The above equalities imply the following inclusion

$$M.M' \subset W_1.W_2 + M.B_t + B_r.M' + B_t.B_r \subset W_1.W_2 + B_{\|M\|t+\|M'\|r+rt}.$$

Hence, keeping in mind the fact that X is a WC–Banach algebra, and in view of the definition of the De Blasi measure of weak noncompactness ω, we obtain the following inequality.

$$\omega(M.M') \leq \|M\|t + \|M'\|r + rt.$$

By letting $r \to \omega(M)$ and $t \to \omega(M')$, we get

$$\omega(M.M') \leq \|M'\|\omega(M) + \|M\|\omega(M') + \omega(M)\omega(M').$$

Q.E.D.

Now, we are ready to state our first fixed point theorem in WC–Banach algebra in order to provide the existence of solutions for the operator equation

$$x = Ax.Bx + Cx.$$

In what follows, let us assume that:

$$(\mathcal{A}_0) \qquad \begin{cases} \text{If } (x_n)_{n\in\mathbb{N}} \subseteq \mathcal{D}(A) \text{ is a weakly convergent sequence in } X, \text{ then} \\ (Ax_n)_{n\in\mathbb{N}} \text{ has a strongly convergent subsequence in } X. \end{cases}$$

Theorem 3.2.1 *Let S be a nonempty, bounded, closed, and convex subset of a WC–Banach algebra X, and let A, $C : X \longrightarrow X$ and $B : S \longrightarrow X$ be three weakly sequentially continuous operators, satisfying the following conditions:*

(i) $\left(\frac{I-C}{A}\right)^{-1}$ *exists on $B(S)$,*

(ii) A satisfies $(\mathcal{A}_0)$, and $A(S)$ is relatively weakly compact,

(iii) B is an ω-β-contraction,

(iv) C is an ω-α-contraction, and

(v) $(x = Ax.By + Cx,\ y \in S) \Longrightarrow x \in S.$

Then, the operator equation $x = Ax.Bx + Cx$ has, at least, a solution in S whenever $\frac{\gamma\beta}{1-\alpha} < 1$, where $\gamma = \|A(S)\|$.

Proof. It is easy to check that the vector $x \in S$ is a solution of the equation

$$x = Ax.Bx + Cx$$

if, and only if, x is a fixed point for the operator

$$T := \left(\frac{I-C}{A}\right)^{-1} B.$$

From assumption (i), it follows that, for each $y \in S$, there is a unique $x_y \in X$ such that

$$\left(\frac{I-C}{A}\right) x_y = By,$$

or, in an equivalent way,

$$Ax_y.By + Cx_y = x_y.$$

Since the hypothesis (v) holds, then $x_y \in S$. Hence, the map $T : S \longrightarrow S$ is well defined. In order to achieve the proof, we will apply Theorem 2.3.4. Hence, we only have to prove that the operator $T : S \longrightarrow S$ is weakly sequentially continuous and ω-condensing. Indeed, let us consider $(x_n)_{n\in\mathbb{N}}$ as a sequence in S which is weakly convergent to x. In this case, the set $\{x_n : n \in \mathbb{N}\}$ is relatively weakly compact, and since B is weakly sequentially continuous, then $\{Bx_n : n \in \mathbb{N}\}$ is also relatively weakly compact. By using the following equality

$$T = AT.B + CT, \tag{3.22}$$

combined with the facts that $A(S)$ is relatively weakly compact and C is a ω-α-contraction, we obtain the following:

$$\begin{aligned}
\omega\Big(\{Tx_n : n \in \mathbb{N}\}\Big) &\leq \omega\Big(\{A(Tx_n)Bx_n : n \in \mathbb{N}\}\Big) + \omega\Big(\{C(Tx_n) : n \in \mathbb{N}\}\Big) \\
&\leq \alpha\omega\Big(\{Tx_n : n \in \mathbb{N}\}\Big) \\
&< \omega\Big(\{Tx_n : n \in \mathbb{N}\}\Big).
\end{aligned}$$

Hence, $\{Tx_n : n \in \mathbb{N}\}$ is relatively weakly compact. Consequently, there exists a subsequence $(x_{n_i})_{i\in\mathbb{N}}$ of $(x_n)_{n\in\mathbb{N}}$ such that $Tx_{n_i} \rightharpoonup y$. Going back to Eq. (3.22), to the weak sequential continuity of A, B, and C, and to the fact that A verifies $(\mathcal{A}_0)$, shows that there exists a subsequence $(x_{n_{i_j}})_{j\in\mathbb{N}}$ of $(x_{n_i})_{i\in\mathbb{N}}$ such that $Tx_{n_{i_j}} = A(Tx_{n_{i_j}})Bx_{n_{i_j}} + C(Tx_{n_{i_j}})$ and then, $y = Tx$. Consequently, $Tx_{n_{i_j}} \rightharpoonup Tx$. Now, we claim that $Tx_n \rightharpoonup Tx$. Let us suppose the contrary. Then, there exist a subsequence $(x_{n_i})_{i\in\mathbb{N}}$ of $(x_n)_{n\in\mathbb{N}}$ and a weak neighborhood V^w of Tx, such that $Tx_{n_i} \notin V^w$ for all $i \in \mathbb{N}$. Since $(x_{n_i})_{i\in\mathbb{N}}$ converges weakly to x, then arguing as before, we may extract a subsequence $(x_{n_{i_{j_k}}})_{k\in\mathbb{N}}$ of $(x_{n_i})_{i\in\mathbb{N}}$, such that $Tx_{n_{i_{j_k}}} \rightharpoonup Tx$, which is absurd, since $Tx_{n_{i_{j_k}}} \notin V^w$, for all $k \in \mathbb{N}$. As a result, T is weakly sequentially continuous. Next, we will prove that T is ω-condensing. For this purpose, let $\mathcal{M}$ be a subset of S with $\omega(\mathcal{M}) > 0$. By using Eq. (3.22), we infer that

$$\omega(T(\mathcal{M})) \leq \omega(A(T(\mathcal{M}))B(\mathcal{M}) + C(T(\mathcal{M}))).$$

The properties of ω in Lemmas 1.4.1 and 3.2.1, when combined with the assumptions (ii), (iii), and (iv) on A, B, and C, allow us to show that

$$\begin{aligned}\omega(T(\mathcal{M})) &\leq \omega(A(T(\mathcal{M}))B(\mathcal{M})) + \omega(C(T(\mathcal{M}))) \\ &\leq \gamma\beta\omega(\mathcal{M}) + \alpha\omega(T(\mathcal{M})),\end{aligned}$$

and then,

$$\omega(T(\mathcal{M})) \leq \frac{\gamma\beta}{1-\alpha}\omega(\mathcal{M}).$$

This inequality means, in particular, that T is ω-condensing. Q.E.D.

Corollary 3.2.1 *Let S be a nonempty, bounded, closed, and convex subset of a WC–Banach algebra X, and let A, $C : X \longrightarrow X$ and $B : S \longrightarrow X$ be three weakly sequentially continuous operators satisfying the following conditions:*

(i) $\left(\frac{I-C}{A}\right)^{-1}$ exists on $B(S)$,

(ii) A satisfies $(\mathcal{A}_0)$,

(iii) $A(S)$, $B(S)$, and $C(S)$ are relatively weakly compact, and

(iv) $(x = Ax.By + Cx,\ y \in S) \Longrightarrow x \in S$.

Then, the operator equation $x = Ax.Bx + Cx$ has, at least, one solution in S.

Proof. According to Theorem 3.2.1, it is sufficient to show that

$$T(S) := \left(\frac{I-C}{A}\right)^{-1} B(S)$$

is relatively weakly compact. By using both Lemmas 1.4.1 and 3.2.1, and knowing the weak compactness of $A(S)$, $B(S)$, and $C(S)$, we infer that

$$\omega(T(S)) \leq \omega(A(T(S))B(S)) + \omega(C(T(S))).$$

This shows that $\omega(T(S)) = 0$. Hence, $T(S)$ is relatively weakly compact. The use of Theorem 3.2.1 achieves the proof. Q.E.D.

In the following theorem, we will use the notion of $\mathcal{D}$-Lipschitzian operators.

Theorem 3.2.2 *Let S be a nonempty, bounded, closed, and convex subset of a WC–Banach algebra X, and let A, $C : X \longrightarrow X$ and $B : S \longrightarrow X$ be three weakly sequentially continuous operators satisfying the following conditions:*

(i) B is a ω-δ-contraction,

(ii) A and C are $\mathcal{D}$-Lipschitzian with the $\mathcal{D}$-functions ϕ_A and ϕ_C, respectively, where $\phi_C(r) < (1 - \delta Q)r$ for $r > 0$ and $Q = \|A(S)\|$,

(iii) A is regular on X, verifies $(\mathcal{A}_0)$, and $A(S)$ is relatively weakly compact, and

(iv) $(x = Ax.By + Cx,\ y \in S) \Longrightarrow x \in S$.

Then, the operator equation $x = Ax.Bx + Cx$ has, at least, one solution in S whenever $L\phi_A(r) + \phi_C(r) < r$, where $L = \|B(S)\|$.

Proof. Let y be fixed in S, and let us define the mapping

$$\begin{cases} \varphi_y : X \longrightarrow X, \\ \qquad x \longrightarrow \varphi_y(x) = Ax.By + Cx. \end{cases}$$

Let $x_1, x_2 \in X$. The use of assumption (i) leads to the following inequality:

$$\begin{aligned} \|\varphi_y(x_1) - \varphi_y(x_2)\| &\leq \|Ax_1.B_y - Ax_2.B_y\| + \|Cx_1 - Cx_2\| \\ &\leq \|Ax_1 - Ax_2\|\|By\| + \|Cx_1 - Cx_2\| \\ &\leq L\phi_A(\|x_1 - x_2\|) + \phi_C(\|x_1 - x_2\|). \end{aligned}$$

Now, an application of Boyd's and Wong's fixed point theorem (see Theorem 1.6.10) shows the existence of a unique point $x_y \in X$, such that

$$\varphi_y(x_y) = x_y.$$

Hence, the operator

$$T := \left(\frac{I - C}{A}\right)^{-1} B : S \longrightarrow X$$

is well defined. Moreover, the use of assumption (iv) allows us to have $T(S) \subset S$. By using arguments similar to those used in the proof of Theorem 3.2.1, we can deduce that the operator is weakly sequentially continuous. By applying Theorem 2.3.4, it is sufficient to check that T is ω-condensing. In order to achieve this, let $\mathcal{M}$ be a subset of S with $\omega(\mathcal{M}) > 0$. By using Eq. (3.22), we have

$$T(\mathcal{M}) \subset A(T(\mathcal{M}))B(\mathcal{M}) + C(T(\mathcal{M})).$$

Making use of Lemmas 1.4.1 and 3.2.1 together with the assumptions on A, B, and C, enables us to have

$$\begin{aligned} \omega(T(\mathcal{M})) &\leq \omega(A(T(\mathcal{M}))B(\mathcal{M})) + \omega(C(T(\mathcal{M}))) \\ &\leq \delta\|Q\|\omega(\mathcal{M}) + \phi_C(\omega(T(\mathcal{M}))). \end{aligned} \tag{3.23}$$

Now, if $\delta = 0$, the inequality (3.23) becomes $\omega(T(\mathcal{M})) \leq \phi_C(\omega(T(\mathcal{M})))$, which implies that $\omega(T(\mathcal{M})) = 0$. Otherwise, by using the inequality $\phi_C(r) < (1 - \delta Q)r$ for $r > 0$, we have

$$\omega(T(\mathcal{M})) < \omega(\mathcal{M}).$$

In both cases, T is shown to be ω-condensing. The use of Theorem 2.3.4 achieves the proof. Q.E.D.

If we only take $\delta = 0$ in the above theorem, we obtain the following corollary:

Corollary 3.2.2 *Let S be a nonempty, bounded, closed, and convex subset of a Banach algebra X, and let $C : X \longrightarrow X$ and $B : S \longrightarrow X$ be two weakly sequentially continuous operators satisfying the following conditions:*

(i) C is a nonlinear contraction,

(ii) $B(S)$ is relatively weakly compact, and

(iii) $(x = By + Cx,\ y \in S) \Longrightarrow x \in S$.

Then, $B + C$ has, at least, a fixed point in S.

If we only take the function $\phi_C(r) = \zeta r$, where $\zeta \in [0, 1 - \delta)$ in the above theorem, we obtain the following corollary:

Corollary 3.2.3 *Let S be a nonempty, bounded, closed, and convex subset of a Banach algebra X, and let $C : X \longrightarrow X$ and $B : S \longrightarrow X$ be two weakly sequentially continuous operators satisfying the following conditions:*

(i) C is a strict contraction with a constant $\zeta \in [0, 1 - \delta)$,

(ii) B is a ω-δ-contraction, and

(iii) $(x = By + Cx,\ y \in S) \Longrightarrow x \in S$.

Then, $B + C$ has, at least, a fixed point in S.

3.3 Leray–Schauder's Alternatives in Banach Algebras Involving Three Operators

In what follows, we are going to give some nonlinear alternatives of the Leray–Schauder type in a Banach algebra involving three operators.

Theorem 3.3.1 *Let Ω be a closed, and convex subset in a Banach algebra X, let $U \subset \Omega$ be a weakly open set (with respect to the weak topology of Ω) such that $0 \in U$, and also let $\overline{U^w}$ be a weakly compact subset of Ω. Let A, $C : X \longrightarrow X$ and $B : \overline{U^w} \longrightarrow X$ be three operators satisfying the following conditions:*

(i) A and C are $\mathcal{D}$-Lipschitzian with the $\mathcal{D}$-functions Φ_A and Φ_C, respectively,

(ii) A is regular on X, i.e., A maps X into the set of all invertible elements of X,

(iii) B is weakly sequentially continuous on $\overline{U^w}$,

(iv) $M\Phi_A(r) + \Phi_C(r) < r$ for $r > 0$, with $M = \|B(\overline{U^w})\|$,

(v) $x = Ax.By + Cx \Rightarrow x \in \Omega$ for all $y \in \overline{U^w}$, and

(vi) $(\frac{I-C}{A})^{-1}$ is weakly sequentially continuous on $B(\overline{U^w})$.

Then, either

(a) the equation $\lambda A(\frac{x}{\lambda}).Bx + \lambda C(\frac{x}{\lambda}) = x$ has a solution for $\lambda = 1$, or

(b) there is an element $u \in \partial_{\Omega}^{w}(U)$ such that $\lambda A(\frac{u}{\lambda}).Bu + \lambda C(\frac{u}{\lambda}) = u$ for some $0 < \lambda < 1$.

Proof. Let $y \in \overline{U^w}$ be fixed and let us define the mapping $\phi_y : X \longrightarrow X$ by:

$$\phi_y(x) = Ax.By + Cx$$

for $x \in X$. Then, for any $x_1, x_2 \in X$, we have

$$\begin{aligned}\|\phi_y(x_1) - \phi_y(x_2)\| &\leq \|Ax_1.By - Ax_2.By\| + \|Cx_1 - Cx_2\| \\ &\leq \|Ax_1 - Ax_2\|\|By\| + \|Cx_1 - Cx_2\| \\ &\leq M\Phi_A(\|x_1 - x_2\|) + \Phi_C(\|x_1 - x_2\|).\end{aligned}$$

In view of the hypothesis (iv), we deduce that ϕ_y is a nonlinear contraction on X. Therefore, an application of Theorem 1.6.10 shows that ϕ_y has a unique fixed point, say x in X. This means that there exists a unique $x \in X$, such that

$$Ax.By + Cx = x.$$

By using the hypothesis (v), it is clear that $x \in \Omega$. So, there exists a unique $x \in \Omega$ such that

$$Ax.By + Cx = x.$$

By virtue of the hypothesis (ii), there exists a unique $x \in \Omega$ such that

$$\left(\frac{I-C}{A}\right) x = By$$

and then,

$$x = \left(\frac{I-C}{A}\right)^{-1} By.$$

Hence, $(\frac{I-C}{A})^{-1}B : \overline{U^w} \longrightarrow \Omega$ is well defined. Since B is weakly sequentially continuous on $\overline{U^w}$, and since $(\frac{I-C}{A})^{-1}$ is weakly sequentially continuous on $B(\overline{U^w})$, so by composition, we have $(\frac{I-C}{A})^{-1}B$ is weakly sequentially continuous on $\overline{U^w}$. Now, an application of Theorem 2.3.7 implies that either

(c) $(\frac{I-C}{A})^{-1}B$ has, at least, a fixed point, or

(d) there is a point $u \in \partial_\Omega U$ and $\lambda \in]0,1[$, with $u = \lambda(\frac{I-C}{A})^{-1}Bu$.

First, let us assume that $x \in U$ is a fixed point of the operator $(\frac{I-C}{A})^{-1}B$. Then, $x = \left(\frac{I-C}{A}\right)^{-1} Bx$, which implies that

$$Ax.Bx + Cx = x.$$

Next, let us suppose that there exist an element $u \in \partial_\Omega(U)$ and a real number $\lambda \in]0,1[$ such that $u = \lambda\left(\frac{I-C}{A}\right)^{-1} Bu$. Then,

$$\left(\frac{I-C}{A}\right)^{-1} Bu = \frac{u}{\lambda},$$

so that

$$\lambda A\left(\frac{u}{\lambda}\right).Bu + \lambda C\left(\frac{u}{\lambda}\right) = u.$$

This completes the proof. Q.E.D.

Notice that this result remains true even when $C \equiv 0$, and we get a nonlinear alternative of the Leray–Schauder type in a Banach algebra for the product of two operators.

Corollary 3.3.1 *Let Ω be a closed and convex subset in a Banach algebra X, let $U \subset \Omega$ be a weakly open set (with respect to the weak topology of Ω) such that $0 \in U$ and also let $\overline{U^w}$ be a weakly compact subset of Ω. Let $A : X \longrightarrow X$ and $B : \overline{U^w} \longrightarrow X$ be two operators satisfying the following conditions:*

(i) A is $\mathcal{D}$-Lipschitzian with the $\mathcal{D}$-function Φ_A,

(ii) A is regular on X,

(iii) B is weakly sequentially continuous on $\overline{U^w}$,

(iv) $M\Phi_A(r) < r$ for $r > 0$, with $M = \|B(\overline{U^w})\|$,

(v) $x = Ax.By \Longrightarrow x \in \Omega$ for all $y \in \overline{U^w}$, and

(vi) $(\frac{I}{A})^{-1}$ is weakly sequentially continuous on $B(\overline{U^w})$.

Then, either

(a) the equation $\lambda A(\frac{x}{\lambda}).Bx = x$ has a solution for $\lambda = 1$, or

(b) there is an element $u \in \partial_\Omega(U)$ such that $\lambda A(\frac{u}{\lambda})Bu = u$ for some $0 < \lambda < 1$.

Now, we may give some Leray–Schauder results for maps acting on Banach algebras satisfying condition $(\mathcal{P})$.

Theorem 3.3.2 *Let Ω be a closed, and convex subset in a Banach algebra X satisfying the condition $(\mathcal{P})$ and let $U \subset \Omega$ be a weakly open set (with respect to the weak topology of Ω) such that $0 \in U$. Moreover, let A, $C : X \longrightarrow X$ and $B : \overline{U^w} \longrightarrow X$ be three operators satisfying the following:*

(i) A and C are $\mathcal{D}$-Lipschitzian with the $\mathcal{D}$-functions Φ_A and Φ_C, respectively,

(ii) A is regular on X,

(iii) $B(\overline{U^w})$ is bounded with a bound M,

(iv) $M\Phi_A(r) + \Phi_C(r) < r$ for $r > 0$,

(v) $x = Ax.By + Cx \Longrightarrow x \in \Omega$ for all $y \in \overline{U^w}$,

(vi) A and C are weakly sequentially continuous on Ω, and B is weakly sequentially continuous on $\overline{U^w}$, and

(vii) $(\frac{I-C}{A})^{-1}B(\overline{U^w})$ is relatively weakly compact.

Then, either

(a) the equation $\lambda A(\frac{x}{\lambda}).Bx + \lambda C(\frac{x}{\lambda}) = x$ has a solution for $\lambda = 1$, or

(b) there is an element $u \in \partial_\Omega(U)$ *such that* $\lambda A(\frac{u}{\lambda}).Bu + \lambda C(\frac{u}{\lambda}) = u$ *for some* $0 < \lambda < 1$.

Proof. Similarly to the proof of Theorem 3.3.1, in order to show that $(\frac{I-C}{A})^{-1}B$ is well defined from $\overline{U^w}$ to Ω, it suffices to establish that $(\frac{I-C}{A})^{-1}B$ is weakly sequentially continuous on $\overline{U^w}$. To see this, Let $\{u_n\}$ be a weakly convergent sequence of $\overline{U^w}$ to a point u in $\overline{U^w}$. Now, let us define the sequence $\{v_n\}$ of the subset Ω by:

$$v_n = \left(\frac{I-C}{A}\right)^{-1} Bu_n.$$

Since $(\frac{I-C}{A})^{-1}B(\overline{U^w})$ is relatively weakly compact, so there is a renamed subsequence such that

$$v_n = \left(\frac{I-C}{A}\right)^{-1} Bu_n \rightharpoonup v.$$

However, the subsequence $\{v_n\}$ verifies $v_n - Cv_n = Av_n.Bu_n$. Therefore, from assumption (vi) and in view of condition $(\mathcal{P})$, we deduce that v verifies the following equation

$$v - Cv = Av.Bu,$$

or equivalently,

$$v = \left(\frac{I-C}{A}\right)^{-1} Bu.$$

Next, we claim that the whole sequence $\{u_n\}$ verifies

$$\left(\frac{I-C}{A}\right)^{-1} Bu_n = v_n \rightharpoonup v.$$

Indeed, let us suppose that this is not the case. So, there is V^w, a weakly neighborhood of v satisfying, for all $n \in \mathbb{N}$, the existence of an $N \geq n$ such that $v_N \notin V^w$. Hence, there is a renamed subsequence $\{v_n\}$ verifying the property:

$$\text{for all } n \in \mathbb{N}, \{\, v_n\} \notin V^w. \tag{3.24}$$

However, for all $n \in \mathbb{N}, v_n \in (\frac{I-C}{A})^{-1}B(\overline{U^w})$. Again, there is a renamed subsequence such that $v_n \rightharpoonup v'$. According to the preceding, we have $v' = (\frac{I-C}{A})^{-1}Bu$ and consequently, $v' = v$, which is in contradiction with Property 3.24. This shows that $(\frac{I-C}{A})^{-1}B$ is weakly sequentially continuous. In view of Remark 2.3.4, an application of Theorem 2.3.7 implies that either

(e) $(\frac{I-C}{A})^{-1}B$ has, at least, a fixed point, or

(f) there exist a point $u \in \partial_\Omega U$ and $\lambda \in]0,1[$, with $u = \lambda(\frac{I-C}{A})^{-1}Bu$.

First, let us assume that $x \in \overline{U^w}$ is a fixed point of the operator $(\frac{I-C}{A})^{-1}B$. Then, $x = (\frac{I-C}{A})^{-1}Bx$, which implies that

$$Ax.Bx + Cx = x.$$

Second, let us suppose that there exist an element $u \in \partial_\Omega^w(U)$ and a real number $\lambda \in]0,1[$ such that $u = \lambda(\frac{I-C}{A})^{-1}Bu$. Then,

$$\left(\frac{I-C}{A}\right)^{-1} Bu = \frac{u}{\lambda},$$

so that

$$\lambda A\left(\frac{u}{\lambda}\right).Bu + \lambda C\left(\frac{u}{\lambda}\right) = u.$$

This completes the proof. Q.E.D.

Notice that this result remains true when $C \equiv 0$, and we get the following results :

Corollary 3.3.2 *Let Ω be a closed, and convex subset in a Banach algebra X satisfying the condition $(\mathcal{P})$, and let $U \subset \Omega$ be a weakly open set (with respect to the weak topology of Ω) such that $0 \in U$. Moreover, let $A : X \longrightarrow X$ and $B : \overline{U^w} \longrightarrow X$ be two operators satisfying the following:*

(i) A is $\mathcal{D}$-Lipschitzian with the $\mathcal{D}$-function Φ_A,

(ii) A is regular on X,

(iii) $B(\overline{U^w})$ is bounded with a bound M,

(iv) $M\Phi_A(r) < r$ for $r > 0$,

(v) $x = Ax.By \Longrightarrow x \in \Omega$ for all $y \in \overline{U^w}$,

(vi) A is weakly sequentially continuous on Ω, and B is weakly sequentially continuous on $\overline{U^w}$, and

(vii) $(\frac{I}{A})^{-1}B(\overline{U^w})$ is relatively weakly compact.

Then, either

(a) the equation $\lambda A(\frac{x}{\lambda}).Bx = x$ has a solution for $\lambda = 1$, or

(b) there is $u \in \partial_\Omega^w(U)$ such that $\lambda A(\frac{u}{\lambda}).Bu = u$ for some $0 < \lambda < 1$.

Theorem 3.3.3 *Let Ω be a closed, and convex subset in a Banach algebra X satisfying the condition $(\mathcal{P})$, and let $U \subset \Omega$ be a weakly open set (with respect to the weak topology of Ω) such that $0 \in U$. Besides, let A, $C : X \longrightarrow X$ and $B : \overline{U^w} \longrightarrow X$ be three operators satisfying the following:*

(i) A and C are $\mathcal{D}$-Lipschitzian with the $\mathcal{D}$-functions Φ_A and Φ_C, respectively,

(ii) A is regular on X,

(iii) $M\Phi_A(r) + \Phi_C(r) < r$ for $r > 0$,

(iv) $x = Ax.By + Cx \Longrightarrow x \in \Omega$ for all $y \in \overline{U^w}$,

(v) A and C are weakly sequentially continuous on Ω and B is weakly sequentially continuous on $\overline{U^w}$, and

(vi) $A(\Omega)$, $B(\overline{U^w})$, and $C(\Omega)$ are relatively weakly compact.

Then, either

(a) the equation $\lambda A(\frac{x}{\lambda}).Bx + \lambda C(\frac{x}{\lambda}) = x$ has a solution for $\lambda = 1$, or

(b) there is $u \in \partial_\Omega(U)$ such that $\lambda A(\frac{u}{\lambda}).Bu + \lambda C(\frac{u}{\lambda}) = u$ for some $0 < \lambda < 1$.

Proof. In view of Theorem 3.3.2, it is sufficient to prove that $(\frac{I-C}{A})^{-1}B(\overline{U^w})$ is relatively weakly compact. To do this, let $\{u_n\}$ be any sequence in $(\overline{U^w})$, and let

$$v_n = \left(\frac{I-C}{A}\right)^{-1} Bu_n. \tag{3.25}$$

Since $B(\overline{U^w})$ is relatively weakly compact, then there is a renamed subsequence $\{Bu_n\}$ which is weakly converging to an element w. Moreover, by using Eq. (3.25), we obtain

$$v_n = Av_n.Bu_n + Cv_n.$$

Since $\{v_n\}$ is a sequence in $(\frac{I-C}{A})^{-1}B(\overline{U^w})$, so by using the assumption (vi), there is a renamed subsequence such that $Av_n \rightharpoonup x$ and $Cv_n \rightharpoonup y$. Hence, in view of both condition $(\mathcal{P})$ and the last equation, we obtain

$$v_n \rightharpoonup x.w + y.$$

This shows that $(\frac{I-C}{A})^{-1}B(\overline{U^w})$ is relatively weakly sequentially compact. An application of the Eberlein–Šmulian theorem shows that $(\frac{I-C}{A})^{-1}B(\overline{U^w})$ is relatively weakly compact. Q.E.D.

When $C \equiv 0$, we get the following results :

Corollary 3.3.3 *Let Ω be a closed, and convex subset in a Banach algebra X satisfying the condition $(\mathcal{P})$, and let $U \subset \Omega$ be a weakly open set (with respect to the weak topology of Ω) such that $0 \in U$. Moreover, let $A : X \longrightarrow X$ and $B : \overline{U^w} \longrightarrow X$ be two operators satisfying the following:*

(i) A is $\mathcal{D}$-Lipschitzian with the $\mathcal{D}$-function Φ_A ,

(ii) A is regular on X,

(iii) $M\Phi_A(r) < r$ for $r > 0$,

(iv) $x = Ax.By \Longrightarrow x \in \Omega$ for all $y \in \overline{U^w}$,

(v) A is weakly sequentially continuous on Ω and B is weakly sequentially continuous on $\overline{U^w}$, and

(vi) $A(\Omega)$ and $B(\overline{U^w})$ are relatively weakly compact.

Then, either

(a) the equation $\lambda A(\frac{x}{\lambda}).Bx = x$ has a solution for $\lambda = 1$, or

(b) there is $u \in \partial_\Omega(U)$ such that $\lambda A(\frac{u}{\lambda}).Bu = u$ for some $0 < \lambda < 1$.

3.4 Convex-Power Condensing Operators

In this section, our main interest is dealing with some fixed point results for convex-power condensing operators. Throughout this section, S is a nonempty, closed, and convex subset of X. Let us first recall the following result proved by R. P. Agarwal, D. O'Regan, and M. A. Taoudi in [5].

Theorem 3.4.1 *Let $F : S \longrightarrow S$ be a weakly sequentially continuous operator and convex-power condensing with respect to ω. If $F(S)$ is bounded, then F has, at least, one fixed point in S.*

As a consequence, we have the following:

Theorem 3.4.2 *Let $F : S \longrightarrow S$ be a weakly sequentially continuous operator and convex-power condensing with respect to ω. If $F(S)$ is bounded, then the set $\mathcal{A}$ of fixed points of F is nonempty and weakly compact in S.*

Proof. Using the preceding Theorem 3.4.1, the set $\mathcal{A}$ is nonempty. Now, let us prove that $\mathcal{A}$ is weakly compact in S. Indeed, since $\mathcal{A} = F(\mathcal{A}) \subset F(S)$, then $\mathcal{A}$ is bounded. Moreover,

$$\mathcal{A} \subset F^{(2,x_0)}(\mathcal{A}) = F\left(\overline{\text{co}}\ \{\mathcal{A} \cup x_0\}\right),$$

which implies that

$$\mathcal{A} \subset \overline{\text{co}}\{F^{(2,x_0)}(\mathcal{A}), x_0\}.$$

Consequently,

$$\mathcal{A} \subset F^{(3,x_0)}(\mathcal{A}).$$

The continuation of this procedure leads to:

$$\mathcal{A} \subset F^{(n_0,x_0)}(\mathcal{A}).$$

Hence, by using Remark 1.4.5, $\mathcal{A}$ is relatively weakly compact. Knowing the sequentially weak continuity of F, $\mathcal{A}$ is sequentially weakly closed. By using the Eberlein–Šmulian theorem (see Theorem 1.3.3), it follows that $\mathcal{A}$ is a weakly closed subset of S. As a result, $\mathcal{A}$ is weakly compact in S. Q.E.D.

Corollary 3.4.1 *Let $F : S \longrightarrow S$ be a weakly sequentially continuous operator. If there exist $x_0 \in S$, $k \in [0, 1)$, and a positive integer n_0 ($n_0 \geq 1$) such that, for any bounded subset $V \subset S$, we have*

$$\omega\left(F^{(n_0,x_0)}(V)\right) \leq k\omega(V)$$

and if we assume that $F(S)$ is bounded, then F has, at least, one fixed point in S.

In [4], R. P. Agarwal et al. proved Theorem 3.4.1 under the condition of ws-compactness of the operator F instead of weakly sequentially continuous. Moreover, they generalize this result as follows:

Theorem 3.4.3 *[4] Let S be a nonempty, closed, and convex subset of X with $0 \in int\ S$. Let $F : S \longrightarrow X$ be a ws-compact operator and convex-power condensing about 0 and n_0 with respect to ω. If $F(S)$ is bounded and $F(\partial S) \subset S$, then F has, at least, a fixed point in S.*

Theorem 3.4.4 *Let X be a Banach space, and let ψ be a regular and set additive measure of weak noncompactness on X. Let C be a nonempty, closed, and convex subset of X, $x_0 \in C$, and let n_0 be a positive integer. Suppose that $F : C \longrightarrow C$ is ψ-convex-power condensing about x_0 and n_0. If F is ws-compact and if $F(C)$ is bounded, then F has, at least, a fixed point in C.*

Proof. Let

$$\mathcal{F} = \{A \subset C \text{ such that } \overline{co}(A) = A,\ x_0 \in A \text{ and } F(A) \subset A\}.$$

The set $\mathcal{F}$ is nonempty since $C \in \mathcal{F}$. Set $M = \bigcap_{A \in \mathcal{F}} A$. Now, let us show that, for any positive integer n, we have

$$M = \overline{co}\left(F^{(n,x_0)}(M) \cup \{x_0\}\right). \qquad \mathcal{P}(n)$$

In order to do this, we proceed by induction. Clearly, M is a closed and convex subset of C and $F(M) \subset M$. Thus, $M \in \mathcal{F}$. This implies that

$$\overline{co}\,(F(M) \cup \{x_0\}) \subset M.$$

Hence,

$$F\left(\overline{co}\,(F(M) \cup \{x_0\})\right) \subset F(M) \subset \overline{co}\,(F(M) \cup \{x_0\}).$$

Consequently,

$$\overline{co}\,(F(M) \cup \{x_0\}) \in \mathcal{F}.$$

Hence,

$$M \subset \overline{co}\,(F(M) \cup \{x_0\}).$$

As a result, $\overline{co}\,(F(M) \cup \{x_0\}) = M$. This shows that $\mathcal{P}(1)$ holds. Let n be fixed, and suppose that $\mathcal{P}(n)$ holds. This implies that

$$F^{(n+1,x_0)}(M) = F(\overline{co}\left(F^{(n,x_0)}(M) \cup \{x_0\}\right) = F(M).$$

Consequently,

$$\overline{co}\left(F^{(n+1,x_0)}(M) \cup \{x_0\}\right) = \overline{co}\,(F(M) \cup \{x_0\}) = M.$$

As a result,

$$\overline{co}\left(F^{(n_0,x_0)}(M) \cup \{x_0\}\right) = M.$$

Knowing that $F(C)$ is bounded also implies that M is bounded. By using the properties of the measure of weak noncompactness, we get

$$\psi(M) = \psi\left(\overline{co}\left(F^{(n_0,x_0)}(M) \cup \{x_0\}\right)\right) = \psi\left(F^{(n_0,x_0)}(M)\right),$$

which implies that M is weakly compact. Now, let us show that $F(M)$ is relatively compact. To do this, let us consider a sequence $(y_n)_n \in F(M)$. For each $n \in \mathbb{N}$, there exists $x_n \in M$ with $y_n = Fx_n$. Now, the Eberlein–Šmulian theorem (Theorem 1.3.3) guarantees the existence of a subsequence S of $\mathbb{N}$ so that $(x_n)_{n\in S}$ is a weakly convergent sequence. Since F is ws-compact, then $(Fx_n)_{n\in S}$ has a strongly convergent subsequence. Thus, $F(M)$ is relatively compact. Keeping in mind that $F(M) \subset M$, the result follows from Schauder's fixed point theorem. Q.E.D.

Theorem 3.4.5 *Let X be a Banach space, and let ψ be a regular set additive measure of weak noncompactness on X. Let Q be a closed and convex subset of X with $0 \in Q$, and let n_0 be a positive integer. Assume that $F : X \longrightarrow X$ is ws-compact and ψ-convex-power condensing about 0 and n_0 and that $F(Q)$ is bounded, and*

$$\begin{cases} \text{if } \{(x_j, \lambda_j)\} \text{ is a sequence in } \partial Q \times [0,1] \text{ converging to } (x, \lambda) \text{ with} \\ x = \lambda F(x) \text{ and } 0 < \lambda < 1, \text{ then } \lambda_j F(x_j) \in Q \text{ for a sufficiently large } j. \end{cases}$$

Also, suppose that the following condition holds:

$$\begin{cases} \text{there exists a continuous retraction } r : X \longrightarrow Q \text{ with } r(z) \in \partial Q \\ \text{for } z \in X \backslash Q \text{ and } r(D) \subset co(D \cup \{0\}) \text{ for any bounded subset } D \text{ of } X. \end{cases} \tag{3.26}$$

Then, F has, at least, a fixed point.

Proof. Let $r : X \longrightarrow Q$ be as described in (3.26). Let us consider

$$B = \Big\{x \in X \text{ such that } x = Fr(x)\Big\}.$$

First, let us show that $B \neq \emptyset$. To see this, consider $rF : Q \longrightarrow Q$. Notice that $rF(Q)$ is bounded since $F(Q)$ is bounded and $r(F(Q)) \subset co(F(Q) \cup \{0\})$. Clearly, rF is continuous since F and r are continuous. Now, let us show that rF is ws-compact. For this purpose, let $(x_n)_n$ be a sequence in Q which converges weakly to some $x \in Q$. Since F is ws-compact, then there exists a subsequence S of $\mathbb{N}$, so that $(Fx_n)_{n\in S}$ converges strongly to some $y \in X$. The continuity of r guarantees that the sequence $(rFx_n)_{n\in S}$ converges strongly to ry. This proves that rF is ws-compact. Our next task is to show that rF is ψ-convex-power condensing about 0 and n_0. To do so, let A be a subset of Q. In view of Eq. (3.26), we have

$$(rF)^{(1,0)}(A) = rF(A) = rF^{(1,0)}(A) \subset \overline{co}(F^{(1,0)}(A) \cup \{0\}).$$

Hence,

$$\begin{aligned}(rF)^{(2,0)}(A) &= rF\left(\overline{co}((rF)^{(1,0)}(A)\cup\{0\})\right)\\ &= rF\left(\overline{co}(rF^{(1,0)}(A)\cup\{0\})\right)\\ &\subset rF\left(\overline{co}(F^{(1,0)}(A)\cup\{0\})\right)\\ &= rF^{(2,0)}(A),\end{aligned}$$

and by induction, we have

$$\begin{aligned}(rF)^{(n_0,0)}(A) &= rF\left(\overline{co}((rF)^{(n_0-1,0)}(A)\cup\{0\})\right)\\ &\subset rF\left(\overline{co}(rF^{(n_0-1,0)}(A)\cup\{0\})\right)\\ &\subset rF\left(\overline{co}(F^{(n_0-1,0)}(A)\cup\{0\})\right)\\ &= rF^{(n_0,0)}(A).\end{aligned}$$

Taking into account the fact that F is ψ-convex-power condensing about 0 and n_0 and by using the condition (3.26), we get

$$\begin{aligned}\psi\left((rF)^{(n_0,0)}(A)\right) &\leq \psi\left(rF^{(n_0,0)}(A)\right)\\ &\leq \psi\left(co\left(F^{(n_0,0)}(A)\cup\{0\}\right)\right)\\ &\leq \psi\left(F^{(n_0,0)}(A)\right)\\ &< \psi(A),\end{aligned}$$

whenever $\psi(A) > 0$. Invoking Theorem 3.4.4, we infer that there exists $y \in Q$ with $rF(y) = y$. Let $z = F(y)$. So, $Fr(z) = Fr(F(y)) = F(y) = z$. Thus, $z \in B$ and $B \neq \emptyset$. In addition, B is closed, since Fr is continuous. Moreover, we claim that B is compact. Q.E.D.

Next, we will also generalize this result as follows:

Theorem 3.4.6 *Let S be a nonempty, closed, and convex subset of X with $int\ S \neq \emptyset$ and fix $x_0 \in int\ S$. Let $F : S \longrightarrow X$ be a ws-compact operator and convex-power condensing about x_0 and n_0 with respect to ω. If $F(S)$ is bounded and $F(\partial S) \subset S$ (the condition of Rothe type), then F has, at least, one fixed point in S.*

Proof. Set $S_1 = S - x_0 := \{x - x_0 : \ x \in S\}$ and let us define the following mapping

$$\begin{cases} F_1 : S_1 \longrightarrow X \\ \qquad x \longrightarrow F_1(x) = F(x + x_0) - x_0. \end{cases}$$

Clearly, S_1 is a closed and convex subset of E and $0 \in \text{int } S_1$. Moreover,

$$F_1(\partial S_1) = F(\partial(S - x_0) + x_0) - x_0 = F(\partial S) - x_0 \subset S - x_0 = S_1.$$

Since $F_1(S_1) = F(S) - x_0$, then $F_1(S_1)$ is bounded. Since F is ws-compact, then F_1 is ws-compact. Our next task is to prove that F_1 is convex-power condensing about 0 and n_0 with respect to the measure of weak noncompactness ω. To do so, let V be an arbitrary bounded subset of S_1 with $\omega(V) > 0$. Then,

$$F_1^{(1,0)}(V) = F_1(V) = F(V + x_0) - x_0 = F^{(1,x_0)}(V + x_0) - x_0.$$

Making an inductive assumption, we have

$$F_1^{(n-1,0)}(V) \subset F^{(n-1,x_0)}(V + x_0) - x_0.$$

Then,

$$F_1^{(n-1,0)}(V) \subset \overline{\text{co}}\ \{F^{(n-1,x_0)}(V + x_0), x_0\} - x_0.$$

Consequently,

$$F_1^{(n-1,0)}(V) \cup \{\theta\} \subset \overline{\text{co}}\ \{F^{(n-1,x_0)}(V + x_0), x_0\} - x_0.$$

Hence,

$$\overline{\text{co}}\ \{F_1^{(n-1,0)}(V), 0\} \subset \overline{\text{co}}\ \{F^{(n-1,x_0)}(V + x_0), x_0\} - x_0.$$

Thus,

$$\begin{aligned} F_1^{(n,0)}(V) &= F_1\left(\overline{\text{co}}\ \{F_1^{(n-1,0)}(V), 0\}\right) \\ &\subset F_1\left(\overline{\text{co}}\ \{F^{(n-1,x_0)}(V + x_0), x_0\} - x_0\right) \\ &= F\left(\overline{\text{co}}\ \{F^{(n-1,x_0)}(V + x_0), x_0\}\right) - x_0 \\ &= F^{(n,x_0)}(V + x_0) - x_0. \end{aligned}$$

Now, The use of the properties (i) and (vii) of ω leads to

$$\begin{aligned}\omega\left(F_1^{(n_0,0)}(V)\right) &\leq \omega\left(F^{(n_0,x_0)}(V+x_0)-x_0\right)\\ &= \omega\left(F^{(n_0,x_0)}(V+x_0)\right)\\ &< \omega(V+x_0)=\omega(V).\end{aligned}$$

That is, F_1 is convex-power condensing about 0 and n_0. Now, by applying Theorem 3.4.3, we infer that F_1 has, at least, a fixed point x_1 in S_1, i.e.,

$$x_1 \in S_1 \text{ and } F_1(x_1)=x_1,$$

or equivalently,

$$F(x_1+x_0)=x_1+x_0, \text{ where } x=x_1+x_0\in S.$$

Q.E.D.

3.5 ws-Compact and ω-Convex-Power Condensing Maps

Theorem 3.5.1 *Let $F: S \longrightarrow S$ be a weakly sequentially continuous operator and ω-convex-power condensing. If $F(S)$ is bounded, then the set $\mathcal{A}$ of fixed points of F is nonempty and weakly compact in S.*

Proof. By using the preceding Theorem 3.4.1, the set $\mathcal{A}$ is nonempty. Now, let us prove that $\mathcal{A}$ is weakly compact in S. Since $\mathcal{A}=F(\mathcal{A})\subset F(S)$, then $\mathcal{A}$ is bounded. Moreover,

$$\mathcal{A}\subset F^{(2,x_0)}(\mathcal{A})=F\left(\overline{\text{co}}\left\{\mathcal{A}\cup x_0\right\}\right),$$

which implies that

$$\mathcal{A}\subset \overline{\text{co}}\{F^{(2,x_0)}(\mathcal{A}),x_0\},$$

and consequently,

$$\mathcal{A}\subset F^{(3,x_0)}(\mathcal{A}).$$

The continuation of this procedure leads to:

$$\mathcal{A}\subset F^{(n_0,x_0)}(\mathcal{A}).$$

Hence, $\mathcal{A}$ is relatively weakly compact. Knowing the sequentially weak continuity of F, we deduce that $\mathcal{A}$ is weakly sequentially closed. By using the Eberlein–Šmulian theorem (Theorem 1.3.3), it follows that $\mathcal{A}$ is a weakly closed subset of S. As a result, $\mathcal{A}$ is weakly compact in S. Q.E.D.

Corollary 3.5.1 *Let $F : S \longrightarrow S$ be a weakly sequentially continuous operator. If there exist $x_0 \in S$, $k \in [0,1)$, and a positive integer n_0 ($n_0 \geq 1$) such that, for any bounded subset $V \subset S$, we have*

$$\omega\left(F^{(n_0,x_0)}(V)\right) \leq k\omega(V),$$

and if we assume that $F(S)$ is bounded, then F has, at least, one fixed point in S.

Theorem 3.5.2 *Let S be a nonempty, closed, and convex subset of X with int $S \neq \emptyset$ and fix $x_0 \in$ int S. Let $F : S \longrightarrow X$ be a ws-compact operator and ω-convex-power condensing about x_0 and n_0. If $F(S)$ is bounded and $F(\partial S) \subset S$ (the condition of Rothe type), then F has, at least, a fixed point in S.*

Proof. Set $S_1 = S - x_0 := \{x - x_0 : \ x \in S\}$ and let us define the following mapping

$$\begin{cases} F_1 : S_1 \longrightarrow X \\ \qquad x \longrightarrow F_1(x) = F(x + x_0) - x_0. \end{cases}$$

Clearly, S_1 is a closed convex subset of E and $0 \in$ int S_1. Moreover,

$$F_1(\partial S_1) = F(\partial(S - x_0) + x_0) - x_0 = F(\partial S) - x_0 \subset S - x_0 = S_1.$$

Since $F_1(S_1) = F(S) - x_0$, then $F_1(S_1)$ is bounded. Since F is ws-compact, then F_1 is ws-compact. Our next task is to prove that F_1 is ω-convex-power condensing about 0 and n_0. To do so, let V be an arbitrary bounded subset of S_1 with $\omega(V) > 0$. Then,

$$F_1^{(1,0)}(V) = F_1(V) = F(V + x_0) - x_0 = F^{(1,x_0)}(V + x_0) - x_0.$$

By making an inductive assumption, we have

$$F_1^{(n-1,0)}(V) \subset F^{(n-1,x_0)}(V + x_0) - x_0,$$

and we deduce that,

$$F_1^{(n-1,0)}(V) \subset \overline{\text{co}}\left\{F^{(n-1,x_0)}(V + x_0), x_0\right\} - x_0.$$

Consequently,

$$F_1^{(n-1,0)}(V) \cup \{\theta\} \subset \overline{\text{co}}\,\{F^{(n-1,x_0)}(V+x_0), x_0\} - x_0.$$

Hence,

$$\overline{\text{co}}\,\{F_1^{(n-1,0)}(V), 0\} \subset \overline{\text{co}}\,\{F^{(n-1,x_0)}(V+x_0), x_0\} - x_0.$$

Thus,

$$\begin{aligned} F_1^{(n,0)}(V) &= F_1\left(\overline{\text{co}}\,\{F_1^{(n-1,0)}(V), 0\}\right) \\ &\subset F_1\left(\overline{\text{co}}\,\{F^{(n-1,x_0)}(V+x_0), x_0\} - x_0\right) \\ &= F\left(\overline{\text{co}}\,\{F^{(n-1,x_0)}(V+x_0), x_0\}\right) - x_0 \\ &= F^{(n,x_0)}(V+x_0) - x_0. \end{aligned}$$

Now, by using the properties of ω, we infer that

$$\begin{aligned} \omega\left(F_1^{(n_0,0)}(V)\right) &\leq \omega\left(F^{(n_0,x_0)}(V+x_0) - x_0\right) \\ &= \omega\left(F^{(n_0,x_0)}(V+x_0)\right) \\ &< \omega(V+x_0) = \omega(V). \end{aligned}$$

That is, F_1 is convex-power condensing about 0 and n_0. Now, by applying the preceding Theorem 3.4.3, we deduce that F_1 has, at least, a fixed point x_1 in S_1, i.e.,

$$x_1 \in S_1 \text{ and } F_1(x_1) = x_1,$$

or equivalently,

$$F(x_1 + x_0) = x_1 + x_0,$$

where $x = x_1 + x_0 \in S$. Q.E.D.

Next, we have:

Theorem 3.5.3 *Let U be an open subset of S with $x_0 \in U$. Let $F : \overline{U} \longrightarrow X$ be a ws-compact operator and ω-convex-power condensing about x_0 and n_0. If $F(\overline{U})$ is bounded and $F(\overline{U}) \subset S$, then either*

(i) F has, at least, a fixed point in $\overline{U}$, or

(ii) there exist an $x \in \partial U$ and $\lambda \in (0,1)$, such that $x = \lambda Fx + (1-\lambda)x_0$.

Proof. By using Theorem 3.5.2, and by replacing F, S, and U by F_1, $S - x_0$, and $U - x_0$, respectively, we may assume that $0 \in U$ and that F is ω-convex-power condensing about 0 and n_0. Now, let us suppose that (ii) does not hold and that F has no fixed points in ∂U (otherwise, we have finished). Then, $x \neq \lambda F(x)$, for all $x \in \partial U$ and $\lambda \in [0,1]$. Now, let us consider

$$Q := \{x \in \overline{U} : x = tF(x) \text{ for some } t \in [0,1]\}.$$

The set Q is nonempty, since $0 \in U$. Moreover, since F is continuous, then Q is closed. By hypothesis, $Q \cap \partial U = \emptyset$. Therefore, by using Urysohn's lemma, there exists a continuous map $\rho : \overline{U} \longrightarrow [0,1]$, with $\rho(Q) = 1$ and $\rho(\partial U) = 0$. Let us define T by:

$$T(x) = \begin{cases} \rho(x)F(x), & if\ x \in \overline{U} \\ 0, & if\ x \in S \backslash \overline{U}. \end{cases}$$

Since $\partial \overline{U} \subset \partial U$ and since ρ, F are continuous, then T is continuous. Now, let us show that $T : S \longrightarrow S$ is ws-compact. For this purpose, let $(x_n)_{n\in\mathbb{N}}$ be any sequence in S which converges weakly to some $x \in S$. Without loss of generality, we may assume that $(x_n)_{n\in\mathbb{N}}$ in $\overline{U}$ and $T(x_n) = \rho(x_n)F(x_n)$, $n \in \mathbb{N}$. Then, $\rho(x_n)_{n\in\mathbb{N}}$ is a sequence in $[0,1]$. Hence, there exists a subsequence $\rho(x_{\varphi(n)})_{n\in\mathbb{N}}$ which converges to some $\alpha \in [0,1]$. Moreover, since F is ws-compact, then there exists a subsequence $F(x_{\varphi o \psi(n)})_{n\in\mathbb{N}}$ which converges to some $y \in S$. Thus, the sequence $T(x_{\varphi o \psi(n)})_{n\in\mathbb{N}}$ converges to $\alpha y \in S$. This proves that T is ws-compact. Next, let us show that T is ω-convex-power condensing about 0 and n_0. First, we note that $T(S) \subset \overline{co}\,(F(\overline{U}) \cup \{\theta\})$. Then, $T(S)$ is bounded. Let V be an arbitrary bounded subset of S with $\omega(V) > 0$. We show that, according to the mathematical induction method, for all positive integers n, we have

$$T^{(n,0)}(V) \subset \overline{co}\, \left\{ \widetilde{F}^{(n,0)}(V \cap \overline{U}), \{0\} \right\}. \tag{3.27}$$

Notice that

$$T(V) \subset \overline{co}\,(F(V \cap \overline{U}), \{0\}), \tag{3.28}$$

which implies that

$$T^{(2,0)}(V) \subset T\left(\overline{co}\{F(V \cap \overline{U}), \{0\}\}\right).$$

Using (3.28) leads to

$$T^{(2,0)}(V) \subset \overline{co} \left\{ \widetilde{F}^{(2,0)}(V \cap \overline{U}), \{0\} \right\}.$$

Now, suppose that

$$T^{(n,0)}(V) \subset \overline{\text{co}}\left\{\widetilde{F}^{(n,0)}(V \cap \overline{U}), \{0\}\right\}.$$

Then,

$$T^{(n+1,0)}(V) \subset T\left(\overline{\text{co}}\left\{\widetilde{F}^{(n,0)}(V \cap \overline{U}), \{0\}\right\}\right).$$

By using (3.28), it follows that

$$T^{(n+1,0)}(V) \subset \overline{\text{co}}\ \left\{\widetilde{F}^{(n+1,0)}(V \cap \overline{U}), \{0\}\right\}.$$

This proves the claim. Hence, by using both (3.27) and the defining properties of ω, we have

$$\omega\left(T^{(n_0,0)}(V)\right) \leq \omega\left(\widetilde{F}^{(n_0,0)}(V \cap \overline{U})\right).$$

Case 1: If $\omega(V \cap \overline{U}) = 0$, then $V \cap \overline{U}$ is relatively weakly compact. Since F is ws-compact, then $F(V \cap \overline{U})$ is relatively compact. Therefore, by using Mazur's theorem (see Theorem 1.3.5), $\overline{\text{co}}\left\{F(V \cap \overline{U}), \{0\}\right\}$ is compact. Thus, $\left(\overline{\text{co}}\left\{F(V \cap \overline{U}), \{0\}\right\} \cap \overline{U}\right)$ is compact. Since F is continuous, then $\widetilde{F}^{(2,0)}(V \cap \overline{U})$ is compact. By induction, for all positive integers n, we get $\widetilde{F}^{(n,0)}(V \cap \overline{U})$ is relatively compact. Consequently, $\omega\left(\widetilde{F}^{(n_0,0)}(V \cap \overline{U})\right) = 0$. As a result,

$$\omega\left(T^{(n_0,0)}(V)\right) < \omega(V).$$

Case 2: If $\omega(V \cap \overline{U}) > 0$, then

$$\omega\left(T^{(n_0,0)}(V)\right) < \omega(V \cap \overline{U}) \leq \omega(V).$$

Now, we may invoke Theorem 3.4.3 in order to conclude that T has, at least, a fixed point x in S, i.e., $T(x) = x$. If $x \in Q$, then $x \in \overline{U}$ and $T(x) = F(x) = x$. If $x \notin Q$, then $T(x) = 0$. (Otherwise, $x = \rho(x)F(x) \in Q$). As a result,

$$T(x) = 0 = x = T(0) = \rho(0)F(0),\ (0 \in \overline{U}),$$

which is in contradiction with $x = 0 \notin Q$. Q.E.D.

Remark 3.5.1 *The set Q is relatively weakly compact. Indeed, since*

$$Q \subset \overline{co}(F(Q) \cup \{0\}) \subset \overline{co}(F(\overline{U}) \cup \{0\}),$$

then Q is bounded. It follows that

$$F(Q) \subset F^{(2,0)}(Q).$$

Thus,

$$Q \subset \overline{co}\left(F^{(2,0)}(Q) \cup \{0\}\right).$$

By induction, for all positive integers n, we get

$$Q \subset \overline{co}\left(F^{(n,0)}(Q) \cup \{0\}\right).$$

Now, from the definition of a ω-convex-power condensing about 0 *and n_0, we deduce the desired result.*

Question 3:

If $\left(\frac{I-C}{A}\right)$ is not invertible, $\left(\frac{I-C}{A}\right)^{-1}$ could be seen as a multi-valued mapping. This case is not discussed in all the theorems of this chapter. To our knowledge, this question is still open.

Chapter 4

Fixed Point Theory for BOM on Banach Spaces and Banach Algebras

In this chapter, we are concerned with fixed point theorems for a 2×2 block operator matrix with nonlinear entries (in short BOM) acting on a product of two Banach spaces or Banach algebras. We are also interested with the case where these entries are assumed nonlinear multi-valued operators.

4.1 Some Variants of Schauder's and Krasnosel'skii's Fixed Point Theorems for BOM

Let Ω (resp. Ω') be a nonempty, closed, and convex subset of a Banach space X (resp. Y). We consider the 2×2 block operator matrix (in short BOM)

$$\mathcal{L} = \begin{pmatrix} A & B \\ C & D \end{pmatrix}, \tag{4.1}$$

in the space $X \times Y$, that is, the nonlinear operator A maps Ω into X, B from Ω' into X, C from Ω into Y and D from Ω' into Y. The aim of this section is to discuss the existence of fixed points for the block operator matrix (4.1) by imposing some conditions on the entries which are, in general, nonlinear operators. This discussion is based on the invertibility or not of the diagonal terms of $\mathcal{I} - \mathcal{L}$.

4.1.1 One of the diagonal entries of $\mathcal{I} - \mathcal{L}$ is invertible

Assume that:

$(\mathcal{H}_1)$ The operator $I - A$ is invertible and $(I - A)^{-1}B(\Omega') \subset \Omega$,

$(\mathcal{H}_2)$ $S := C(I - A)^{-1}B$ is an operator with a closed graph and the subset

$S(\Omega')$ is relatively compact in Y,

$(\mathcal{H}_3)$ the operator $I-D$ is invertible and $(I-D)^{-1}$ is continuous on $(I-D)(\Omega')$, and

$(\mathcal{H}_4)$ $(I-D)^{-1}S(\Omega') \subset \Omega'$.

Theorem 4.1.1 *Let f be a function from a space X into a space Y. Then, the following conditions are equivalent.*

(i) f has a closed graph.

(ii) If $x \in X$, $y \in Y$ and $y \neq f(x)$, then there exists an open neighborhood U of x and there exists an open neighborhood V of y, such that $f(U) \cap V = \emptyset$.

(iii) If K is a compact subset of X then,

$$f(K) = \bigcap \{\overline{f(U)} \text{ such that } U \text{ is an open neighborhood of } K\}.$$

(iv) If C is a compact subset of Y, then

$$f^{-l}(C) = \bigcap \{\overline{f^{-1}(V)} \text{ such that } V \text{ is an open neighborhood of } C\}.$$

Proof. $(i) \Leftrightarrow (ii)$ is trivial and follows immediately from the definition of the product topology. Note that (ii) can be rewritten in the following forms:

$$\{f(x)\} = \bigcap \{\overline{f(U)} \text{ such that } U \text{ is an open neighborhood of } x\}$$

for each $x \in X$; or

$$f^{-1}(y) = \bigcap \{\overline{f^{-1}(V)} \text{ such that } V \text{ is an open neighborhood of } y\}$$

for each $y \in Y$. Since compact sets " behave" like points, we obtain $(ii) \Leftrightarrow (iii)$ and $(ii) \Leftrightarrow (iv)$. Q.E.D.

Corollary 4.1.1 *Let $f : X \longrightarrow Y$ have a closed graph. Then,*

(i) If K is a compact subset of X, then $f(K)$ is closed.

(ii) If C is a compact subset of Y, then $f^{-l}(C)$ is closed.

(iii) If K is a compact subset of X, C is a compact subset of Y and $f(K) \cap C = \emptyset$, then there is an open neighborhood U of K and also an open neighborhood V of C, such that $\overline{f^{-1}(V)} \cap U = \emptyset = V \cap \overline{f(U)}$.

Proof. For (i), use (iii) from Theorem 4.1.1; for (ii), use (iv) from Theorem 4.1.1; for (iii), use both (iii) and (iv) from Theorem 4.1.1. Q.E.D.

Corollary 4.1.2 *Let $f : X \longrightarrow Y$ have a closed graph.*

(i) If X is compact, then f is a closed function.

(ii) If Y is compact, then f is a continuous function.

Proof. Because closed subsets of a compact space are compact, the result follows immediately from Corollary 4.1.1. Q.E.D.

Corollary 4.1.3 *Let X and Y be two metric spaces. Assume that $J : X \longrightarrow Y$ has a closed graph and $\overline{J(X)}$ is a compact set of Y. Then, J is continuous.*

Proof. Corollary 4.1.3 is an immediate consequence of Corollary 4.1.2. Q.E.D.

Theorem 4.1.2 *Let $\mathcal{K}$ be a closed, convex, and nonempty subset of a Banach space X. Suppose that J maps $\mathcal{K}$ into $\mathcal{K}$, and that*

(i) J has a closed graph, and

(ii) $\overline{J(\mathcal{K})}$ is compact.

Then, J has, at least, a fixed point in $\mathcal{K}$.

Proof. The proof of Theorem 4.1.2 follows from both Corollary 4.1.3 and the Schauder's fixed point theorem. Q.E.D.

Theorem 4.1.3 *Under assumptions $(\mathcal{H}_1)$–$(\mathcal{H}_4)$, the block operator matrix (4.1) has, at least, a fixed point in $\Omega \times \Omega'$.*

Proof. Since S has a closed graph and $S(\Omega')$ is relatively compact in Y and using Corollary 4.1.3, it follows that S is continuous on Ω'. Now, let M be a bounded subset of Ω'. Obviously from $(\mathcal{H}_2)$, the set $S(M)$ is relatively compact. Then, by hypothesis $(\mathcal{H}_3)$, $(I - D)^{-1}S(M)$ is relatively compact. According to Schauder's fixed point theorem, there exists $y_0 \in \Omega'$, such that

$$(I - D)^{-1}Sy_0 = y_0.$$

Let $x_0 := (I - A)^{-1}By_0$. Hence, $\mathcal{L}\begin{pmatrix} x_0 \\ y_0 \end{pmatrix} = \begin{pmatrix} x_0 \\ y_0 \end{pmatrix}$. Q.E.D.

Corollary 4.1.4 *If assumptions $(\mathcal{H}_1)$, $(\mathcal{H}_2)$, and $(\mathcal{H}_4)$ hold, and if D is a separate contraction mapping satisfying $C(\Omega) \subset (I - D)(\Omega')$, then the block operator matrix (4.1) has, at least, a fixed point in $\Omega \times \Omega'$.*

Proof. In Lemma 1.2.2, it was shown that $I - D$ is a homeomorphism from Ω' onto $(I - D)(\Omega')$. Then, $(\mathcal{H}_3)$ is satisfied and the result follows from Theorem 4.1.2. Q.E.D.

Corollary 4.1.5 *If assumptions $(\mathcal{H}_1)$, $(\mathcal{H}_2)$, and $(\mathcal{H}_4)$ hold, and if $I - D$ is a semi-expansive mapping satisfying $C(\Omega) \subset (I - D)(\Omega')$, then the block operator matrix (4.1) has, at least, a fixed point in $\Omega \times \Omega'$.*

Proof. From Definition 1.2.7 (i), it is clear that $I - D$ is one-to-one. Now, we show that $(I - D)^{-1} : (I - D)(\Omega') \longrightarrow \Omega'$ is continuous. Let $(y_n)_{n\in\mathbb{N}}$ and y in $(I - D)(\Omega')$, such that $y_n \to y$. Then, there exist $\alpha_n, \alpha \in \Omega'$ such that $y_n = (I - D)(\alpha_n)$ and $y = (I - D)(\alpha)$. Now,

$$\begin{aligned} \|y_n - y\| &= \|(I - D)(\alpha_n) - (I - D)(\alpha)\| \\ &\geq \Phi(\alpha_n, \alpha). \end{aligned}$$

It follows that $\Phi(\alpha_n, \alpha) \to 0$ and so, by Definition 1.2.7 (iii), $\alpha_n \to \alpha$. Hence, the assumption $(\mathcal{H}_3)$ is satisfied and the result follows from Theorem 4.1.2. Q.E.D.

At the end of this section, we will treat only the case of invertibility of $I - A$. The other case is similar, just simply exchanging the roles of A and D, and B and C.

An immediate application of the Krasnosel'skii theorem (see Theorem 1.6.8) for the operator $S + D$ allows us to get the following result:

Theorem 4.1.4 *If the entries satisfy the following assumptions:*

(i) The operator $I - A$ is invertible and $(I - A)^{-1}B(\Omega') \subset \Omega$,

(ii) $S := C(I - A)^{-1}B$ is a contraction map,

(iii) the operator D is completely continuous, and

(iv) $Sy + Dy' \in \Omega'$ for every y, y' in Ω'.

Then, the block operator matrix (4.1) has, at least, a fixed point in $\Omega \times \Omega'$.

4.1.2 None of the diagonal entries of $\mathcal{I} - \mathcal{L}$ is invertible

In this subsection, we discuss the existence of fixed points for the following perturbed block operator matrix by laying down some conditions on the entries, using Krasnosel'skii's theorem.

$$\widetilde{\mathcal{L}} = \begin{pmatrix} A_1 & B \\ C & D_1 \end{pmatrix} + \begin{pmatrix} P_1 & 0 \\ 0 & P_2 \end{pmatrix}. \tag{4.2}$$

Assume that the nonlinear operators A_1 and P_1 maps Ω into X, B from Ω' into X, C from Ω into Y, and D and P_2 from Ω' into Y. Suppose that the operator (4.2) fulfills the following assumptions:

($\mathcal{H}_5$) The operator $I - A_1$ (resp. $I - D_1$) is invertible from Ω into X, (resp. from Ω' into Y),

($\mathcal{H}_6$) $(I - A_1)^{-1}B$ and $(I - D_1)^{-1}C$ are completely continuous maps,

($\mathcal{H}_7$) $(I - A_1)^{-1}P_1$ and $(I - D_1)^{-1}P_2$ are contractions maps, and

($\mathcal{H}_8$) For every $x_1, x_2 \in \Omega$, $y_1, y_2 \in \Omega'$, $(I - A_1)^{-1}P_1x_1 + (I - A_1)^{-1}By_2 \in \Omega$ and $(I - D_1)^{-1}Cx_2 + (I - D_1)^{-1}P_2y_1 \in \Omega'$.

Theorem 4.1.5 *Under the previous four assumptions ($\mathcal{H}_5$)–($\mathcal{H}_8$), the block operator matrix (4.2) has, at least, a fixed point in $\Omega \times \Omega'$.*

Proof. Let us consider the following matrix equation

$$\begin{pmatrix} A_1 + P_1 & B \\ C & D_1 + P_2 \end{pmatrix} \begin{pmatrix} x \\ y \end{pmatrix} = \begin{pmatrix} x \\ y \end{pmatrix}. \tag{4.3}$$

Observe that Eq. (4.3) is equivalent to

$$\begin{pmatrix} P_1 & B \\ C & P_2 \end{pmatrix} \begin{pmatrix} x \\ y \end{pmatrix} = \begin{pmatrix} I - A_1 & 0 \\ 0 & I - D_1 \end{pmatrix} \begin{pmatrix} x \\ y \end{pmatrix}.$$

By using the assumption ($\mathcal{H}_5$), we deduce that

$$\begin{pmatrix} (I - A_1)^{-1} & 0 \\ 0 & (I - D_1)^{-1} \end{pmatrix} \begin{pmatrix} P_1 & B \\ C & P_2 \end{pmatrix} \begin{pmatrix} x \\ y \end{pmatrix} = \begin{pmatrix} x \\ y \end{pmatrix}$$

or, equivalently

$$\begin{pmatrix} (I - A_1)^{-1}P_1 & (I - A_1)^{-1}B \\ (I - D_1)^{-1}C & (I - D_1)^{-1}P_2 \end{pmatrix} \begin{pmatrix} x \\ y \end{pmatrix} = \begin{pmatrix} x \\ y \end{pmatrix}.$$

Hence, Eq. (4.3) may be transformed into

$$\mathcal{Z}_1 \begin{pmatrix} x \\ y \end{pmatrix} + \mathcal{Z}_2 \begin{pmatrix} x \\ y \end{pmatrix} = \begin{pmatrix} x \\ y \end{pmatrix},$$

where

$$\mathcal{Z}_1 = \begin{pmatrix} (I - A_1)^{-1}P_1 & 0 \\ 0 & (I - D_1)^{-1}P_2 \end{pmatrix}$$

and

$$\mathcal{Z}_2 = \begin{pmatrix} 0 & (I - A_1)^{-1}B \\ (I - D_1)^{-1}C & 0 \end{pmatrix}.$$

Obviously, from $(\mathcal{H}_6)$ the operator matrix $\mathcal{Z}_2$ is completely continuous, and from $(\mathcal{H}_7)$ the map $\mathcal{Z}_1$ is a contraction. Using $(\mathcal{H}_8)$ and Krasnosel'skii's theorem, it follows that the operator matrix (4.2) has, at least, a fixed point in $\Omega \times \Omega'$. Q.E.D.

4.2 Fixed Point Theory under Weak Topology Features

Let A be a nonlinear operator from X into itself. Let us denote by ω the measure of weak noncompactness of De Blasi (see Lemma 1.4.1). We introduce the following conditions:

$$(\mathcal{A}1) \quad \begin{cases} \text{If } (x_n)_{n\in\mathbb{N}} \text{ is a weakly convergent sequence in } X, \text{ then} \\ (Ax_n)_{n\in\mathbb{N}} \text{ has a strongly convergent subsequence in } X, \end{cases}$$

and

$$(\mathcal{A}2) \quad \begin{cases} \text{if } (x_n)_{n\in\mathbb{N}} \text{ is a weakly convergent sequence in } X, \text{ then} \\ (Ax_n)_{n\in\mathbb{N}} \text{ has a weakly convergent subsequence in } X. \end{cases}$$

Regarding these two conditions, we notice that:

Remark 4.2.1 *(i) Operators satisfying $(\mathcal{A}1)$ or $(\mathcal{A}2)$ are not necessarily weakly continuous.*

(ii) The map A satisfies $(\mathcal{A}1)$ if, and only if, it maps relatively weakly compact sets into relatively compact ones.

(iii) The assumption $(\mathcal{A}1)$ is weaker than the weakly-strongly sequentially continuity of the operator A.

(iv) Every ω-contractive map satisfies $(\mathcal{A}2)$.

(v) The map A satisfies $(\mathcal{A}2)$ if, and only if, it maps relatively weakly compact sets into relatively weakly compact ones (use the Eberlien-Šmulian theorem 1.3.3).

(vi) The condition $(\mathcal{A}2)$ holds true for every bounded linear operator.

Now, we recall the following well-known results in [150].

Theorem 4.2.1 *Let $\mathcal{M}$ be a nonempty, closed, and convex subset of a Banach space X. Assume that $A : \mathcal{M} \longrightarrow \mathcal{M}$ is a continuous map which verifies $(\mathcal{A}1)$. If $A(\mathcal{M})$ is relatively weakly compact, then there exists $x \in \mathcal{M}$ such that $Ax = x$.*

Proof. Let $\mathcal{C} = \overline{co}(A(\mathcal{M}))$ (the closed convex hull of $A(\mathcal{M})$). Since $\mathcal{M}$ is a closed convex subset of X satisfying $A(\mathcal{M}) \subset \mathcal{M}$, we obtain $\mathcal{C} \subset \mathcal{M}$ and therefore, $A(\mathcal{C}) \subset A(\mathcal{M}) \subset \overline{co}(A(\mathcal{M}))$. This shows that A maps $\mathcal{C}$ into itself. By hypothesis, $A(\mathcal{M})$ is relatively weakly compact. Hence, applying the Krein–Šmulian theorem (see Theorem 1.3.5), one sees that $\mathcal{C}$ is weakly compact too. Let $(\theta_n)_n$ be a sequence in $\mathcal{C}$. Then, it has a weakly convergent subsequence, say $(\theta_{n_k})_k$. By hypothesis, $(A\theta_{n_k})_k$ has a strongly convergent subsequence and therefore, $A(\mathcal{C})$ is relatively compact. Now, the use of Schauder's fixed point theorem concludes the proof. Q.E.D.

Theorem 4.2.2 *Let $\mathcal{M}$ be a nonempty, bounded, closed, and convex subset of a Banach space X. Assume that $A : \mathcal{M} \longrightarrow \mathcal{M}$ is a continuous map satisfying $(\mathcal{A}1)$. If A is ω-contractive, then there exists $x \in \mathcal{M}$ such that $Ax = x$.*

Proof. Let $M_1 = \mathcal{M}$ and $M_{n+1} = \overline{co}(A(M_n))$. It is clear that the sequence $(M_n)_n$ consists of nonempty, closed, convex decreasing subsets of $\mathcal{M}$. Since A is ω-contractive, then, for some $\beta \in [0, 1)$, we have

$$\omega(M_2) = \omega(\overline{co}(A(M_1))) = \omega(A(M_1)) \leq \beta\omega(M_1).$$

Proceeding by induction, we get $\omega(M_{n+1}) \leq \beta^n\omega(\mathcal{M})$ and therefore, $\lim_{n\to\infty}\omega(M_n) = 0$. Using the property (8) of Lemma 1.4.1, we infer that $\mathcal{N} := \cap_{n=1}^{\infty} M_n$ is a nonempty, closed, convex, and weakly compact subset of $\mathcal{M}$. Moreover, we can easily verify that $A(\mathcal{N}) \subset \mathcal{N}$. Accordingly, $A(\mathcal{N})$ is relatively weakly compact. Now, the use of Theorem 4.2.1 concludes the proof. Q.E.D.

Theorem 4.2.3 *Let $\mathcal{M}$ be a nonempty, closed, bounded, and convex subset of a Banach space X. Suppose that $A : \mathcal{M} \longrightarrow X$ and $B : X \longrightarrow X$ such that:*

(i) A is continuous, $A(\mathcal{M})$ is relatively weakly compact and A satisfies $(\mathcal{A}1)$,

(ii) B is a contraction satisfying $(\mathcal{A}2)$, and

(iii) $A(\mathcal{M}) + B(\mathcal{M}) \subset \mathcal{M}$.

Then, there is $x \in \mathcal{M}$ such that $Ax + Bx = x$.

Proof. Since B is a contraction, the operator $I - B$ is continuous. Moreover,

$$\|(I-B)x - (I-B)y\| \geq \|x-y\| - \|Bx - By\| \geq (1-\tau)\|x-y\|$$

for some $\tau \in (0,1)$. This shows that $(I-B)^{-1}$ exists and is continuous on X. Let y be fixed in $\mathcal{M}$, the map which assigns to each $x \in X$ the value $Bx + Ay$ defines a contraction from X into X. So, according to assumption (iii), the equation $x = Bx + Ay$ has a unique solution $x = (I-B)^{-1}Ay$ in $\mathcal{M}$. Therefore,

$$(I-B)^{-1}A(\mathcal{M}) \subset \mathcal{M}. \tag{4.4}$$

Now, let us define the sequence $(M_n)_n$ of subsets of $\mathcal{M}$ by $M_1 = \mathcal{M}$ and $M_{n+1} = \overline{co}((I-B)^{-1}A(M_n))$. We claim that the sequence $(M_n)_n$ satisfies the conditions of property (8) of Lemma 1.4.1. Indeed, it is clear that the sequence $(M_n)_n$ consists of nonempty, closed, and convex subsets of $\mathcal{M}$. Using Eq. (4.4) one sees that it is also decreasing. Moreover, since

$$(I-B)^{-1}A(M_n) \subset A(M_n) + B(I-B)^{-1}A(M_n) \subset A(M_n) + B(M_{n+1}),$$

it follows, from the monotonicity of M_n, that

$$(I-B)^{-1}A(M_n) \subset A(M_n) + B(M_n).$$

Accordingly,

$$\omega((I-B)^{-1}A(M_n)) \leq \omega(A(M_n)) + \omega(B(M_n))$$

(use the Properties (1) and (7) of Lemma 1.4.1). Since $A(\mathcal{M})$ is relatively weakly compact, then we get

$$\omega((I-B)^{-1}A(M_n)) \leq \omega(B(M_n)).$$

Next, let $t > 0$ and $Y \in \mathcal{K}^w$ such that $M_n \subset Y + B_t$. Since B is a contraction with a constant τ, one sees that

$$B(M_n) \subset B(Y) + B_{t\tau} \subset \overline{B(Y)}^w + B_{t\tau}.$$

Moreover, since B satisfies hypothesis $(\mathcal{A}_2)$, it follows that $B(Y)$ is relatively weakly compact. Accordingly,

$$\omega(B(M_n)) \leq \omega(M_n)$$

and so

$$\omega(M_{n+1}) \leq \tau\omega(M_n).$$

By induction, it follows that

$$\omega(M_n) \leq \tau^{n-1}\omega(\mathcal{M})$$

and therefore, $\lim_{n\to\infty} \omega(M_n) = 0$ because $\tau \in (0, 1)$. This proves the claim. Now, making use of the Property (8) of Lemma 1.4.1, we infer that $\mathcal{N} := \cap_{n=1}^{\infty} M_n$ is a nonempty, closed, convex, and weakly compact subset of $\mathcal{M}$. Moreover, it is easily seen that

$$(I - B)^{-1}A(\mathcal{N}) \subset \mathcal{N}.$$

Consequently,

$$(I - B)^{-1}A(\mathcal{N})$$

is relatively weakly compact. Finally, from the properties of A and the continuity of $(I - B)^{-1}$, it follows that the map $(I - B)^{-1}A$ verifies the condition $(\mathcal{A}_1)$. Now, the use of Theorem 4.2.1 achieves the proof. Q.E.D.

Remark 4.2.2 *From the proof of Theorem 4.2.3, it follows that if a mapping* $A : X \longrightarrow X$ *is a contraction and satisfies* $(\mathcal{A}2)$*, then* A *is* ω*-contractive.*

In the remaining part of this section, we will use the notations and results of Section 4.1 in order to develop a general matrix fixed point theory under weak topology. As above, the discussion will be based on the invertibility or not of the diagonal terms of $\mathcal{I} - \mathcal{L}$.

4.2.1 One of the diagonal entries of $\mathcal{I} - \mathcal{L}$ is invertible

In this subsection, we assume that:

$(\mathcal{H}_9)$ The operator $I - A$ is invertible and $(I - A)^{-1}B(\Omega') \subset \Omega$,

($\mathcal{H}_{10}$) $(I-A)^{-1}B$ is a continuous operator satisfying ($\mathcal{A}1$) and $(I-A)^{-1}B(\Omega')$ is relatively weakly compact,

($\mathcal{H}_{11}$) C is a continuous operator satisfying ($\mathcal{A}2$),

($\mathcal{H}_{12}$) the operator $I-D$ is invertible and its inverse $(I-D)^{-1}$ is continuous on $(I-D)(\Omega')$ and satisfies ($\mathcal{A}2$), and

($\mathcal{H}_{13}$) $(I-D)^{-1}C(I-A)^{-1}B(\Omega') \subset \Omega'$.

Theorem 4.2.4 *Under the assumptions* ($\mathcal{H}_9$)–($\mathcal{H}_{13}$), *the block operator matrix* (4.1) *has, at least, a fixed point in* $\Omega \times \Omega'$.

Proof. Let Γ be the operator defined by:

$$\Gamma := (I-D)^{-1}C(I-A)^{-1}B : \Omega' \longrightarrow \Omega'.$$

In order to prove the theorem, we have to check that:

1. $\Gamma(\Omega')$ is relatively weakly compact.

For this, let $(y_n)_{n\in\mathbb{N}}$ be a sequence in $\Gamma(\Omega')$. There exists a sequence $(x_n)_{n\in\mathbb{N}} \subset \Omega'$ such that $y_n = \Gamma x_n$ for all $n \in \mathbb{N}$, because $((I-A)^{-1}Bx_n)_{n\in\mathbb{N}} \subset (I-A)^{-1}B(\Omega')$ and $(I-A)^{-1}B(\Omega')$ is relatively weakly compact. Then, according to the Eberlien-Šmulian theorem (see Theorem 1.3.3), $((I-A)^{-1}Bx_n)_{n\in\mathbb{N}}$ has a weakly convergent subsequence. Moreover, using the fact that C and $(I-D)^{-1}$ verify ($\mathcal{A}_2$), the sequence $(y_n)_{n\in\mathbb{N}}$ also has a weak converging subsequence. Consequently, $\Gamma(\Omega')$ is relatively weakly compact.

2. Γ satisfies the condition ($\mathcal{A}_1$).

Let $(x_n)_{n\in\mathbb{N}}$ be a weakly convergent sequence of Ω'. Since $(I-A)^{-1}B$ satisfies ($\mathcal{A}_1$), it follows that $((I-A)^{-1}Bx_n)_{n\in\mathbb{N}}$ has a strongly convergent subsequence. By the continuity of the operator C and $(I-D)^{-1}$, $(\Gamma x_n)_{n\in\mathbb{N}}$ also has a strongly convergent subsequence, that is, Γ satisfies ($\mathcal{A}_1$).

Clearly, Γ is continuous. Hence, Γ satisfies the hypotheses of Theorem 4.2.1 as we claimed, and there exists $y_0 \in \Omega'$ such that

$$\Gamma y_0 = y_0.$$

Let $x_0 := (I-A)^{-1}By_0$, hence $\mathcal{L}\begin{pmatrix} x_0 \\ y_0 \end{pmatrix} = \begin{pmatrix} x_0 \\ y_0 \end{pmatrix}$. Q.E.D.

In these other cases, we will also assume that Ω and Ω' are bounded. As above, we will treat only the case of invertibility of $I - A$. The other case is similar, just simply exchanging the roles of A and D, and B and C.

Let us assume that:

($\mathcal{H}_{14}$) The operator $I - A$ is invertible and $(I-A)^{-1}B(\Omega') \subset \Omega$,

($\mathcal{H}_{15}$) $S := C(I-A)^{-1}B$ is a contraction satisfying ($\mathcal{A}2$) with a constant k,

($\mathcal{H}_{16}$) D is a continuous operator satisfying ($\mathcal{A}1$) and ω-α-contractive for some $\alpha \in [0, 1-k)$, and

($\mathcal{H}_{17}$) $S(\Omega') + D(\Omega') \subset \Omega'$.

Theorem 4.2.5 *Under the assumptions* ($\mathcal{H}_{14}$)–($\mathcal{H}_{17}$), *the block operator matrix* (4.1) *has, at least, a fixed point in* $\Omega \times \Omega'$.

Proof. Since S is a contraction with a constant $k \in (0,1)$, it follows from Remark 1.2.2 and Lemma 1.2.2 that the mapping $I - S$ is a homeomorphism from Ω' into $(I-S)(\Omega')$. Let y' be fixed in Ω', the map which assigns to each $y \in \Omega'$ the value $Sy + Dy'$ defines a contraction from Ω' into Ω'. So, by the Banach fixed point theorem, the equation $y = Sy + Dy'$ has a unique solution $y = (I-S)^{-1}Dy'$ in Ω'. Therefore,

$$(I-S)^{-1}D(\Omega') \subset \Omega'.$$

Next, we will prove that $T := (I-S)^{-1}D$ satisfies the conditions of Theorem 4.2.2. It is clear that T is continuous and satisfies ($\mathcal{A}1$). Now, let us check that T is ω-β-contractive for some $\beta \in [0,1)$. To do this, let $\mathcal{N}$ be a subset of Ω'. Using the following equality

$$(I-S)^{-1}D = D + S(I-S)^{-1}D,$$

we infer that

$$\omega(T(\mathcal{N})) \leq \omega(D(\mathcal{N}) + ST(\mathcal{N})).$$

The properties of $\omega(.)$ in Lemma 1.4.1 and the assumptions on S and D imply that

$$\omega(T(\mathcal{N})) \leq \omega(D(\mathcal{N})) + \omega(ST(\mathcal{N})) \leq \alpha\omega(\mathcal{N}) + k\omega(T(\mathcal{N})),$$

and therefore,

$$\omega(T(\mathcal{N})) \leq \frac{\alpha}{1-k}\omega(\mathcal{N}).$$

This inequality means that T is ω-β-contractive with $\beta := \frac{\alpha}{1-k}$. Consequently, T satisfies the hypotheses of Theorem 4.2.2 as we claimed. Hence, such an operator has, at least, a fixed point in Ω', and the block operator matrix (4.1) has, at least, a fixed point in $\Omega \times \Omega'$. Q.E.D.

4.2.2 None of the diagonal entries of $\mathcal{I} - \mathcal{L}$ is invertible

In this subsection, we discuss the existence of fixed points for the following perturbed block operator matrix by laying down some conditions on the entries.

$$\widetilde{\mathcal{L}} = \begin{pmatrix} A_1 & B \\ C & D_1 \end{pmatrix} + \begin{pmatrix} P_1 & 0 \\ 0 & P_2 \end{pmatrix}. \tag{4.5}$$

Assume that the nonlinear operators A_1 and P_1 map Ω into X, B from Ω' into X, C from Ω into Y, and D_1 and P_2 from Ω' into Y. Suppose that the operator (4.5) respects the following assumptions:

($\mathcal{H}_{18}$) The operator $I - A_1$ (resp. $I - D_1$) is invertible from Ω into X (resp. from Ω' into Y),

($\mathcal{H}_{19}$) $(I - A_1)^{-1}B$ and $(I - A_1)^{-1}C$ are continuous, weakly compact maps and verify ($\mathcal{A}_1$),

($\mathcal{H}_{20}$) $(I - A_1)^{-1}P_1$ and $(I - D_1)^{-1}P_2$ are contraction maps and verify ($\mathcal{A}_2$), and

($\mathcal{H}_{21}$) $(I - A_1)^{-1}P_1(\Omega) + (I - A_1)^{-1}B(\Omega') \subset \Omega$ and $(I - D_1)^{-1}C(\Omega) + (I - D_1)^{-1}P_2(\Omega') \subset \Omega'$.

Theorem 4.2.6 *Under the above assumptions* ($\mathcal{H}_{18}$)–($\mathcal{H}_{21}$)*, the block operator matrix* (4.5) *has, at least, a fixed point in* $\Omega \times \Omega'$*.*

Proof. By using the assumption ($\mathcal{H}_{18}$) and the same decomposition used in the proof of Theorem 4.1.5, the following equation

$$\begin{pmatrix} A_1 + P_1 & B \\ C & D_1 + P_2 \end{pmatrix} \begin{pmatrix} x \\ y \end{pmatrix} = \begin{pmatrix} x \\ y \end{pmatrix}$$

may be transformed into

$$\mathcal{Z}_1 \begin{pmatrix} x \\ y \end{pmatrix} + \mathcal{Z}_2 \begin{pmatrix} x \\ y \end{pmatrix} = \begin{pmatrix} x \\ y \end{pmatrix},$$

where

$$\mathcal{Z}_1 = \begin{pmatrix} (I - A_1)^{-1} P_1 & 0 \\ 0 & (I - D_1)^{-1} P_2 \end{pmatrix},$$

and

$$\mathcal{Z}_2 = \begin{pmatrix} 0 & (I - A_1)^{-1} B \\ (I - D_1)^{-1} C & 0 \end{pmatrix}.$$

Obviously, the operator matrix $\mathcal{Z}_2$ is continuous. Now, let us check that $\mathcal{Z}_2$ is a weakly compact operator and satisfies $\mathcal{A}_1$. To see this, let $\big((x_n, y_n)\big)_{n\in\mathbb{N}}$ be a sequence in $\Omega \times \Omega'$. Since $(I - A_1)^{-1}B$ is weakly compact, the sequence $\big((I - A_1)^{-1}By_n\big)_{n\in\mathbb{N}}$ has a weakly convergent subsequence, say $\big((I - A_1)^{-1}By_{n_k}\big)_{k\in\mathbb{N}}$. Moreover, the sequence $\big((I - D_1)^{-1}Cx_{n_k}\big)_{k\in\mathbb{N}}$ has a weak converging subsequence say $\big((I-D_1)^{-1}Cx_{n_{k_j}}\big)_{j\in\mathbb{N}}$. Then, $\big(\mathcal{Z}_2(x_{n_{k_j}}, y_{n_{k_j}})\big)_{j\in\mathbb{N}}$ is a weakly convergent subsequence of $\big(\mathcal{Z}_2(x_n, y_n)\big)_{n\in\mathbb{N}}$, hence $\mathcal{Z}_2$ is a weakly compact operator. We also show that the operator matrix $\mathcal{Z}_2$ verifies $(\mathcal{A}_1)$ and from $(\mathcal{H}_{20})$ the operator matrix $\mathcal{Z}_1$ is a contraction which satisfies $(\mathcal{A}_2)$. From the assumption $(\mathcal{H}_{21})$ and Theorem 4.2.3, it follows that the operator matrix (4.5) has, at least, a fixed point in $\Omega \times \Omega'$. Q.E.D.

4.3 Fixed Point Theorems for BOM in Banach Algebras

In this section, we are concerned with fixed point results on Banach algebras of operators defined by a 2×2 block operator matrix

$$\mathcal{L} = \begin{pmatrix} A & B.B' \\ C & D \end{pmatrix}, \tag{4.6}$$

where the entries of the matrix are generally nonlinear operators defined on a Banach algebra. The operators occurring in the representation (4.6) are nonlinear, and our assumptions are as follows: A maps a bounded, closed, convex, and nonempty subset S of a Banach algebra X into X, B, and B' from another bounded, closed, convex, and nonempty subset S' of a Banach algebra Y into X, C from S into Y, and D from S' into Y.

Theorem 4.3.1 *Let S be a convex, closed, and bounded subset of a Banach algebra X and let A, B, B', C, $D : S \longrightarrow X$, be five operators such that:*

(i) A, B, and C are Lipschitzian with Lipschitz constants α, β, and γ respectively,

(ii) $I - D$ is one-to-one such that $(I - D)^{-1}$ is Lipschitzian with a Lipschitz constant δ on $C(S)$,

(iii) B' is continuous and C is a compact operator, and

(iv) $Ax + Tx.T'z \in S$ for all x, $z \in S$, where $T = B(I - D)^{-1}C$ and $T' = B'(I - D)^{-1}C$.

Then, the block operator matrix (4.6) has, at least, a fixed point in $S \times S$ whenever $\alpha + \beta\delta\gamma M < 1$, where $M = \|T'(S)\|$.

Proof. Let $y \in S$ be any fixed point and let us define a mapping φ_y on S into itself by:

$$\varphi_y(x) = Ax + Tx.T'y.$$

Notice that φ_y is a contraction on S, since we have

$$\|\varphi_y(x_1) - \varphi_y(x_2)\| \leq (\alpha + \beta\gamma\delta M)\|x_1 - x_2\|$$

whenever x_1, $x_2 \in S$. From the Banach contraction principle, it follows that there is a unique point x^* in S such that

$$\varphi_y(x^*) = x^*$$

or,

$$x^* = Ax^* + Tx^*.T'y.$$

Let us define the operator

$$\begin{cases} N : S \longrightarrow S \\ \qquad y \longrightarrow N(y) = z, \end{cases}$$

where z is the unique solution of the operator equation

$$z = Az + Tz.T'y.$$

The operator N is continuous on S. To see this, let $\{y_n\}$ be any sequence in S converging to a point y. Then, we have

$$\begin{aligned} \|Ny_n - Ny\| &= \|Az_n + Tz_n.T'y_n - Az - Tz.T'y\| \\ &\leq \alpha\|z_n - z\| + \|Tz_n.T'y_n - Tz.T'y\| \\ &\leq (\alpha + \beta\gamma\delta M)\,\|z_n - z\| + \|Tz\|\|T'y_n - T'y\|. \end{aligned}$$

Hence,

$$\|Ny_n - Ny\| \leq \big(\alpha + \beta\gamma\delta M\big)\|Ny_n - Ny\| + \|T(Ny)\|\|T'y_n - T'y\|.$$

Taking the supremum limit in the above inequality yields

$$\lim_{n\to+\infty}\|Ny_n - Ny\| = 0.$$

This proves that N is a continuous operator on S. Next, we will show that N is a compact operator on S. For this purpose let $z \in S$, then we have

$$\|Tz\| \leq \|Ta\| + M\beta\gamma\delta\|z - a\| \leq c,$$

where $c = \|Ta\| + M\beta\gamma\delta \operatorname{diam}(S)$ for some fixed point a in S. Let $\varepsilon > 0$ be given. Since $C(S)$ is a totally bounded subset and $B'(I-D)^{-1}$ is a continuous operator then, $T'(S)$ is a totally bounded subset. Consequently, there exists a subset $Y = \{y_1, \ldots, y_n\}$ of S, such that

$$T'(S) \subset \bigcup_{i=1}^{n} B_r(w_i),$$

where $r = \left(\frac{1-\big(\alpha+M\beta\gamma\delta\big)}{c}\right)\varepsilon$ and $w_i = T'y_i$. So, for any $y \in S$, we have $y_k \in Y$ such that

$$\|T'y_k - T'y\| < \left(\frac{1-(\alpha+M\beta\gamma\delta)}{c}\right)\varepsilon.$$

Therefore,

$$\begin{aligned}\|Ny_k - Ny\| &= \|z_k - z\|\\ &\leq \|Tz_k.T'y_k - Tz.T'y\| + \|Az_k - Az\|\\ &\leq \|Tz_k - Tz\|\|T'y_k\| + \|Tz\|\|T'y_k - T'y\| + \|Az_k - Az\|\\ &\leq (\alpha + M\beta\gamma\delta)\|z_k - z\| + c\|T'y_k - T'y\|.\end{aligned}$$

Then,

$$\|Ny_k - Ny\| \leq \left(\frac{c}{1-(\alpha+M\beta\gamma\delta)}\right)\|T'y_k - T'y\| < \varepsilon.$$

Since $y \in S$ was arbitrary,

$$N(S) \subset \bigcup_{i=1}^{n} B_\varepsilon(k_i),$$

where $k_i = N(y_i)$. As a result, $N(S)$ is totally bounded in X. Hence, N is completely continuous on S into itself. Now, an application of Schauder's fixed point theorem shows that N has, at least, a fixed point in S. Consequently, we deduce that the operator equation

$$x = Ax + Tx.T'x$$

has a solution in S. Now, the use of the vector $y = (I - D)^{-1}Cx$ allows us to achieve the proof of Theorem 4.3.1. Q.E.D.

Theorem 4.3.2 *Let S be a convex, closed, and bounded subset of a Banach algebra X and let A, B, B', C, and $D : S \longrightarrow S$, be five operators satisfying:*

(i) A, B, and C are $\mathcal{D}$-Lipschitzian with $\mathcal{D}$-functions ϕ_A, ϕ_B, and ϕ_C, respectively,

(ii) $(I - D)^{-1}$ exists and $\mathcal{D}$-Lipschitzian on $(I - D)(S)$ with the $\mathcal{D}$-function ϕ_ψ,

(iii) B' is continuous and C is compact,

(iv) $\left(\frac{I}{T}\right)^{-1}$ exists on $T'(S)$, and

(v) $Ax + Tx.T'z \in S$ for all x, $z \in S$, where $T = B(I - D)^{-1}C$ and $T' = B'(I - D)^{-1}C$.

Then, the block operator matrix (4.6) *has, at least, a fixed point in $S \times S$ whenever $M\phi_B \circ \phi_\psi \circ \phi_C(r) + \phi_A(r) < r$, where $M = \|T'(S)\|$.*

Proof. Arguing as in the proof of Theorem 3.1.7, we infer that

$$x = Ax + Tx.T'x$$

has a solution in S. Now, the use of equation $y = (I - D)^{-1}Cx$ leads the block operator matrix (4.6) to have a fixed point in $S \times S$. Q.E.D.

Remark 4.3.1 *If the operator D is a contraction on S into itself with a constant contraction k and C is Lipschitzian with constant γ and $C(S) \subset (I - D)(X)$, then the inverse operator $(I - D)^{-1}C$ exists and is Lipschitzian with constant $\frac{\gamma}{1-k}$.*

Theorem 4.3.3 *Let S be a nonempty, convex, closed, and bounded subset of a Banach algebra X and let S' be a nonempty, convex, closed, and bounded subset of a Banach space Y. Let $A : S \longrightarrow X$, B, $B' : S' \longrightarrow X$, $C : S \longrightarrow Y$,* *and $D : S' \longrightarrow S'$ be five operators such that:*

(i) The operator B is Lipschitzian with a constant β, and A and C are $\mathcal{D}$-Lipschitzian with $\mathcal{D}$-functions ϕ_A and ϕ_C, respectively,

(ii) $C(S) \subset (I - D)(S')$,

(iii) D is a contraction with a constant k,

(iv) B' is continuous and C is compact, and

(v) $Ax + Tx.T'z \in S$ for all $x, z \in S$, where $T = B(I-D)^{-1}C$ and $T' = B'(I-D)^{-1}C$.

Then, the operator matrix (4.6) has, at least, a fixed point in $S \times S'$ whenever $\frac{\beta M}{1-k}\phi_C(r) + \phi_A(r) < r$, where $M = \|T'(S)\|$.

Proof. It is easy to verify that the inverse operator $(I - D)^{-1}$ exists and is Lipschitzian with the Lipschitz constant $\frac{1}{1-k}$ on $(I - D)(S')$ in view of hypothesis (ii) and hypothesis (iii). Now, since C is $\mathcal{D}$-Lipschitzian, then it is continuous on S and from hypothesis (iv), it follows that $T' = B'(I - D)^{-1}C$ is completely continuous. Let $y \in S$ be fixed and let us define a mapping

$$\begin{cases} \varphi_y : S \longrightarrow S \\ \qquad x \longrightarrow Ax + Tx.T'y. \end{cases}$$

Notice that φ_y is a nonlinear contraction on S into itself with function $\phi_A + \frac{\beta M}{1-k}\phi_C$. Hence, by the fixed point theorem of Boyd and Wong (see Theorem 1.6.10), it follows that there is a unique point x^* in S such that

$$x^* = Ax^* + Tx^*.T'y.$$

Define the operator N by:

$$\begin{cases} N : S \longrightarrow S \\ \qquad y \longrightarrow Ny = z \end{cases}$$

where z is the unique solution of the equation $z = Az + Tz.T'y$. Arguing as in the proof of Theorem 3.1.6, we get N is a continuous and compact operator on S. Hence, an application of Schauder's fixed point theorem shows that N has, at least, a fixed point x in S. Then, by the definition of N, we have

$$x = Nx = A(Nx) + T(Nx).T'x = Ax + Tx.T'x.$$

To achieve the proof, it is sufficient to take $y = (I - D)^{-1}Cx$. Q.E.D.

In what follows, we will combine Theorems 3.1.7 and 4.3.3 in order to get the following fixed point theorem in a Banach algebra.

Theorem 4.3.4 *Let S (resp. S') be a nonempty, convex, closed, and bounded subset of a Banach algebra X (resp. Y) and let $R = S \cap S'$ (assumed nonempty). Let A, B, C, D, $B' : R \longrightarrow R$ be five operators such that:*

(i) The operator B is Lipschitzian with a constant β, and A and C are $\mathcal{D}$-Lipschitzian with the functions ϕ_A and ϕ_C, respectively,

(ii) $C(R) \subset (I - D)(R)$,

(iii) D is a contraction with a constant k,

(iv) B' is continuous and C is compact,

(v) $\left(\frac{I}{T}\right)^{-1}$ exists on R, and

(vi) $Ax + Tx.T'z \in R$, $\forall\ x,\ z \in R$, where $T = B(I - D)^{-1}C$ and $T' = B'(I - D)^{-1}C$.

Then, the block operator matrix (4.6) has at, least, a fixed point in $R \times R$ whenever $\left(\frac{\beta M}{1-k}\right)\phi_C(r) + \phi_A(r) < r$ if $r > 0$, where $M = \|T'(S)\|$.

Proof. The operator $D : R \longrightarrow R$ is a nonlinear contraction on R. By using Theorem 1.6.10, we infer that $(I - D)^{-1}$ exists and is continuous on R. In view of Theorem 4.3.3, it is sufficient to prove that the equation $x = Ax + Tx.T'x$ has a solution in R. To do this, the operator T is $\mathcal{D}$-Lipschitzian with the $\mathcal{D}$-function $\left(\frac{\beta}{1-k}\right)\phi_C$. Let

$$\begin{cases} N : R \longrightarrow R \\ \qquad y \longrightarrow \left(\dfrac{I - A}{T}\right)^{-1} T'y. \end{cases} \tag{4.7}$$

From the hypothesis $\phi_A(r) < r$, for all $r > 0$, it follows that the operator $(I - A)^{-1}$ exists and is continuous on R. Therefore, the operator

$$\left(\frac{I - A}{T}\right)^{-1} = \left(\frac{I}{T}\right)^{-1}(I - A)^{-1}$$

exists on R. We show that the operator N given by Eq. (4.7) is well defined. We claim that

$$\left(\frac{I - A}{T}\right)^{-1} T' : R \longrightarrow R.$$

It is sufficient to prove that

$$T'(R) \subset \left(\frac{I-A}{T}\right)(R).$$

Let $y \in R$ be a fixed point. Let us define a mapping $\varphi_y : R \longrightarrow R$ by:

$$\varphi_y(x) = Ax + Tx.T'y.$$

Let $x_1,\ x_2 \in R$. Then, by hypothesis (i), we have

$$\begin{aligned}
\|\varphi_y(x_1) - \varphi_y(x_2)\| &= \|Ax_1 + Tx_1.T'y - Ax_2 - Tx_2.T'y\| \\
&\leq \|Ax_1 - Ax_2\| + \|Tx_1 - Tx_2\|\|T'y\| \\
&\leq \psi(\|x_1 - x_2\|),
\end{aligned}$$

where $\psi(r) = \frac{\beta M}{1-k}\phi_C(r) + \phi_A(r) < r$, if $r > 0$. From Theorem 1.6.10, it follows that there is a unique point $x^* \in R$ such that

$$\varphi_y(x^*) = x^*.$$

Or, also

$$\left(\frac{I-A}{T}\right) x^* = T'y.$$

Hence, $\left(\frac{I-A}{T}\right)^{-1} T'$ defines a mapping

$$\left(\frac{I-A}{T}\right)^{-1} T' : R \longrightarrow R.$$

Next, we show that N is completely continuous. In fact, let $\{x_n\}$ be any sequence in R such that $x_n \to x$ and let

$$\begin{cases}
y_n = T'(x_n) & \text{and} \quad y = T'(x) \\
z_n = \left(\frac{I-A}{T}\right)^{-1}(y_n) & \text{and} \quad z = \left(\frac{I-A}{T}\right)^{-1}(y).
\end{cases}$$

Then, it is easy to show that $y_n \to y$, and we have

$$\begin{cases}
z_n = Az_n + Tz_n.y_n \\
z = Az + Tz.y.
\end{cases}$$

So,

$$\|z_n - z\| \leq \psi(\|z_n - z\|) + \|Tz\|\|y_n - y\|,$$

where $\psi(r) = \frac{\beta}{1-k}\phi_C(r) + \phi_A(r) < r$, for $r > 0$. Taking the supremum limit in the above inequality shows that N is continuous. Therefore, the operator $\left(\frac{I-A}{T}\right)^{-1}$ is continuous on $T'(S)$. Moreover, from hypothesis (iv), it follows that N is a compact operator on R, and consequently it is completely continuous. Now, use the vector $y = (I-D)^{-1}Cx$ to solve the problem. Q.E.D.

Note that the hypothesis (iv) of Theorem 4.3.4 is very strong and can be replaced by a milder one. We state the following result.

Theorem 4.3.5 *Let S be a nonempty, convex, closed, and bounded subset of a Banach algebra X and, let A, $C : S \longrightarrow X$ and B, B', $D : X \longrightarrow X$ be five operators satisfying:*

(i) A, B, and C are $\mathcal{D}$-Lipschitzian with the $\mathcal{D}$-functions ϕ_A, ϕ_B, and ϕ_C, respectively,

(ii) B' is completely continuous on S,

(iii) $\left(\frac{I-A}{T}\right)^{-1}$ exists on $T'(S)$ and $C(S) \subset (I-D)(S)$,

(iv) D is a contraction with a constant k, and

(v) $x = Ax + Tx.T'z \Longrightarrow x \in S$, $\forall\ z \in S$, where $T = B(I-D)^{-1}C$ and $T' = B'(I-D)^{-1}C$.

Then, the block operator matrix (4.6) has, at least, a fixed point in $S \times S$ whenever $M\phi_B \circ L_{\frac{1}{1-k}} \circ \phi_C(r) + \phi_A(r) < r$ for $r > 0$, where $M = \|T'(S)\|$.

Proof. Clearly, $(I-D)^{-1}$ exists and is continuous on X and the operator T is $\mathcal{D}$-Lipschitzian with the $\mathcal{D}$-function $\phi_B \circ L_{\frac{1}{1-k}} \circ \phi_C$. From the assumption (iii), it follows that, for each $y \in S$, there exists a unique point $x \in S$, such that

$$\left(\frac{I-A}{T}\right)x = T'y,$$

or also, $x = Ax + Tx.T'y$. Since the hypothesis (v) holds, then $x \in S$. Therefore, we can define $N : S \longrightarrow S$ by:

$$N(x) = \left(\frac{I-A}{T}\right)^{-1} T'(x).$$

This map is continuous. To see this, let $\{x_n\}$ be any sequence in S such that $x_n \to x$ and let

$$\begin{cases} y_n = T'x_n & \text{and} \qquad y = T'x \\ z_n = \left(\frac{I-A}{T}\right)^{-1}(y_n) & \text{and} \quad z = \left(\frac{I-A}{T}\right)^{-1}(y). \end{cases}$$

Then,

$$\begin{cases} y_n = T'x_n & \text{and} \qquad y = T'x \\ z_n = Az_n + Tz_n.y_n & \text{and} \quad z = Az + Tz.y. \end{cases}$$

Hence, $y_n \to y$ and $z_n \to z$. Indeed, it is easy to verify that

$$\|z_n - z\| \leq \psi(\|z_n - z\|) + \|T(z)\|\|y_n - y\|,$$

where $\psi(r) = M\phi_B \circ L_{\frac{1}{1-k}} \circ \phi_C(r) + \phi_A(r) < r$. Now, taking the supremum limit in the above inequality shows that N is continuous. Moreover,

$$N(S) \subset \left(\frac{I-A}{T}\right)^{-1} B'(S).$$

Now, the continuity of the operator $\left(\frac{I-A}{T}\right)^{-1}$ combined with hypothesis (ii), allow $N(S)$ to be a relatively compact subset. Again, an application of Schauder's fixed point theorem completes this proof. Q.E.D.

An interesting corollary of Theorem 4.3.5 is as follows:

Corollary 4.3.1 *Let S be a nonempty, convex, and compact subset of a Banach algebra X and let A, $C : S \longrightarrow X$ and B, B', $D : X \longrightarrow X$ be five operators such that:*

(i) A, B, and C are $\mathcal{D}$-Lipschitzian with $\mathcal{D}$-functions ϕ_A, ϕ_B, and ϕ_C, respectively,

(ii) $C(S) \subset (I-D)(X)$,

(iii) D is a contraction with a constant k,

(iv) B' is continuous,

(v) $\left(\frac{I-A}{T}\right)^{-1}$ exists on $B'(X)$, and

(vi) $x = Ax + Tx.T'z \Longrightarrow x \in S,\ \forall\ z \in S$, where $T = B(I-D)^{-1}C$ and $T' = B'(I-D)^{-1}C$.

Then, the block operator matrix (4.6) has, at least, a fixed point in $S \times X$ whenever $\phi_B \circ L_{\frac{1}{1-k}} \circ \phi_C(r) + \phi_A(r) < r$ if $r > 0$, where $M = \|T'(S)\|$.

Proof. The proof follows from both Theorem 4.3.5 and Schauder's fixed point theorem. Q.E.D.

4.3.1 Banach algebras satisfying the condition $(\mathcal{P})$

At the beginning of this section, we are going to discuss a fixed point theorem for the operator matrix (4.6) involving the concept of the De Blasi measure of weak noncompactness in Banach algebras satisfying condition $(\mathcal{P})$.

The following result gives sufficient conditions to the block operator matrix (4.6) acting on a product of Banach algebras satisfying condition $(\mathcal{P})$ to have a fixed point.

Theorem 4.3.6 *Let S be a nonempty, convex, closed, and bounded subset of X, and let A, $C : S \longrightarrow X$, and B, B', $D : X \longrightarrow X$ be five operators satisfying:*

(i) A and C are weakly compact,

(ii) D is linear, bounded, and there exists a strictly positive integer p such that D^p is a separate contraction on X,

(iii) A, B, C, and B' are weakly sequentially continuous, and

(iv) $Ax + B(I - D)^{-1}Cx \cdot B'(I - D)^{-1}Cx \in S$ for all $x \in S$.

Then, the block operator matrix (4.6) has, at least, one fixed point in $S \times X$.

Proof. From the assumption *(ii)* and Lemma 1.2.2, we infer that the operator $(I - D^p)^{-1}$ exists on X and

$$(I - D)^{-1} = (I - D^p)^{-1}\sum_{k=0}^{p-1} D^k.$$

Define the mapping

$$\begin{cases} F : S \longrightarrow S \\ \quad x \longrightarrow Ax + B(I - D)^{-1}Cx \cdot B'(I - D)^{-1}Cx. \end{cases} \tag{4.8}$$

Notice that $(I - D)^{-1}$ is weakly continuous. Moreover, taking into account that X is a Banach algebra satisfying condition $(\mathcal{P})$ and using assumption *(iii)*, we show that F is weakly sequentially continuous on S. Besides, since

$$F(S) \subset A(S) + B(I - D)^{-1}C(S) \cdot B'(I - D)^{-1}C(S),$$

it follows from assumption (i) that $F(S)$ is relatively weakly compact. Accordingly, the operator F has, at least, a fixed point x in S in view of Theorem 2.2.1. So, the vector $y = (I - D)^{-1}Cx$ solves the problem. Q.E.D.

Remark 4.3.2 *Theorem 4.3.6 remains true if we suppose that there exists a strictly positive integer p such that D^p is a nonlinear contraction.*

Notice that the proof of Theorem 4.3.6 is based on the linearity of the operator D. Hence, it would be interesting to investigate the case when D is not linear.

Theorem 4.3.7 *Let S be a nonempty, convex, closed, and bounded subset of X, and let $A,\ C : S \longrightarrow X$, and $B,\ B',\ D : X \longrightarrow X$ be five weakly sequentially continuous operators satisfying:*

(i) C is weakly compact and A is condensing,

(ii) D is a ϕ-nonlinear contraction and $(I - D)^{-1}C(S)$ is bounded, and

(iii) $Ax + B(I - D)^{-1}Cx \cdot B'(I - D)^{-1}Cx \in S$ for all $x \in S$.

Then, the block operator matrix (4.6) *has, at least, one fixed point in $S \times X$.*

Proof. Since D is a ϕ-nonlinear contraction, the operator $(I - D)^{-1}$ exists on X. We first claim that $(I - D)^{-1}C(S)$ is relatively weakly compact. If it is not the case, then $d = \beta\left((I - D)^{-1}C(S)\right) > 0$. Using

$$(I - D)^{-1}C = C + D(I - D)^{-1}C, \tag{4.9}$$

and taking into account that $C(S)$ is a relatively weakly compact subset of X, we get

$$\beta\left((I - D)^{-1}C(S)\right) \leq \beta\left(D(I - D)^{-1}C(S)\right). \tag{4.10}$$

Let $\varepsilon > 0$, then there exists a $K \in \mathcal{W}(X)$ satisfying $(I - D)^{-1}C(S) \subset K + B_{d+\varepsilon}$. From assumption (ii), it follows that

$$D(I - D)^{-1}C(S) \subset D(K) + B_{\phi(d+\varepsilon)}.$$

Moreover, since D is weakly sequentially continuous, $D(K)$ is a weakly compact subset. Consequently, $\beta\left(D(I - D)^{-1}C(S)\right) \leq \phi(d + \varepsilon)$. Since $\varepsilon > 0$ is arbitrary,

$$\beta\left(D(I - D)^{-1}C(S)\right) \leq \phi(d) < d = \beta\left((I - D)^{-1}C(S)\right). \tag{4.11}$$

Now, combining Eqs. (4.10) and (4.11) allows us to get

$$\beta\left((I-D)^{-1}C(S)\right) \leq \beta\left(D(I-D)^{-1}C(S)\right) < \beta\left((I-D)^{-1}C(S)\right)$$

which is a contradiction and the claim is proved. Consequently,

$$(I-D)^{-1}C$$

is sequentially weakly continuous and so is F, where F is defined in Eq. (4.8). One can show that

$$B(I-D)^{-1}C(S)$$

and

$$B'(I-D)^{-1}C(S)$$

are relatively weakly compact. It follows that the product

$$B(I-D)^{-1}C(S).B'(I-D)^{-1}C(S)$$

is also a relatively weakly compact subset of X in view of Lemma 1.5.1. Clearly, the operator $(I-D)^{-1}C$ is weakly sequentially continuous which is also true for the operator F already defined in Eq. (4.8). Indeed, let $(\theta_n)_n$ be a sequence in S which converges weakly to θ. Since

$$(I-D)^{-1}C(S)$$

is relatively weakly compact, there is a subsequence $(\theta_{n_k})_k$ of $(\theta_n)_n$ such that

$$(I-D)^{-1}C\theta_{n_k} \rightharpoonup \rho.$$

Therefore, the use of equality Eq. (4.9) combined with the weak sequential continuity of C and D yields $\rho = C\theta + D\rho$. Consequently, $\rho = (I-D)^{-1}C\theta$. Hence,

$$(I-D)^{-1}C\theta_{n_k} \rightharpoonup (I-D)^{-1}C\theta.$$

Moreover, taking into account that X satisfies the condition $(\mathcal{P})$, and using the weak sequential continuity of B and B', we get

$$B(I-D)^{-1}C\theta_{n_k}.B'(I-D)^{-1}C\theta_{n_k} \rightharpoonup B(I-D)^{-1}C\theta.B'(I-D)^{-1}C\theta.$$

As a result, $F(\theta_{n_k}) \rightharpoonup F(\theta)$. Now, we show that

$$(I-D)^{-1}C(\theta_n) \rightharpoonup (I-D)^{-1}C(\theta).$$

Let us suppose the contrary. Then, there exists a weak neighborhood $\overline{U^w}$ of

$(I-D)^{-1}C\theta$ and a subsequence $(\theta_{n_j})_j$ of $(\theta_n)_n$ such that $(I-D)^{-1}C\theta_{n_j} \notin \overline{U}^w$, for all $j \geq 1$. Since $(\theta_{n_j})_j$ converges weakly to θ, and arguing as before, we may extract a subsequence $(\theta_{n_{j_k}})_k$ of $(\theta_{n_j})_j$ such that

$$(I-D)^{-1}C\theta_{n_{j_k}} \rightharpoonup (I-D)^{-1}C\theta.$$

This is impossible since $(I-D)^{-1}Cx_{n_{j_k}} \notin \overline{U}^w$, $\forall k \geq 1$. As a result, the operator $(I-D)^{-1}C$ is weakly sequentially continuous which is also valid for F. In addition, the use of both Lemma 1.5.1 and assumption (i) leads to the situation where the operator F is β-condensing. Now, applying Theorem 2.3.4 shows that F has a fixed point x in S. Hence, the vector $y = (I-D)^{-1}Cx$ solves the problem. Q.E.D.

Now, we may combine Theorem 4.3.7 and Lemma 3.1.3 to obtain the following fixed point theorem.

Corollary 4.3.2 *Let S be a nonempty, convex, closed, and bounded subset of X, and let A, $C : S \longrightarrow X$, and B, B', $D : X \longrightarrow X$ be five weakly sequentially continuous operators satisfying:*

(i) C is weakly compact,

(ii) A is a contraction with a constant k,

(iii) D is a ϕ-nonlinear contraction and $(I-D)^{-1}C(S)$ is bounded, and

(iv) $Ax + B(I-D)^{-1}Cx \cdot B'(I-D)^{-1}Cx \in S$, for all $x \in S$.

Then, the block operator matrix (4.6) *has, at least, one fixed point in $S \times X$.*

By the same arguments used in the proof of Theorem 4.3.7, we have the following result:

Theorem 4.3.8 *Let S be a nonempty, convex, closed, and bounded subset of X, and let A, $C : S \longrightarrow X$, and B, B', $D : X \longrightarrow X$ be five operators satisfying:*

(i) B and C are Lipschitzian with the Lipschitz constants α and γ, respectively,

(ii) A and B' are weakly compact,

(iii) D is expansive with a constant $h > \gamma + 1$ and $C(S) \subset (I-D)(S)$,

(iv) A, B, and B' are weakly sequentially continuous and C is strongly continuous, and

(*v*) $Ax + B(I - D)^{-1}Cx \cdot B'(I - D)^{-1}Cx \in S$, *for all* $x \in S$.

Then, the block operator matrix (4.6) *has, at least, one fixed point in* $S \times S$ *provided* $0 \leq \alpha M < 1$, *where* $M = \|B'(I - D)^{-1}C(S)\|$.

Proof. Since D is expansive, by Proposition 1.2.1, the operator $(I - D)^{-1}$ exists on $(I - D)(X)$, and for all $x, y \in (I - D)(X)$, we have

$$\|(I - D)^{-1}x - (I - D)^{-1}y\| \leq \frac{1}{h-1}\|x - y\|.$$

Then, $(I - D)^{-1}$ is continuous, and by using assumption (*iv*), the operator $B(I - D)^{-1}C$ is sequentially weakly continuous on S. Moreover, the mapping $(I - D)^{-1}C$ is a contraction on S in view of assumption (*iii*). Thus, $(I - D)^{-1}C(S)$ is bounded. Now, the use of assumption (*ii*) and Eberlein–Šmulian's theorem (see Theorem 1.3.3) shows that $B'(I - D)^{-1}C(S)$ is a relatively weakly compact subset of X. Similarly to the proof of Theorem 4.3.7, we show that the operator F defined in Eq. (4.8) is weakly sequentially continuous and β-condensing. Indeed,

$$\beta\left(F(S)\right) \leq \beta(A(S)) + \beta\left(B(I - D)^{-1}C(S) \cdot B'(I - D)^{-1}C(S)\right).$$

Taking into account that $A(S)$ is relatively weakly compact, and using Lemma 1.5.2, we get

$$\beta\left(F(S)\right) \leq M\beta\left(B(I - D)^{-1}C(S)\right).$$

So, if $\beta(S) \neq 0$, we have

$$\beta\left(F(S)\right) \leq \frac{\alpha\gamma}{h-1}M\beta(S) < \beta(S).$$

Hence, the operator F is β-condensing. Now, the result follows from Theorem 2.3.4. Q.E.D.

Next, we can modify some assumptions of Theorem 4.3.8 in order to study the same problem.

Theorem 4.3.9 *Let S be a nonempty, convex, closed, and bounded subset of a Banach algebra X satisfying condition* ($\mathcal{P}$), *and let* $A, C : S \longrightarrow X$, B, B', $D : X \longrightarrow X$ *be five weakly sequentially continuous operators satisfying:*

(*i*) *B and C are Lipschitzian with the Lipschitz constants α and γ, respectively,*

(*ii*) *C is weakly compact and* $C(S) \subset (I - D)(S)$,

(*iii*) *A and D are two contractions with constants k and k', respectively, and* *(iv)* $Ax + B(I-D)^{-1}Cx \cdot B'(I-D)^{-1}Cx \in S$, *for all* $x \in S$.

Then, the block operator matrix (4.6) *has, at least, a fixed point in* $S \times S$ *whenever* $0 \leq k + \frac{\alpha\gamma}{1-k'}M < 1$, *where* $M = \|\overline{B'(I-D)^{-1}C(S)^w}\| > 1$.

Proof. Notice that the map $(I-D)^{-1}$ exists and is continuous. Our next task is to show that the mapping F defined in Eq. (4.8) respects all conditions of Lemma 1.4.2. We first claim that $(I-D)^{-1}C(S)$ is relatively weakly compact. If not, then $\beta\left((I-D)^{-1}C(S)\right) > 0$. It is easy, in view of Eq. (4.9), to see that

$$\beta((I-D)^{-1}C(S)) \leq \beta(\overline{C(S)}) + \beta(D(I-D)^{-1}C(S)) \leq \beta(D(I-D)^{-1}C(S)).$$

Let $r > \beta((I-D)^{-1}C(S))$. Then, there exists $0 \leq r_0 < r$ and $K \in \mathcal{W}(X)$ such that

$$(I-D)^{-1}C(S) \subset K + B_{r_0}.$$

Arguing as above, we get

$$D(I-D)^{-1}C(S) \subset D(K) + B_{k'r_0}.$$

Now, since D is sequentially weakly continuous, $D(K) \in \mathcal{W}(X)$ and

$$\beta(D(I-D)^{-1}C(S)) \leq k'r_0 < r.$$

Letting $r \to \beta((I-D)^{-1}C(S))$, we get

$$\beta((I-D)^{-1}C(S)) \leq \beta(D(I-D)^{-1}C(S)) < \beta((I-D)^{-1}C(S)),$$

which is a contradiction, and the claim is proved. An argument similar to the one in the proof of Theorem 4.3.7 leads to the weak sequential continuity of the maps $B(I-D)^{-1}C$ and $B'(I-D)^{-1}C$, which is also true for F. Moreover, the operator F is convex-power condensing. Indeed, it is easy to see that

$$F(S) \subset A(S) + B(I-D)^{-1}C(S) \cdot B'(I-D)^{-1}C(S).$$

Keeping in mind the relatively weak compactness of $B'(I-D)^{-1}C(S)$, and using the sub-additivity of the De Blasi measure of weak noncompactness, we get

$$\beta\left(F(S)\right) \leq \beta\left(A(S)\right) + \beta\left(B(I-D)^{-1}C(S) \cdot \overline{B'(I-D)^{-1}C(S)^w}\right).$$

The use of assumption *(iv)*, Lemma 3.1.3 and Lemma 1.5.2, leads to

$$\beta\left(F(S)\right) \leq k\beta\left(S\right) + \|\overline{B'(I-D)^{-1}C(S)^w}\|\beta\left(B(I-D)^{-1}C(S)\right).$$

Since the operator $B(I-D)^{-1}C$ is Lipschitzian with a Lipschitz constant $\frac{\alpha\gamma}{1-k'}$,

$$\beta(F(S)) \leq \left(k + M\frac{\alpha\gamma}{1-k'}\right)\beta(S).$$

Letting $x_0 \in S$ and a positive integer $n \geq 1$, then

$$\begin{aligned}\beta\left(F^{(n,x_0)}(S)\right) &= \beta\left(F\left(\overline{co}\left\{F^{(n-1,x_0)}(S), \{x_0\}\right\}\right)\right)\\ &\leq \left(k + M\tfrac{\alpha\gamma}{1-k'}\right)\beta\left(\overline{co}\left\{F^{(n-1,x_0)}(S), \{x_0\}\right\}\right)\\ &\leq \left(k + M\tfrac{\alpha\gamma}{1-k'}\right)^n \beta(S).\end{aligned}$$

Since $0 < k + M\frac{\alpha\gamma}{1-k'} < 1$, F is a convex power condensing operator. Now, we may apply Lemma 1.4.2 to infer that F has, at least, one fixed point x in S. Consequently, the vector $y = (I-D)^{-1}Cx$ solves the problem. Q.E.D.

Remark 4.3.3 *It should be noted that if K is convex, then by Theorem 1.6.9, the operator D has, at least, a fixed point.*

4.4 Fixed Point Results in a Regular Case

In what follows, we will study a fixed point for the block operator matrix (4.6) in the case where X is a commutative Banach algebra satisfying the condition $(\mathcal{P})$. Before stating the main result, we need the following lemma.

Lemma 4.4.1 *Let S be a nonempty, convex, and closed subset of X, and let A, $C : S \longrightarrow X$ and B, B', $D : X \longrightarrow X$ be five operators such that:*

(i) A, B, and C are $\mathcal{D}$-Lipschitzian with the $\mathcal{D}$-functions ϕ_A, ϕ_B, and ϕ_C, respectively,

(ii) D is a contraction with a constant k and $C(S) \subset (I-D)(S)$,

(iii) T is regular on $T'(S)$, i.e., T maps S into the set of all invertible elements of X included in $T'(S)$, where $T = B(I-D)^{-1}C$ and $T' = B'(I-D)^{-1}C$,

(iv) $T'(S)$ is bounded with a bound M, and

(v) $Ax + Tx \cdot T'y \in S$, for all x, $y \in S$.

Then, $\left(\frac{I-A}{T}\right)^{-1}$ *exists on* $T'(S)$*, whenever* $M\phi_B \circ \left(\frac{1}{1-k}\phi_C\right)(r) + \phi_A(r) < r$ *for* $r > 0$.

Proof. From hypothesis (ii), it follows that $(I-D)^{-1}$ exists on $(I-D)(X)$, and for any $x,\, y \in S$, we have

$$\begin{aligned} \|Tx - Ty\| &\leq \phi_B\left(\|(I-D)^{-1}Cx - (I-D)^{-1}Cy\|\right) \\ &\leq \phi_B \circ \left(\tfrac{1}{1-k}\phi_C\right)(\|x-y\|). \end{aligned}$$

Then, $B(I-D)^{-1}C$ is $\mathcal{D}$-Lipschitzian with $\mathcal{D}$-function $\phi_B \circ \varphi$, where $\varphi = \frac{1}{1-k}\phi_C$. Now, let y be a fixed point in S, and let us define a mapping

$$\begin{cases} \varphi_y : S \longrightarrow S \\ \qquad x \longrightarrow Ax + Tx \cdot T'y. \end{cases}$$

Notice that this operator is $\mathcal{D}$-Lipschitzian with the $\mathcal{D}$-function $\psi = M\phi_B \circ \varphi + \phi_A$. Hence, an application of Theorem 1.6.10 shows that there is a unique point $x_y \in S$ such that $\varphi_y(x_y) = x_y$. Or in an equivalent way:

$$Ax_y + Tx_y \cdot T'y = x_y.$$

Consequently, in view of assumption (iii), we have

$$\left(\frac{I-A}{T}\right)x_y = T'y.$$

Thus, the mapping $\left(\frac{I-A}{T}\right)^{-1}$ is well defined and the desired result is deduced. Q.E.D.

In the following result, we will combine Theorem 2.3.4 and Lemma 4.4.1.

Theorem 4.4.1 *Let* S *be a nonempty, convex, closed, and bounded subset of* X*, and let* $A,\, C : S \longrightarrow X$ *and* $B,\, B',\, D : X \longrightarrow X$ *be five operators satisfying:*

(i) $A, B,$ *and* C *are* $\mathcal{D}$*-Lipschitzian with* $\mathcal{D}$*-functions* ϕ_A, ϕ_B, *and* ϕ_C*, respectively.*

(ii) B' *is Lipschitzian with a constant* α,

(iii) D *is a contraction with a constant* $k \in \left[0, \frac{3-\sqrt{5}}{2}\right[$,

(iv) T *is regular on* $T'(S)$*, where* $T = B(I-D)^{-1}C$ *and* $T' = B'(I-D)^{-1}C$,

(v) C is strongly continuous and $C(S) \subset (I-D)(S)$,

(vi) $x = Ax + Tx \cdot T'y$; $y \in S \Longrightarrow x \in S$, and

(vii) $\max\left\{\left(M_1\phi_B \circ \left(\frac{1}{1-k}\phi_C\right)\right)(r) + \phi_A(r), M_2\alpha\phi_C(r)\right\} \leq kr$, *for* $r > 0$, *where* $M_1 = \|T'(S)\| > k$ *and* $M_2 = \|T(S)\| > 1$.

Then, the block operator matrix (4.6) has, at least, a fixed point in $S \times S$.

Proof. By using our assumptions, it is easy to verify that $B(I-D)^{-1}C$ and $B'(I-D)^{-1}C$ are two contractions on S, and consequently $T(S)$ and $T'(S)$ are two bounded subsets. The use of Lemma 4.4.1 and the Browder's fixed point theorem [44] shows that the operator $\left(\frac{I-A}{T}\right)^{-1}$ exists on $T'(S)$. Define a mapping

$$\begin{cases} N : S \longrightarrow X \\ \quad x \longrightarrow \left(\dfrac{I-A}{T}\right)^{-1} T'x. \end{cases} \tag{4.12}$$

Now, in view of Lemma 1.4.2, it is sufficient to prove that the operator N is weakly sequentially continuous and convex-power condensing and that $N(S)$ is bounded.

Step 1: N is weakly sequentially continuous.
We will show that N is weakly sequentially continuous on S. To do so, let $(x_n)_n$ be any sequence in S weakly converging to a point x in S. In view of hypothesis (v), one has

$$Cx_n \to Cx.$$

Keeping in mind the continuity of $(I-D)^{-1}$ and B', we get

$$B'(I-D)^{-1}Cx_n \to B'(I-D)^{-1}Cx.$$

Moreover, the operator $\left(\frac{I-A}{T}\right)^{-1}$ is continuous on $T'(S)$. Indeed, let $(x_n)_n$ be any sequence in $T'(S)$ converging to a point x, and let

$$\begin{cases} y_n = \left(\dfrac{I-A}{T}\right)^{-1} x_n \\ y = \left(\dfrac{I-A}{T}\right)^{-1} x. \end{cases}$$

Or, which is equivalent:

$$\begin{cases} y_n = Ay_n + Ty_n \cdot x_n \\ y = Ay + Ty \cdot x. \end{cases}$$

Then,

$$\begin{aligned}
\|y_n - y\| &\leq \|Ay_n - Ay\| + \|Ty_n \cdot x_n - Ty \cdot x\| \\
&\leq \|Ay_n - Ay\| + \|Ty_n \cdot x_n - Ty \cdot x_n\| + \|Ty \cdot x_n - Ty \cdot x\| \\
&\leq \|x_n\| \left(\phi_B \circ \left(\tfrac{1}{1-k}\phi_C\right) + \phi_A\right)(\|y_n - y\|) + \|Ty\|\|x_n - x\| \\
&\leq \left(M_1\phi_B \circ \left(\tfrac{1}{1-k}\phi_C\right) + \phi_A\right)(\|y_n - y\|) + M_2\|x_n - x\|.
\end{aligned}$$

Consequently,

$$\limsup_n \|y_n - y\| \leq \left(M_1\phi_B \circ \left(\frac{1}{1-k}\phi_C\right) + \phi_A\right)(\limsup_n \|y_n - y\|).$$

If $\limsup_n \|y_n - y\| \neq 0$, then we get a contradiction and so, the operator $\left(\frac{I-A}{T}\right)^{-1}$ is continuous on $T'(S)$. Accordingly,

$$\left(\frac{I-A}{T}\right)^{-1} B'(I-D)^{-1}Cx_n \to \left(\frac{I-A}{T}\right)^{-1} B'(I-D)^{-1}Cx \text{ in } X.$$

Then, $Nx_n \rightharpoonup Nx$. As a result, the operator N is weakly sequentially continuous.

Step 2: N is convex-power condensing.
Let $x_1, x_2 \in S$ and $y_1, y_2 \in S$ such that $y_1 = Nx_1$ and $y_2 = Nx_2$. Then,

$$\begin{cases} y_1 = Ay_1 + Ty_1 \cdot T'x_1 \\ y_2 = Ay_2 + Ty_2 \cdot T'x_2 \end{cases}$$

and so, by assumption (vii), we have

$$\begin{aligned}
\|y_1 - y_2\| &\leq \|Ay_1 - Ay_2\| + \|Ty_1 \cdot T'x_1 - Ty_2 \cdot T'x_2\| \\
&\leq \left(\phi_A + M_1\phi_B \circ \left(\tfrac{1}{1-k}\phi_C\right)\right)(\|y_1 - y_2\|) + \frac{M_2\alpha}{1-k}\phi_C(\|x_1 - x_2\|) \\
&\leq k\|y_1 - y_2\| + \frac{k}{1-k}\|x_1 - x_2\|.
\end{aligned}$$

This implies that

$$\|Nx_1 - Nx_2\| \leq \frac{k}{(1-k)^2}\|x_1 - x_2\|.$$

Now, the use of both Lemma 3.1.3 and Step 1 yields

$$\beta(N(S)) \leq \frac{k}{(1-k)^2}\beta(S).$$

From assumption (*iii*), it follows that $0 \leq \frac{k}{(1-k)^2} < 1$. Consequently, N is β-condensing and so, is convex-power condensing. The result follows from Lemma 1.4.2. Q.E.D.

Now, we may combine Theorem 2.2.1 and Lemma 4.4.1 in order to obtain the following fixed point theorem in Banach algebra.

Theorem 4.4.2 *Let S be a nonempty, weakly compact, and convex subset of X, and let A, $C : S \longrightarrow X$ and B, B', $D : X \longrightarrow X$ be five operators satisfying:*

(i) A, B, and C are $\mathcal{D}$-Lipschitzian with the $\mathcal{D}$-functions ϕ_A, ϕ_B, and ϕ_C, respectively,

(ii) B' is continuous on S,

(iii) T is regular on $B'(S)$, where $T = B(I-D)^{-1}C$,

(iv) D is a contraction with a constant k,

(v) C is strongly continuous and $C(S) \subset (I-D)(S)$, and

(vi) $(x = Ax + Tx \cdot T'y\ ,\ y \in S) \Longrightarrow x \in S$, where $T' = B'(I-D)^{-1}C$.

Then, the block operator matrix (4.6) *has, at least, a fixed point in S^2 whenever $M\phi_B \circ \left(\frac{1}{1-k}\phi_C\right)(r) + \phi_A(r) < r$ for $r > 0$, where $M = \|T'(S)\|$.*

Proof. Similarly to the proof of Theorem 4.4.1, we show that the operator N defined in Eq. (4.12) is weakly sequentially continuous. Moreover, taking into account that S is weakly compact and using the Eberlein–Šmulian theorem (see Theorem 1.3.3), we deduce that $N(S)$ is relatively weakly compact. Hence, from Theorem 2.2.1, we prove that the equation $Nx = x$ has, at least, one solution in S. Consequently, the use of the vector $y = (I-D)^{-1}Cx$ solves the problem. Q.E.D.

4.5 BOM with Multi-Valued Inputs

In this section, we are concerned with the fixed point results on Banach algebras of the block operator matrix Eq. (4.6), where the entries of the matrix are assumed to be nonlinear multi-valued operators defined on Banach algebras. Our assumptions are as follows: A maps a bounded, closed, convex, and nonempty subset S of a Banach algebra X into the classes of all closed, convex, and bounded subsets of X, denoted by $\mathcal{P}_{cl,\ cv,\ bd}(X)$, B, B' from X into $\mathcal{P}_{cl,\ cv,\ bd}(X)$, C from S into X and D from X into X. Let us introduce the following definition.

Definition 4.5.1 *A point* $(u, v) \in X \times X$ *is called a fixed point of the multi-valued block operator matrix* (4.6), *if*

$$\begin{cases} u \in Au + Bv \cdot B'v, \ \textit{and} \\ v = Cu + Dv. \end{cases} \tag{4.13}$$

In what follows, let X be a Banach algebra and let $\mathcal{P}_p(X)$ denote the class of all nonempty subsets of X with the property p. Thus, $\mathcal{P}_{cl}(X), \mathcal{P}_{bd}(X)$, $\mathcal{P}_{cp}(X)$, and $\mathcal{P}_{cv}(X)$ denote, respectively, the classes of all closed, bounded, compact, and convex subsets of X. Similarly, $\mathcal{P}_{cl,\ bd}(X)$ and $\mathcal{P}_{cp,\ cv}(X)$ denote, respectively, the classes of all closed-bounded and compact-convex subsets of X. Recall that a correspondence $Q : X \longrightarrow \mathcal{P}_p(X)$ is called a multi-valued operator or multi-valued mapping on X into X. A point $x \in X$ is called a fixed point of Q if, $x \in Qx$ and the set of all fixed points of Q in X is denoted by $\mathcal{F}_Q$. For the sake of convenience, we denote $Q(A) = \cup_{x \in A} Qx$ for a subset A of X. For $x \in X$ and A, $B \in \mathcal{P}_{cl}(X)$ we denote by:

$$D(x, A) := \inf\{\|x - y\| \ ; \ y \in A\}.$$

Let us define a function $d_H : \mathcal{P}_{cl}(X) \times \mathcal{P}_{cl}(X) \longrightarrow \mathbb{R}_+$ by:

$$d_H(A, B) = \max\{\sup_{a \in A} D(a, B), \ \sup_{b \in B} D(A, b)\}.$$

The function d_H is called a Hausdorff metric on X. Note that $\|A\|_{\mathcal{P}} = d_H(A, \{0\})$. The concept of Hausdorff metric was used by many authors in order to prove the fixed point and the coincidence point results in the setting of metric spaces. Let $Q : X \longrightarrow \mathcal{P}_p(X)$ be a multi-valued map. For any subset A of X, we define

$$Q^-(A) = \{x \in X \ ; \ Q(x) \cap A \neq \emptyset\} \quad \text{and} \quad Q^+(A) = \{x \in X \ ; \ Q(x) \subset A\}.$$

Definition 4.5.2 *A multi-valued operator Q is called lower semi-continuous (resp. upper semi-continuous) if $Q^{-}(A)$ (resp. $Q^{+}(A)$) is an open set in X for every open subset U of X.*

Lemma 4.5.1 *(See [65]) A multi-valued operator Q is lower semi-continuous at a point $x \in X$ if, and only if, for every sequence $\{x_n\}_{n=0}^{\infty}$ in X which converges to x, and for each $y \in Q(x)$, there exists a sequence $y_n \in Q(x_n)$ such that (y_n) converges to y.*

Definition 4.5.3 *A multi-valued operator $Q : X \longrightarrow \mathcal{P}_p(X)$ is called H-lower semi-continuous at x_0 if, and only if, for $\varepsilon > 0$, there exists $\eta > 0$ such that $Q(x_0) \subset V(Q(x), \varepsilon)$ for all $x \in B_\eta(x_0)$, where $V(Q(x), \varepsilon)$ is a closed neighborhood of $Q(x)$ in X. Q is called H-lower semi-continuous on X if it is H-lower semi-continuous at each point x_0 of X. Similarly, Q is called H-upper semi-continuous at $x_0 \in X$ if, and only if, for $\varepsilon > 0$, there exists $\eta > 0$ such that $Q(x) \subset V(Q(x_0), \varepsilon)$ for all $x \in B_\eta(x_0)$. Q is called H-upper semi-continuous on X, if it is H-upper semi-continuous at each point x_0 of X.*

Remark 4.5.1 *Note that every contraction multi-valued operator is H-lower semi-continuous as well as H-upper semi-continuous on X.*

Remark 4.5.2 *Notice that every upper semi-continuous operator is H-upper semi-continuous, but the converse may not be true.*

4.5.1 Fixed point theorems of multi-valued mappings

Let us recall the following definition.

Definition 4.5.4 *A multi-valued mapping $Q : X \longrightarrow \mathcal{P}_p(X)$ is called totally bounded if $Q(S)$ is a totally bounded subset of X for all bounded subsets S of X. Again, Q is called completely continuous on X, if it is upper semi-continuous and totally bounded on X.*

Before reaching the major part of the main fixed point results for this section, let us state the useful lemmas for the sequel.

Lemma 4.5.2 *[69] If A, $B \in \mathcal{P}_{bd,\ cl}(X)$, then*

$$d_H(AC, BC) \leq d_H(0, C) d_H(A, B).$$

Lemma 4.5.3 *[74] Let $Q : X \longrightarrow \mathcal{P}_{bd}(X)$ be a multi-valued Lipschitzian operator. Then, for any bounded subset S of X, we deduce that $Q(S)$ is bounded.*

Using the notion of Hausdorff metric, in 1969, H. Covitz and S. B. Nadler, Jr. [58] proved a set-valued version of the Banach contraction principle. We now consider some known results which will be used in the following sections.

Theorem 4.5.1 *[58] Let (X, d) be a complete metric space, and let $T : X \longrightarrow \mathcal{P}_{cl}(X)$ be a multi-valued contraction. Then, the fixed point set $\mathcal{F}_T$ of T is a nonempty and closed subset of X.*

The following result is due to L. Rybinski [143] and will be useful in the sequel.

Theorem 4.5.2 *Let S be a nonempty and closed subset of a Banach space X, and let Y be a metric space. Assume that the multi-valued operator $Q : S \times Y \longrightarrow \mathcal{P}_{cl,\ cv}(S)$ satisfies the following conditions:*

(i) $d_H(Q(x_1, y), Q(x_2, y)) \leq q\|x_1 - x_2\|$ for all $(x_1, y), (x_2, y) \in S \times Y$, where $q < 1$, and

(ii) for every $x \in S$, $Q(x, \cdot)$ is lower semi-continuous on Y.

Then, there exists a continuous mapping $f : S \times Y \longrightarrow S$ such that $f(x, y) \in F(f(x, y), y)$ for each $(x, y) \in S \times Y$.

Lemma 4.5.4 *Let X be a complete metric space, and let $Q_1,\ Q_2 : X \longrightarrow \mathcal{P}_{bd,\ cl}(X)$ be two multi-valued contractions with the same constant a. Then,*

$$d_H(\mathcal{F}_{Q_1}, \mathcal{F}_{Q_2}) \leq \frac{1}{1-a} \sup \{d_H(Q_1(x), Q_2(x))\ ;\ x \in X\}.$$

Proof. Notice that the fixed point sets $\mathcal{F}_{Q_i}$, $i = 1, 2$ are nonempty and closed (see [75]). Write $K = \sup \{d_H(Q_1(x), Q_2(x))\ ;\ x \in X\}$ and assume that it is finite. Let ε be an arbitrary strictly positive real. Next, choose $c > 0$ such that,

$$c\sum_{i=1}^{\infty} i a^{i-1} < 1$$

and set $\varepsilon_1 = c(1-a)\varepsilon$. Let $x_0 \in \mathcal{F}_{Q_1}$. Since $d_H(Q_1(x_0), Q_2(x_0)) \leq K$, we may choose $x_1 \in Q_2(x_0)$ such that $\|x_1 - x_0\| \leq K + \varepsilon$. As

$$d_H(Q_2(x_0), Q_2(x_1)) \leq a\|x_1 - x_0\|,$$

we may choose $x_2 \in Q_2(x_1)$ with

$$\|x_2 - x_1\| \leq a\|x_1 - x_0\| + \varepsilon_1.$$

Proceeding by induction, we can define a sequence $(x_n)_{n\in\mathbb{N}}$ such that $x_{n+1} \in Q_2(x_n)$ and

$$\|x_{n+1} - x_n\| \leq a^n \|x_1 - x_0\| + na^{n-1}\varepsilon_1.$$

It follows that,

$$\begin{aligned}
\sum_{i=m}^{n+1} \|x_i - x_{i-1}\| &\leq \sum_{i=m}^{n+1} a^i \|x_1 - x_0\| + \sum_{i=m}^{+\infty} ia^{i-1}(\varepsilon_1) \\
&\leq \frac{a^m}{1-a}\|x_1 - x_0\| + \sum_{i=m}^{+\infty} ia^{i-1}(\varepsilon_1) \to 0.
\end{aligned}$$

as $m \to \infty$ and therefore, $\lim_{n\to\infty} x_n = x^*$. Using the fact that $x_{n+1} \in Q_2(x_n)$, we infer that $x^* \in \mathcal{F}_{Q_2}$. Moreover,

$$\begin{aligned}
\|x_0 - x\| &\leq \sum_{i=0}^{\infty} \|x_{i+1} - x_i\| \\
&\leq \frac{1}{1-a}\|x_1 - x_0\| + \sum_{i=1}^{\infty} ia^{i-1}(\varepsilon_1) \\
&\leq \frac{1}{1-a}(K + 2\varepsilon).
\end{aligned}$$

Q.E.D.

We think that this result remains true if we replace the two multi-valued contractions by two multi-valued nonlinear $\mathcal{D}$-contractions with the same subadditive $\mathcal{D}$-function. In fact, we may have:

Question 4:

Let X be a Banach space and $Q_1, Q_2 : X \to \mathcal{P}_{bd,\ cl}(X)$ be two multi-valued nonlinear $\mathcal{D}$-contractions with the same subadditive $\mathcal{D}$-function ϕ. Is there $0 \leq a < 1$ such that:

$$d_H(\mathcal{F}_{Q_1}, \mathcal{F}_{Q_2}) \leq \frac{1}{1-a} \sup\{d_H(Q_1(x), Q_2(x))\ ;\ x \in X\}?$$

The following theorem can be found in [74].

Theorem 4.5.3 *Let S be a closed, convex, and bounded subset of the Banach algebra X, and let $A,\ C : X \longrightarrow \mathcal{P}_{cl,\ cv,\ bd}(X)$ and $B : S \longrightarrow \mathcal{P}_{cp,\ cv}(X)$ be three multi-valued operators such that:*

(i) A and C are multi-valued Lipschitzian operators with the Lipschitz constants q_1 and q_2, respectively,

(ii) B is a lower semi-continuous and compact operator,

(iii) $Ax \cdot By + Cx \in \mathcal{P}_{cl,\ cv}(X)$, for all $x, y \in S$, and

(iv) $q_1 M + q_2 < 1$, where $M = \| \cup B(S)\|_{\mathcal{P}} = \sup\{\|Bx\|_{\mathcal{P}}\ ;\ x \in S\}$.

Then, the operator inclusion $x \in Ax \cdot Bx + Cx$ has a solution in X.

Theorem 4.5.4 *Let S be a closed, convex, and bounded subset of the Banach algebra X, and let $A : S \longrightarrow \mathcal{P}_{cl,\ cv,\ bd}(X)$, $B : X \longrightarrow \mathcal{P}_{cl,\ cv,\ bd}(X)$, $B' : X \longrightarrow \mathcal{P}_{cp,\ cv}(X)$, $C : S \longrightarrow X$, and $D : X \longrightarrow X$ be five multi-valued operators satisfying:*

(i) A and B are $\mathcal{D}$-Lipschitzian with the $\mathcal{D}$-functions ψ_1 and ψ_2, respectively,

(ii) C and D are Lipschitzian with the constants q and k such that $q+k<1$,

(iii) $C(S) \subseteq (I-D)(S)$,

(iv) B' is a lower semi-continuous and totally bounded operator, and

(v) $Ax + B(I-D)^{-1}Cx \cdot B'(I-D)^{-1}Cy \in \mathcal{P}_{cl,\ cv}(S)$, for all $x,\ y \in S$.

Then, the multi-valued block operator matrix (4.6) has, at least, a fixed point whenever, $\psi_1(r) + M\psi_2(r) < \alpha r$, $r > 0$ and $\alpha \in [0,1[$, where $M = \| \cup B'(S)\|_{\mathcal{P}} = \sup\{\|B'x\|_{\mathcal{P}}\ ;\ x \in S\}$.

Proof. As shown in *(ii)*, we notice that $(I-D)^{-1}$ exists on $(I-D)(X)$. Let $y \in S$ and let us define a multi-valued operator $T : S \longrightarrow \mathrm{P}_{cl,\ cv}(S)$ by:

$$T_y(x) = Ax + B(I-D)^{-1}Cx \cdot B'(I-D)^{-1}Cy.$$

We will demonstrate that $T_y(\cdot) = T(\cdot, y)$ is a multi-valued contraction for each fixed $y \in S$. Let $x_1, x_2 \in S$. Then,

$$\begin{aligned} d_H(T_y(x_1), T_y(x_2)) &\leq d_H(Ax_1, Ax_2) + d_H(\Pi x_1 \cdot \Pi' y, \Pi x_2 \cdot \Pi' y) \\ &\leq d_H(Ax_1, Ax_2) + \|\Pi'(S)\|_{\mathcal{P}} d_H(\Pi x_1, \Pi x_2) \\ &\leq \psi_1(\|x_1 - x_2\|) + M\psi_2\left(\tfrac{q}{1-k}\|x_1 - x_2\|\right) \\ &\leq (\psi_1 + M\psi_2)(\|x_1 - x_2\|) \\ &\leq \alpha\|x_1 - x_2\|, \end{aligned}$$

where $\Pi := B(I-D)^{-1}C$ and $\Pi' := B'(I-D)^{-1}C$. This demonstrates that the multi-valued operator $T_y(\cdot)$ is a contraction on S. Hence, when we apply Covitz–Nadler's fixed point theorem [58], we notice that the fixed point set

$$\mathcal{F}_{T_y} = \left\{x \in S \text{ such that } x \in Ax + B(I-D)^{-1}Cx \cdot B'(I-D)^{-1}Cy\right\}$$

is a nonempty and closed subset of S, for each $y \in S$. The application of both hypothesis (iii) and Theorem 4.5.2 shows the existence of a continuous mapping $f : S \times S \longrightarrow S$ such that

$$f(x,y) \in A(f(x,y)) + B(I-D)^{-1}C(f(x,y)) \cdot B'(I-D)^{-1}Cy. \qquad (4.14)$$

Now, let us define

$$\left\{\begin{array}{l} G : S \longrightarrow \mathrm{P}_{cl}(S) \\ \quad y \longrightarrow \mathcal{F}_{T_y} \end{array}\right. \quad \text{and} \quad \left\{\begin{array}{l} g : S \longrightarrow S \\ \quad x \longrightarrow f(x,x). \end{array}\right.$$

It follows that g is a continuous mapping having the following property:

$$g(x) = f(x,x) \in A(f(x,x)) + B(I-D)^{-1}C(f(x,x)) \cdot B'(I-D)^{-1}Cx$$

for each $x \in S$. In order to prove that g is completely continuous on S, we only have to demonstrate that G is continuous and totally bounded on S. In fact, for any $z \in S$, we have

$$\begin{aligned}
\|B(I-D)^{-1}C(z)\|_{\mathcal{P}} &\leq \|B(I-D)^{-1}C(a)\|_{\mathcal{P}} + \psi_2\left(\tfrac{q}{1-k}\|z-a\|\right) \\
&< \|B(I-D)^{-1}C(a)\|_{\mathcal{P}} + \psi_2(\|z-a\|) \\
&\leq \|B(I-D)^{-1}C(a)\|_{\mathcal{P}} + \tfrac{\|z-a\|}{M} \\
&\leq \eta,
\end{aligned}$$

where

$$\eta = \|B(I-D)^{-1}C(a)\|_{\mathcal{P}} + \frac{\text{diam}(S)}{M}, \tag{4.15}$$

for some fixed point a in S. Let $\varepsilon > 0$ be given. Since $B'(I-D)^{-1}C$ is totally bounded on S, then there exists a subset $Y = \{y_1, \ldots, y_n\}$ of points in S, such that

$$\begin{aligned}
B'(I-D)^{-1}C(S) &\subset \{w_1, \ldots, w_n\} + B\left(0, \frac{1-\alpha}{\eta}\varepsilon\right) \\
&\subset \bigcup_{i=1}^{n} B\left(w_i, \frac{1-\alpha}{\eta}\varepsilon\right),
\end{aligned}$$

where $w_i \in B'(I-D)^{-1}C(y_i)$ and $B(w,r)$ is an open ball in X, centered at w with a radius r. Then, for each $y \in S$, there is an element $y_k \in Y$ such that

$$d_H\left(B'(I-D)^{-1}Cy,\ B'(I-D)^{-1}Cy_k\right) < \frac{1-\alpha}{\eta}\varepsilon.$$

This implies that

$$\begin{aligned}
d_H(G(y), G(y_k)) &= d_H(\mathcal{F}_{T_y}, \mathcal{F}_{T_{y_k}}) \\
&\leq \frac{1}{1-\alpha}\sup_{x\in S}\{d_H\left(T_y(x), T_{y_k}(x)\right)\} \\
&\leq \frac{1}{1-\alpha}\|B(I-D)^{-1}Cx\|_{\mathcal{P}} d_H\left(\Pi' y, \Pi' y_k\right) \\
&< \frac{\eta}{1-\alpha}\frac{1-\alpha}{\eta}\varepsilon \\
&\leq \varepsilon.
\end{aligned}$$

Therefore, for each $u \in G(y)$, there exists $u_k \in G(y_k)$ such that $\|u - u_k\| < \varepsilon$, and thereby, for each $y \in Y$, one has

$$G(y) \subseteq \bigcup_{i=1}^{n} B(u_k, \varepsilon),$$

where $u_i \in G(y_i), i = 1, \ldots, n$. Then,

$$g(S) \subset G(S) \subseteq \bigcup_{i=1}^{n} B(u_k, \varepsilon)$$

and so, h is a completely continuous operator on S. Now, all the assumptions of Schauder's fixed point theorem are satisfied by the mapping h. As a result, there exists $u \in S$ such that $u = h(u)$. From Eq. (4.14), it follows that

$$u = h(u) \in A(h(u)) + B(I - D)^{-1}C(h(u)) \cdot B'(I - D)^{-1}Cu.$$

The vector $v = (I - D)^{-1}Cu$ completes this proof. Q.E.D.

An improved version of Theorem 4.5.4 under a weaker hypothesis (iv) is provided in the following multi-valued fixed point theorem.

Theorem 4.5.5 *Let S be a closed, convex, and bounded subset of the Banach algebra X and let A, $B : X \longrightarrow \mathcal{P}_{cl,\ cv,\ bd}(X)$, $B' : X \longrightarrow \mathcal{P}_{cp,\ cv}(X)$ and $C : S \longrightarrow X$ and $D : X \longrightarrow X$ be five multi-valued operators satisfying:*

(i) A and B are $\mathcal{D}$-Lipschitzian with the $\mathcal{D}$-functions ψ_1 and ψ_2, respectively,

(ii) C and D are Lipschitzian with the constants q and k such that $q + k < 1$,

(iii) $C(S) \subseteq (I - D)(X)$,

(iv) B' is a lower semi-continuous and totally bounded operator,

(v) $Ax + B(I - D)^{-1}Cx \cdot B'(I - D)^{-1}Cy \in \mathcal{P}_{cl,\ cv}(X)$, for all $x, y \in S$, and

(vi) $x \in Ax + B(I - D)^{-1}Cx \cdot B'(I - D)^{-1}Cy \ ; \ y \in S \Longrightarrow x \in S$.

Then, the multi-valued block operator matrix (4.6) has, at least, a fixed point whenever $\psi_1(r) + M\psi_2(r) < \alpha r$, where $\alpha \in [0, 1[$ and $M = \| \cup B'(S)\|_{\mathcal{P}}$.

Proof. Let us define a multi-valued operator $T : X \times S \longrightarrow \mathrm{P}_{cl,\ cv}(X)$ by:

$$T(x, y) = Ax + B(I - D)^{-1}Cx \cdot B'(I - D)^{-1}Cy.$$

By following the same procedures as in the proof of Theorem 4.5.4, it can be proved that T_y is a multi-valued contraction on X. When we apply Covitz–Nadler's fixed point theorem, we reach the result that the fixed point set

$$\mathcal{F}_{T_y} = \Big\{x \in X \text{ such that } x \in Ax + B(I - D)^{-1}Cx \cdot B'(I - D)^{-1}Cy\Big\}$$

is a nonempty and closed subset of S for each $y \in S$. As shown in hypothesis (v), we notice that $\mathcal{F}_{T_y} \subset S$ for each $y \in S$. Now, let us define the mappings G and g as in the proof of Theorem 4.5.4. It follows that g is a continuous mapping having the property that

$$g(y) = f(y,y) \in A(f(y,y)) + B(I-D)^{-1}C(f(y,y)) \cdot B'(I-D)^{-1}Cy,$$

for each $y \in S$. Once again, we proceed with the same arguments as in the proof of Theorem 4.5.4. Then, we can show that there exists $u \in S$ such that

$$u \in Au + B(I-D)^{-1}Cu \cdot B'(I-D)^{-1}Cu.$$

The vector $v = (I-D)^{-1}Cu$ completes this proof. Q.E.D.

Let us recall that the Kuratowskii measure of noncompactness α in a Banach space is a nonnegative real number $\alpha(S)$ defined by:

$$\alpha(S) = \inf\left\{r > 0 \ ; \ S \subseteq \bigcup_{i=1}^{n} S_i, \quad \mathrm{diam}(S) \leq r \ \ \forall i\right\}$$

for all bounded subsets S of X. The Hausdorff measure of noncompactness of a bounded subset S of X is a nonnegative real number $\beta(S)$ defined by:

$$\beta(S) = \inf\{r > 0; \ S \subseteq \bigcup_{i=1}^{n} B(x_i, r), \quad \text{for some } x_i \in X\},$$

where $B(x_i, r) = \{x \in X \ ; \ d(x_i, x) < r\}$.

Definition 4.5.5 *A multi-valued mapping $Q : X \longrightarrow \mathcal{P}_{cl,\ bd}(X)$ is called $\mathcal{D}$-set-Lipschitz, if there is a continuous nondecreasing function $\psi : \mathbb{R}^+ \longrightarrow \mathbb{R}^+$ such that $\beta(Q(A)) \leq \psi(\beta(A))$ for all $A \in \mathcal{P}_{cl,\ bd}(X)$ with $Q(A) \in \mathcal{P}_{cl,\ bd}(X)$ where $\psi(0) = 0$. Sometimes, we call the function ψ a $\mathcal{D}$-function of Q on X. In the special case where $\psi(r) = kr, k > 0, Q$ is called a k-set-Lipschitz mapping, and if $k < 1$, then Q is called a k-set-contraction on X. Further, if $\psi(r) < r$ for $r > 0$, then Q is called a nonlinear $\mathcal{D}$-set-contraction on X.*

The following results are needed in the sequel.

Theorem 4.5.6 *[75] Let S be a nonempty, closed, convex, and bounded subset of a Banach space X, and let $Q : S \longrightarrow \mathcal{P}_{cl,\ cv}(S)$ be a closed and nonlinear $\mathcal{D}$-set-contraction. Then, Q has, at least, a fixed point.*

Lemma 4.5.5 *[15] If $S_1,\ S_2 \in \mathcal{P}_{bd}(X)$, then $\beta(S_1 \cdot S_2) \leq \beta(S_1)\|S_2\|_{\mathcal{P}} + \beta(S_2)\|S_1\|_{\mathcal{P}}$.*

Lemma 4.5.6 *[6] Let α and β be, respectively, the Kuratowskii and Hausdorff measure of noncompactness in a Banach space X. Then, for any bounded subset S of X, we have $\alpha(S) \leq 2\beta(S)$.*

Now, we can use Theorem 4.5.6, together with Lemma 4.5.5, in order to reach the following fixed point theorem.

Theorem 4.5.7 *Let S be a closed, convex, and bounded subset of the Banach algebra X, and let A, B, $B' : X \longrightarrow \mathcal{P}_{bd,\ cv}(X)$, and $C, D : S \longrightarrow X$ be five multi-valued operators satisfying:*

(i) A and B are $\mathcal{D}$-Lipschitzian with the $\mathcal{D}$-functions ψ_1 and ψ_2, respectively,

(ii) C and D are Lipschitzian with constants q and k, respectively, such that $q + k < 1$,

(iii) $C(S) \subseteq (I - D)(S)$,

(iv) B' is an upper semi-continuous and totally bounded operator, and

(v) $Ax + B(I-D)^{-1}Cx \cdot B'(I-D)^{-1}Cx \in \mathcal{P}_{cl,\ bd,\ cv}(S)$, for all x, $y \in S$.

Then, the multi-valued block operator matrix (4.6) has, at least, a fixed point whenever $M\psi_2(r) + \psi_1(r) < \alpha r$, where $\alpha \in [0, \frac{1}{2}[$ and $M = \| \cup B'(S)\|_{\mathcal{P}}$.

Proof. Let us define the mapping

$$\begin{cases} T : S \longrightarrow \mathcal{P}_{\mathcal{P}}(S) \\ \qquad x \longrightarrow Ax + B(I-D)^{-1}Cx \cdot B'(I-D)^{-1}Cx. \end{cases}$$

Obviously, Tx is a convex subset of S for each $x \in S$ in view of hypothesis (iv). Therefore, from Lemma 4.5.5, it follows that

$$\begin{aligned} \beta(Tx) &= \beta\left(Ax + B(I-D)^{-1}Cx \cdot B'(I-D)^{-1}Cx\right) \\ &\leq \beta(Ax) + \beta\left(B(I-D)^{-1}Cx \cdot B'(I-D)^{-1}Cx\right) \\ &\leq \beta(Ax) + \beta\left(B(I-D)^{-1}Cx\right) \|B'(I-D)^{-1}Cx\|_{\mathcal{P}} \\ &\leq \alpha(Ax) + \alpha\left(B(I-D)^{-1}Cx\right) \|B'(I-D)^{-1}Cx\|_{\mathcal{P}} \\ &\leq \psi_1\left(\alpha(\{x\}) + \psi_2\left(\alpha(\{\tfrac{q}{1-k}x\})\right) \|B'(I-D)^{-1}Cx\|_{\mathcal{P}}\right) \end{aligned}$$

for every $x \in S$. Then, T defines a multi-valued map $T : S \longrightarrow \mathcal{P}_{cv,\ cp}(S)$. In order to show that the map $T : S \longrightarrow \mathcal{P}_{cv,\ cp}(S)$ is closed, let $\{x_n\}$ be a sequence in S converging to the point $x \in S$. Let $y_n \in Tx_n$ converge to the point y. It is sufficient to prove that $y \in Tx$. Now, for any $x, z \in S$, we have

$$\begin{aligned}
d_H(Tx, Tz) &\leq \delta(x,z) + d_H\left(\Pi x \cdot \Pi' x, \Pi z \cdot B'(I-D)^{-1}Cz\right) \\
&\leq \delta(x,z) + d_H(\Pi x \cdot \Pi' x, \Pi z \cdot \Pi' x) + d_H(\Pi z \cdot \Pi' x, \Pi z \cdot \Pi' z) \\
&\leq \delta(x,z) + d_H(\Pi x, \Pi z)\, d_H(0, \Pi' x) + d_H(0, \Pi z) d_H(\Pi' x, \Pi z) \\
&\leq \psi_1(\|x-z\|) + M\psi_2\left(\tfrac{q}{1-k}\|x-z\|\right) + \eta d_H(\Pi' x, \Pi' z) \\
&\leq (\psi_1 + M\psi_2)(\|x-z\|) + \eta d_H(\Pi' x, \Pi' z),
\end{aligned}$$

where η was defined in Eq. (4.15), $\delta(x,z) := d_H(Ax, Az)$, $\Pi := B(I-D)^{-1}C$, and $\Pi' := B'(I-D)^{-1}C$. Since B' is upper semi-continuous, then

$$d_H(\Pi' x_n, \Pi' x) \to 0, \quad \text{whenever } x_n \to x.$$

Then, $d_H(Tx_n, Tx) \to 0$ as $n \to \infty$. This shows that the multi-valued operator T is H-upper semi-continuous on S. Since the multi-valued map T is compact-valued, then it is upper semi-continuous on S. Consequently, the multi-valued $T : S \longrightarrow \mathcal{P}_{cp,\, cv}(S)$ is closed on S. Finally, let us demonstrate that T is a nonlinear $\mathcal{D}$-set-contraction on S. For this purpose, let $S_1 \subset S$ be arbitrary. Then, S_1 is bounded. Notice that $A(S_1)$ and $(I-D)^{-1}C(S_1)$ are also bounded. Since B' is compact, then the set $B'(I-D)^{-1}C(S_1)$ is relatively compact and hence is bounded in X. Since

$$T(S_1) \subset A(S_1) + B(I-D)^{-1}C(S_1) \cdot B'(I-D)^{-1}C(S_1),$$

this implies that $T(S_1)$ is a bounded set in X. By using the sub-linearity of β, we obtain

$$\begin{aligned}
\beta(T(S_1)) &\leq \beta(A(S_1)) + \beta(\Pi(S_1) \cdot \Pi'(S_1)) \\
&\leq \beta(A(S_1)) + \beta(\Pi(S_1))\, \|\Pi'(S_1)\|_{\mathcal{P}} + \beta(\Pi'(S_1))\, \|\Pi(S_1)\|_{\mathcal{P}} \\
&\leq \alpha(A(S_1)) + \alpha(\Pi(S_1))\, \|\Pi'(S_1)\|_{\mathcal{P}} + \alpha(\Pi'(S_1))\, \|\Pi(S_1)\|_{\mathcal{P}} \\
&\leq (\psi_1 + M\psi_2)(\alpha(S_1)) \\
&\leq 2(\psi_1 + M\psi_2)(\beta(S_1)).
\end{aligned}$$

This demonstrates that T is a nonlinear $\mathcal{D}$-set-contraction multi-valued mapping on S into itself. Applying Theorem 4.5.6, we deduce that T has, at least, a fixed point. Q.E.D.

A special case of Theorem 4.5.7, useful in applications to differential and integral inclusions (see Chapter 7), is introduced in the following theorem.

Theorem 4.5.8 *Let S be a closed, convex, and bounded subset of the Banach algebra X, and let S' be a closed, convex, and bounded subset of the Banach space Y. Let $A : S \longrightarrow X$, $B : S' \longrightarrow X$, $C : S \longrightarrow Y$, $D : S' \longrightarrow Y$ and $B' : S' \longrightarrow \mathcal{P}_{cp,\ cv}(X)$ be five multi-valued operators satisfying:*

(i) A, B, and C are Lipschitzian with the constants q_1, q_2, and q_3, respectively,

(ii) D is a contraction with a constant k and $C(S) \subseteq (I-D)(S')$,

(iii) B' is an upper semi-continuous and compact operator, and

(iv) $Ax + B(I-D)^{-1}Cx \cdot B'(I-D)^{-1}Cx \subseteq S$, for all $x \in S$.

Then, the multi-valued block operator matrix (4.6) *has, at least, a fixed point whenever $\frac{q_2 q_3}{1-k}M + q_1 < \frac{1}{2}$, where $M = \| \cup B'(I-D)^{-1}C(S)\|_{\mathcal{P}}$.*

Proof. Clearly, every single-valued Lipschitzian mapping with a constant k is multi-valued Lipschitzian mapping with the constant $2k$ (see [98]). Once again,

$$Ax + B(I-D)^{-1}Cx \cdot B'(I-D)^{-1}Cx$$

is a convex and closed subset of X for each $x \in S$, and when we apply Theorem 4.5.7, we reach the desired conclusion. Q.E.D.

Part II

Applications in Mathematical Physics and Biology

Chapter 5

Existence of Solutions for Transport Equations

This chapter deals with some open problems carried from [99, 115, 116, 117, 118] concerning the existence of solutions on L_1 spaces to nonlinear boundary value problems derived from three models. The first one deals with nonlinear one-dimensional stationary transport equations arising in the kinetic theory of gas where we must describe the interaction of gas molecules with solid walls bounding the region where the gas follows. The second one, introduced by J. L. Lebowitz and S. I. Rubinow [121] in 1974, models microbial populations by age and cycle length formalism. The third one, introduced by M. Rotenberg [142] in 1983, describes the growth of a cell population. These three models can be transformed into a fixed point problem which has two types of equations. The first type involves a nonlinear weakly compact operator on L_1 spaces. The second type deals with two nonlinear operators depending on the parameter λ, say, $\psi = A_1(\lambda)\psi + A_2(\lambda)\psi$ where $A_1(\lambda)$ is a weakly compact operator on L_1 spaces and $A_2(\lambda)$ is a (strict) contraction mapping for a large enough $Re\lambda$. Consequently, Schauder's (resp. Krasnosel'skii's) fixed point theorem [149] cannot be used in the first (resp. second) type of equation. This is essentially due to the lack of compactness.

5.1 Transport Equations in the Kinetic Theory of Gas

5.1.1 Leakage of energy at the boundary of the slab

The first purpose of this section is to give some existence results for the stationary model presented in [14] on L_1 spaces :

$$v_3 \frac{\partial \psi}{\partial x} - \lambda \psi(x, v) + \mathcal{V}(x, v, \psi(x, v)) = \int_K r(x, v, v', \psi(x, v'))\, dv' \text{ in } D, \quad (5.1)$$

where $D = (0, 1) \times K$, K is the unit sphere of $\mathbb{R}^3$, $x \in (0, 1)$, $v = (v_1, v_2, v_3) \in K$, $r(., ., ., .)$ and $\mathcal{V}(., ., .)$ are nonlinear functions of ψ, and λ is a complex number. The main point in this equation is the nonlinear dependence of the functions $r(x, v, v', \psi(x, v'))$ on ψ. This equation describes the asymptotic behavior of the energy distribution inside the channel in the variables x and v. The unknown of this equation is a scalar function $\psi(x, v)$ which represents the energy density. The boundary conditions are modeled by:

$$\psi_{|D^i} = H(\psi_{|D^0}), \quad (5.2)$$

where D^i (resp. D^0) is the incoming (resp. outgoing) part of the space boundary and which is given by:

$$\begin{aligned} D^i = &\ D_1^i \cup D_2^i = &\ \{0\} \times K^1 \cup \{1\} \times K^0, \\ D^0 = &\ D_1^0 \cup D_2^0 = &\ \{0\} \times K^0 \cup \{1\} \times K^1, \end{aligned}$$

for

$$K^0 = K \cap \{v_3 < 0\},\ K^1 = K \cap \{v_3 > 0\}.$$

We will treat the problem (5.1)–(5.2) in the following functional setting. Let

$$X = L_1(D, dxdv),$$

$$\begin{aligned} X^i &:= L_1(D^i, |v_3|dv) \sim L_1(D_1^i, |v_3|dv) \oplus L_1(D_2^i, |v_3|dv) \\ &:= X_1^i \oplus X_2^i \end{aligned}$$

endowed with the norm :

$$\begin{aligned} \|\psi^i, X^i\| &= (\|\psi_1^i, X_1^i\| + \|\psi_2^i, X_2^i\|) \\ &= \left[\int_{K^1} |\psi(0, v)||v_3|\, dv + \int_{K^0} |\psi(1, v)||v_3|\, dv\right] \end{aligned}$$

and

$$\begin{aligned} X^0 &:= L_1(D^0, |v_3|dv) \sim L_1(D_1^0, |v_3|dv) \oplus L_1(D_2^0, |v_3|dv) \\ &:= X_1^0 \oplus X_2^0 \end{aligned}$$

endowed with the norm :

$$\begin{aligned} \|\psi^0, X^0\| &= (\|\psi_1^0, X_1^0\| + \|\psi_2^0, X_2^0\|) \\ &= \left[\int_{K^0} |\psi(0, v)||v_3|\, dv + \int_{K^1} |\psi(1, v)||v_3|\, dv\right], \end{aligned}$$

where $\sim$ means the natural identification of these spaces. Now, let us introduce the boundary operator H by:

$$\begin{cases} H : X_1^0 \oplus X_2^0 \longrightarrow X_1^i \oplus X_2^i \\ H\begin{pmatrix} u_1 \\ u_2 \end{pmatrix} = \begin{pmatrix} H_{11} & H_{12} \\ H_{21} & H_{22} \end{pmatrix}\begin{pmatrix} u_1 \\ u_2 \end{pmatrix} \end{cases}$$

for $j, k \in \{1, 2\}$, $H_{jk} : X_k^0 \longrightarrow X_j^i$, $H_{jk} \in \mathcal{L}(X_k^0, X_j^i)$, defined such that, on natural identification, the boundary conditions can be written as $\psi^i = H(\psi^0)$.

5.1.2 Case where $\mathcal{V}(x, v, \psi(x, v)) = \sigma(x, v)\psi(x, v)$

Let us define the streaming operator T_H with a domain including the following boundary conditions:

$$\begin{cases} T_H : \mathcal{D}(T_H) \subset X \longrightarrow X \\ \qquad\qquad \psi \longrightarrow T_H\psi(x, v) = v_3 \dfrac{\partial \psi}{\partial x}(x, v) + \sigma(x, v)\psi(x, v) \\ \mathcal{D}(T_H) = \left\{\psi \in X, v_3 \dfrac{\partial \psi}{\partial x} \in X, \psi^i \in X^i, \psi^0 \in X^0 \text{ and } \psi^i = H(\psi^0)\right\}, \end{cases}$$

where $\psi^0 = \psi_{|D^0} = (\psi_1^0, \psi_2^0)^\top$, $\psi^i = \psi_{|D^i} = (\psi_1^i, \psi_2^i)^\top$ and ψ_1^0, ψ_2^0, ψ_1^i, ψ_2^i are given by:

$$\begin{cases} \psi_1^i(v) = \psi(0, v), & v \in K^1 \\ \psi_2^i(v) = \psi(1, v), & v \in K^0 \\ \psi_1^0(v) = \psi(0, v), & v \in K^0 \\ \psi_2^0(v) = \psi(1, v), & v \in K^1. \end{cases}$$

Remark 5.1.1 *The derivative of ψ in the definition of T_H is meant in a distributional sense. Note that, if $\psi \in \mathcal{D}(T_H)$, then it is absolutely continuous with respect to x. Hence, the restrictions of ψ to D^i and D^0 are meaningful. Note also that $\mathcal{D}(T_H)$ is dense in X because it contains $C_0^\infty(\overset{o}{D})$.*

Let $\varphi \in X$, and consider the resolvent equation for T_H

$$(T_H - \lambda)\psi = \varphi, \tag{5.3}$$

where λ is a complex number and the unknown ψ must be sought in $\mathcal{D}(T_H)$. Let $\underline{\sigma}$ be the real defined by:

$$\underline{\sigma} := \text{ess-}\inf\{\sigma(x,v), (x,v) \in D\}.$$

Thus, for $Re\lambda < \underline{\sigma}$, the solution of Eq. (5.3) is formally given by:

$$\psi(x,v) = \psi(0,v)\, e^{-\int_0^x \frac{\sigma(s,v)-\lambda}{|v_3|}\,ds} + \frac{1}{|v_3|}\int_0^x e^{-\int_{x'}^x \frac{\sigma(s,v)-\lambda}{|v_3|}\,ds}\,\varphi(x',v)\,dx', v \in K^1 \tag{5.4}$$

$$\psi(x,v) = \psi(1,v)\, e^{-\int_x^1 \frac{\sigma(s,v)-\lambda}{|v_3|}\,ds} + \frac{1}{|v_3|}\int_x^1 e^{-\int_{x'}^x \frac{\sigma(s,v)-\lambda}{|v_3|}\,ds}\,\varphi(x',v)\,dx', v \in K^0 \tag{5.5}$$

whereas, $\psi(1,v)$ and $\psi(0,v)$ are given by:

$$\psi(1,v) = \psi(0,v)\, e^{-\int_0^1 \frac{\sigma(s,v)-\lambda}{|v_3|}\,ds} + \frac{1}{|v_3|}\int_0^1 e^{-\int_{x'}^1 \frac{\sigma(s,v)-\lambda}{|v_3|}\,ds}\,\varphi(x',v)\,dx', v \in K^1 \tag{5.6}$$

$$\psi(0,v) = \psi(1,v)\, e^{-\int_0^1 \frac{\sigma(s,v)-\lambda}{|v_3|}\,ds} + \frac{1}{|v_3|}\int_0^1 e^{-\int_0^{x'} \frac{\sigma(s,v)-\lambda}{|v_3|}\,ds}\,\varphi(x',v)\,dx', v \in K^0. \tag{5.7}$$

In order to allow the abstract formulation of Eqs. (5.4)–(5.7), let us define the following operators depending on the parameter λ :

$$\begin{cases} M_\lambda : X^i \longrightarrow X^0, M_\lambda u := (M_\lambda^+ u, M_\lambda^- u), & with \\ (M_\lambda^+ u(0,v) := u(1,v)\, e^{-\int_0^1 \frac{\sigma(s,v)-\lambda}{|v_3|}\,ds}, & v \in K^0; \\ (M_\lambda^- u(1,v) := u(0,v)\, e^{-\int_0^1 \frac{\sigma(s,v)-\lambda}{|v_3|}\,ds}, & v \in K^1; \end{cases}$$

$$\begin{cases} B_\lambda : X^i \longrightarrow X, B_\lambda u := \chi_{K^0}(v) B_\lambda^+ u + \chi_{K^1}(v) B_\lambda^- u, \ with \\ (B_\lambda^- u)(x,v) := u(0,v)\, e^{-\int_0^x \frac{\sigma(s,v)-\lambda}{|v_3|}\,ds}, \qquad v \in K^1; \\ (B_\lambda^+ u)(x,v) := u(1,v)\, e^{-\int_x^1 \frac{\sigma(s,v)-\lambda}{|v_3|}\,ds}, \qquad v \in K^0; \end{cases}$$

$$\begin{cases} G_\lambda : X \longrightarrow X^0, G_\lambda u := (G_\lambda^+ \varphi, G_\lambda^- \varphi), & with \\ G_\lambda^- \varphi := \dfrac{1}{|v_3|}\displaystyle\int_0^1 e^{-\int_x^1 \frac{\sigma(s,v)-\lambda}{|v_3|}\,ds}\,\varphi(x,v)\,dx, & v \in K^1; \\ G_\lambda^+ \varphi := \dfrac{1}{|v_3|}\displaystyle\int_0^1 e^{-\int_0^x \frac{\sigma(s,v)-\lambda}{|v_3|}\,ds}\,\varphi(x,v)\,dx, & v \in K^0; \end{cases}$$

and

$$\begin{cases} C_\lambda : X \longrightarrow X, C_\lambda\varphi := \chi_{K^0}(v)C_\lambda^+\varphi + \chi_{K^1}(v)C_\lambda^-\varphi, \ \textit{with} \\ C_\lambda^-\varphi := \dfrac{1}{|v_3|}\displaystyle\int_0^x e^{-\int_{x'}^x \frac{\sigma(s,v)-\lambda}{|v_3|}ds}\varphi(x',v)\,dx', \quad v \in K^1; \\ C_\lambda^+\varphi := \dfrac{1}{|v_3|}\displaystyle\int_x^1 e^{-\int_{x'}^x \frac{\sigma(s,v)-\lambda}{|v_3|}ds}\varphi(x',v)\,dx', \quad v \in K^0, \end{cases}$$

where $\chi_{K^0}(.)$ and $\chi_{K^1}(.)$ denote, respectively, the characteristic functions of the sets K^0 and K^1. Let λ_0 denote the real defined by:

$$\lambda_0 := \begin{cases} \underline{\sigma}, & \text{if } \|H\| \le 1 \\ \underline{\sigma} - \log(\|H\|), & \text{if } \|H\| > 1. \end{cases}$$

A simple calculation shows that the above operators are bounded on their respective spaces. In fact, for $Re\lambda < \underline{\sigma}$, the norms of the operators M_λ, B_λ, C_λ, and G_λ are bounded above, respectively, by $e^{Re\lambda-\underline{\sigma}}$, $\frac{1}{\underline{\sigma}-Re\lambda}$, $\frac{1}{\underline{\sigma}-Re\lambda}$ and 1. By using these operators and the fact that ψ must satisfy the boundary conditions, we deduce that Eqs. (5.6) and (5.7) are written in the space X^0 in the operator form, as follows:

$$\psi^0 = M_\lambda H\psi^0 + G_\lambda\varphi.$$

The solution of this equation reduces to the invertibility of the operator $\mathcal{U}(\lambda) := I - M_\lambda H$ (which is the case if $Re\lambda < \lambda_0$, see the norm estimate of $M_\lambda H$). This gives

$$\psi^0 = \{\mathcal{U}(\lambda)\}^{-1}G_\lambda\varphi = \sum_{n\ge 0}(M_\lambda H)^n G_\lambda\varphi. \tag{5.8}$$

Moreover, Eq. (5.5) can be rewritten as

$$\psi = B_\lambda H\psi^0 + C_\lambda\varphi.$$

By substituting Eq. (5.8) into the above equation, we get

$$\psi = \sum_{n\ge 0} B_\lambda H(M_\lambda H)^n G_\lambda\varphi + C_\lambda\varphi.$$

Proposition 5.1.1 *$\{\lambda \in \mathbb{C}$ such that $Re\lambda < \lambda_0\} \subset \varrho(T_H)$, and for $Re\lambda < \lambda_0$, we have*

$$(T_H - \lambda)^{-1} = \sum_{n\ge 0} B_\lambda H(M_\lambda H)^n G_\lambda + C_\lambda, \tag{5.9}$$

where $\varrho(T_H)$ denotes the resolvent set of T_H.

For our subsequent analysis, we need the following hypothesis

$$(\mathcal{B}_1) \qquad r(x,v,v',\psi(x,v')) = \xi(x,v,v')f(x,v',\psi(x,v')),$$

where f is a measurable function defined by:

$$\begin{cases} f:[0,1]\times K\times\mathbb{C}\longrightarrow\mathbb{C} \\ \qquad\qquad (x,v,u)\longrightarrow f(x,v,u). \end{cases}$$

The function $\xi(.,.,.)$ is measurable from $[0,1]\times K\times K$ into $\mathbb{R}$. It defines a linear operator F by:

$$\begin{cases} F:X\longrightarrow X \\ \quad \psi\longrightarrow \displaystyle\int_K \xi(x,v,v')\psi(x,v')\,dv'\,. \end{cases} \tag{5.10}$$

Let us notice that the operator F acts only on the variables v', so x may be viewed merely as a parameter in $[0,1]$. Hence, we may consider F as a function

$$F(.) : x\in[0,1]\longrightarrow F(x)\in Z,$$

where $Z=\mathcal{L}(L_1(K;\,dv))$ denotes the set of all bounded linear operators on $L_1(K;\,dv)$. In the following, we will make the assumptions:

$$(\mathcal{B}_2) \qquad \begin{cases} \text{– the function } F(.) \text{ is strongly measurable,} \\ \text{– there exists a compact subset } \mathcal{C}\subset\mathcal{L}(L_1(K;\,dv)) \text{ such that:} \\ \quad F(x)\in\mathcal{C} \text{ a.e. on } [0,1], \\ \text{– } F(x)\in\mathcal{K}(L_1(K;\,dv)) \text{ a.e.,} \end{cases}$$

where $\mathcal{K}(L_1(K;\,dv))$ denotes the set of all compact operators on $L_1(K;\,dv)$. Obviously, the second assumption of $(\mathcal{B}_2)$ implies that

$$F(.)\in L_\infty((0,1),Z). \tag{5.11}$$

Let $\psi\in X$. It is easy to show that $(F\psi)(x,v)=F(x)\psi(x,v)$ and then, by using (5.11), we have

$$\int_K |(F\psi)(x,v)|\,dv \le \|F(.)\|_{L_\infty((0,1),Z)}\int_K |\psi(x,v)|\,dv$$

and therefore,

$$\int_0^1 \int_K |(F\psi)(x,v)| \, dvdx \leq \|F(.)\|_{L_\infty((0,1),Z)} \int_0^1 \int_K |\psi(x,v)| \, dvdx.$$

This leads to the estimate

$$\|F(.)\|_{\mathcal{L}(X)} \leq \|F(.)\|_{L_\infty((0,1),Z)}. \tag{5.12}$$

The interest of the operators in the forms which satisfy $(\mathcal{B}_2)$ lies in the following lemma.

Lemma 5.1.1 *Let F be a linear operator given by (5.10) and assume that $(\mathcal{B}_2)$ holds. Then, F can be approximated, in the uniform topology, by a sequence $(F_n)_n$ of linear operators with kernels of the form*

$$\sum_{i=1}^{n} \eta_i(x)\theta_i(v)\beta_i(v'),$$

where $\eta_i \in L_\infty([0,1];dx), \theta_i \in L_1(K;dv)$ and $\beta_i \in L_\infty(K;dv)$.

Proof. Let $\mathcal{C}^* = \mathcal{C} \cap \mathcal{K}(L_1(K;dv))$. By using the second and third assumptions of $(\mathcal{B}_2)$, $\mathcal{C}^*$ is a nonempty and closed subset of $\mathcal{C}$. Then, $\mathcal{C}^*$ is a compact set of Z. Let $\varepsilon > 0$, there exist $F_1, \ldots, F_m$ such that $(F_i)_i \subset \mathcal{C}^*$ and

$$\mathcal{C} \subset \underset{1 \leq i \leq m}{\cup} B(F_i, \varepsilon),$$

where $B(F_i, \varepsilon)$ is the open ball in Z, centered at F_i and with a radius ε. Let $C_1 = B(F_1, \varepsilon)$, and for $m \geq 2$, $C_m = B(F_m, \varepsilon) - \bigcup_{1 \leq i \leq m-1} C_i$. Clearly, $C_i \cap C_j = \emptyset$ if $i \neq j$ and $\mathcal{C}^* \subset \bigcup_{1 \leq i \leq m} C_i$. Let $1 \leq i \leq m$ and let us denote by I_i the set:

$$F^{-1}(C_i) = \{x \in (0,1) \text{ such that } F(x) \in C_i\}.$$

Hence, we have $I_i \cap I_j = \emptyset$ if $i \neq j$ and $(0,1) = \bigcup_{1 \leq i \leq m} I_i$. Now, let us consider the following step function from $(0,1)$ to Z defined by:

$$S(x) = \sum_{i=1}^{m} \chi_{I_i}(x) F_i,$$

where $\chi_{I_i}(.)$ denotes the characteristic function of I_i. Obviously, $S(.)$ satisfies the hypothesis $(\mathcal{B}_2)$. Then, using (5.10), we get $F - S \in L_\infty((0,1),Z)$. Moreover, an easy calculation leads to

$$\|F - S\|_{L_\infty((0,1),Z)} \leq \varepsilon.$$

Now, using Inequation (5.12), we obtain

$$\|F - S\|_{\mathcal{L}(X)} \leq \|F - S\|_{L_\infty((0,1),Z)} \leq \varepsilon\,.$$

Hence, we infer that the operator F may be approximated (in the uniform topology) by operators of the form

$$U(x) = \sum_{i=1}^{n} \eta_i(x) F_i,$$

where $\eta_i \in L_\infty((0,1); dx)$ and $F_i \in \mathcal{K}(L_1(K;\, dv))$. Moreover, each compact operator F_i is a limit (for the norm topology) of a sequence of finite rank operators because $L_1(K;\, dv)$ admits a Schauder's basis. Q.E.D.

The following lemma is fundamental for the sequel.

Lemma 5.1.2 *Assume that* $(\mathcal{B}_2)$ *holds. Then, for any* $\lambda \in \mathbb{C}$ *satisfying* $Re\lambda < \lambda_0$*, the operator* $(T_H - \lambda)^{-1}F$ *is weakly compact on* X.

Proof. Let λ be such that $Re\lambda < \lambda_0$. In view of Eq. (5.9), we have

$$(T_H - \lambda)^{-1} F = \sum_{n \geq 0} B_\lambda H (M_\lambda H)^n G_\lambda F + C_\lambda F\,.$$

In order to conclude, it is sufficient to show that $\sum_{n\geq 0} B_\lambda H(M_\lambda H)^n G_\lambda F$ and $C_\lambda F$ are weakly compact on X. We claim that $G_\lambda F$ and $C_\lambda F$ are weakly compact on X. By using Lemma 5.1.1, we only need to prove the result for an operator whose kernel is in the form:

$$\xi(x, v, v') = \eta(x)\theta(v)\beta(v'),$$

where $\eta \in L_\infty([0,1];\, dx), \theta \in L_1(K;\, dv), \beta \in L_\infty(K;\, dv)$. Let $\varphi \in X$. Then, for $v \in K_1$ we have

$$\begin{aligned}(G_\lambda^- F\varphi)(v) &= \int_K \int_0^1 \frac{1}{|v_3|} \eta(x)\theta(v) e^{-\int_x^1 \frac{\sigma(s,v)-\lambda}{|v_3|}\, ds} \beta(v')\, \varphi(x, v')\, dx dv' \\ &= J_\lambda U_\lambda \varphi,\end{aligned}$$

where U_λ and J_λ denote the following bounded operators

$$\begin{cases} U_\lambda : X \longrightarrow L_1([0,1];\, dx) \\ \qquad \varphi \longrightarrow \displaystyle\int_K \beta(v)\, \varphi(x, v)\, dv \end{cases}$$

and

$$\begin{cases} J_\lambda : L_1([0,1]; dx) \longrightarrow X_2^0 \\ \qquad \psi \longrightarrow \displaystyle\int_0^1 \frac{1}{|v_3|}\eta(x)\theta(v)e^{-\int_x^1 \frac{\sigma(s,v)-\lambda}{|v_3|}\,ds}\,\psi(x)\,dx\,. \end{cases}$$

Now, it is sufficient to show that J_λ is weakly compact. To do so, let $\mathcal{O}$ be a bounded set of $L_1([0,1]; dx)$, and let $\psi \in \mathcal{O}$. We have

$$\int_E |J_\lambda\psi(v)|\,|v_3|\,dv \le \|\eta\|\,\|\psi\| \int_E |\theta(v)|\,dv,$$

for all measurable subsets E of K^1. Next, by applying Theorem 1.3.7, we infer that the set $J_\lambda(\mathcal{O})$ is weakly compact, since $\lim\limits_{|E|\to 0}\displaystyle\int_E |\theta(v)|\,dv = 0$, $(\theta \in L_1(K; dv))$, where $|E|$ is the measure of E. A similar reasoning allows us to reach the same result for the operators $G_\lambda^+ F$ and $C_\lambda F$. Q.E.D.

Now, let us recall some facts concerning the superposition operators required below. A function $f : D \times \mathbb{C} \longrightarrow \mathbb{C}$ is said to satisfy the Carathéodory conditions on $D \times \mathbb{C}$, if

$$\begin{cases} (x,v) \longrightarrow f(x,v,u) \text{ is measurable on } D \text{ for all } u \in \mathbb{C}, \\ \qquad u \longrightarrow f(x,v,u) \text{ is continuous on } \mathbb{C} \text{ a.e. } (x,v) \in D. \end{cases}$$

If f satisfies the Carathéodory conditions, then we can define the operator $\mathcal{N}_f$ on the set of functions $\{\,\psi : D \longrightarrow \mathbb{C}\,\}$, by:

$$(\mathcal{N}_f\psi)(x,v) = f(x,v,\psi(x,v))$$

for every $(x,v) \in D$. The operator $\mathcal{N}_f$ is said to be the Nemytskii's operator generated by f. In L_p spaces, the Nemytskii's operator has been extensively investigated (see, for example, [110] and the bibliography therein). However, it is useful to recall the following result which states a basic fact for these operators in L_p spaces.

Proposition 5.1.2 *[110]. If $\mathcal{N}_f : L_{p_1} \longrightarrow L_{p_2}, p_1, p_2 \ge 1$, then $\mathcal{N}_f$ is continuous and takes bounded sets into bounded sets. Moreover, if $p_2 < \infty$, then there exist a constant $b > 0$ and a positive function $a \in L_{p_2}$ such that*

$$|f(x,y)| \le a(x) + b|y|^{\frac{p_1}{p_2}} \text{ a.e in } x, \text{ for all } y \in \mathbb{R}.$$

We will also assume that

$$(\mathcal{B}_3)\quad \begin{cases} f \text{ satisfies the Carathéodory conditions and there exist a constant} \\ k > 0 \text{ and a positive function } h \in X \text{ such that :} \\ |f(x, v, \psi(x, v))| \leq h(x, v) + k|\psi(x, v)| \text{ a.e. in } x, \text{ for all } \psi \in X. \end{cases}$$

We note that $(\mathcal{B}_3)$ implies that $\mathcal{N}_f$ acts from L_1 into L_1. Then, by using Proposition 5.1.2, we deduce that $\mathcal{N}_f$ is continuous and takes bounded sets into bounded sets. Let r be a positive real and set

$$B_r = \{\psi \in X \text{ such that } \|\psi\|_X \leq r\}.$$

Theorem 5.1.1 *Let $(\mathcal{B}_1)$–$(\mathcal{B}_3)$ be satisfied. Then, for each $r > 0$, there exists a real $\lambda_1 < 0$ such that, for all λ satisfying $Re\lambda < \lambda_1$, the boundary value problem*

$$\begin{cases} v_3 \dfrac{\partial \psi}{\partial x} + \sigma(x, v)\psi(x, v) - \lambda\psi(x, v) = \displaystyle\int_K \xi(x, v, v')f(x, v', \psi(x, v'))\, dv' \text{ in } D, \\ \psi^i = H(\psi^0), \lambda \in \mathbb{C} \end{cases} \tag{5.13}$$

has, at least, one solution on B_r.

Proof. Let λ be a complex number such that $Re\lambda < \lambda_0$. Then, according to Eq. (5.9), $(T_H - \lambda)$ is invertible and therefore, the problem may be transformed into

$$\begin{cases} \psi = \mathcal{F}(\lambda)\psi \\ \psi^i = H(\psi^0), \end{cases}$$

where $\mathcal{F}(\lambda)\psi = (T_H - \lambda)^{-1}F\mathcal{N}_f\psi$. First, let us check that, for a suitable λ, $\mathcal{F}(\lambda)$ is continuous and leaves B_r invariant. It is clear that $\mathcal{F}(\lambda)$ is continuous. By using Eq. (5.9), we have

$$(T_H - \lambda)^{-1} = \sum_{n \geq 0} B_\lambda H(M_\lambda H)^n G_\lambda + C_\lambda.$$

Then,

$$\begin{aligned}
\|(T_H-\lambda)^{-1}\| &\leq \sum_{n\geq 0}\|B_\lambda\|\|H\|\|M_\lambda H\|^n\|G_\lambda\|+\|C_\lambda\| \\
&\leq \|B_\lambda\|\|H\|\|G_\lambda\|\sum_{n\geq 0}\|M_\lambda H\|^n+\|C_\lambda\| \\
&\leq \frac{\|H\|}{\underline{\sigma}-Re\lambda}\quad\frac{1}{1-\|M_\lambda H\|}+\frac{1}{\underline{\sigma}-Re\lambda} \\
&\leq \frac{\|H\|}{\underline{\sigma}-Re\lambda}\quad\frac{1}{1-\|M_\lambda\|\|H\|}+\frac{1}{\underline{\sigma}-Re\lambda} \\
&\leq \frac{1}{\underline{\sigma}-Re\lambda}\left(\frac{\|H\|}{1-e^{Re\lambda-\underline{\sigma}}\|H\|}+1\right). \qquad (5.14)
\end{aligned}$$

Let $\psi \in B_r$. From inequality (5.14) and $(\mathcal{B}_3)$, it follows that

$$\begin{aligned}
\|\mathcal{F}(\lambda)\psi\| &\leq \frac{\|F\|}{\underline{\sigma}-Re\lambda}\left(\frac{\|H\|}{1-e^{Re\lambda-\underline{\sigma}}\|H\|}+1\right)\|\mathcal{N}_f\psi\| \\
&\leq \frac{\|F\|}{\underline{\sigma}-Re\lambda}\left(\frac{\|H\|}{1-e^{Re\lambda-\underline{\sigma}}\|H\|}+1\right)(\|h\|+kr).
\end{aligned}$$

Let $\varepsilon < \min(0, \lambda_0)$. For $Re\lambda < \varepsilon$, we have

$$\frac{\|H\|}{1-e^{Re\lambda-\underline{\sigma}}\|H\|} < \frac{\|H\|}{1-e^{\varepsilon-\underline{\sigma}}\|H\|}$$

and therefore,

$$\begin{aligned}
\|\mathcal{F}(\lambda)\psi\| &\leq \frac{1}{\underline{\sigma}-Re\lambda}\left(\frac{\|H\|}{1-e^{\varepsilon-\underline{\sigma}}\|H\|}+1\right)\|F\|(\|h\|+kr) \\
&= Q(Re\lambda),
\end{aligned}$$

where

$$Q(t) = \frac{1}{\underline{\sigma}-t}\left(\frac{\|H\|}{1-e^{\varepsilon-\underline{\sigma}}\|H\|}+1\right)\|F\|(\|h\|+kr).$$

Clearly, $Q(.)$ is continuous, strictly increasing in t, $t < 0$, and satisfies $\lim_{t\to-\infty} Q(t) = 0$. Hence, there exists λ_1 such that $Q(\lambda_1) \leq r$ and therefore, $\|\mathcal{F}(\lambda)\psi\| \leq r$ for $Re\lambda \leq \lambda_1$. This shows that $\mathcal{F}_\lambda$ leaves B_r invariant. Now, let $\mathcal{O}$ be a weakly compact subset of X, and we claim that $\mathcal{N}_f(\mathcal{O})$ is weakly compact in X. To see this, let $(\psi_n)_n$ be a sequence in $\mathcal{O}$. Then, $(\psi_n)_n$ has a weak converging subsequence $(\psi_{n_p})_p$. In particular, the set $\Omega = \{(\psi_{n_p})_p\}$ is weakly compact in X and consequently, $\{(|\psi_{n_p}|)_p\}$ is weakly compact in X (use Theorem 1.3.6). Next, by applying Theorem 1.3.7, we notice that

$$\lim_{|E|\to 0} \int_E |\psi(x,v)|\, dxdv = 0$$

uniformly for ψ in Ω. From $(\mathcal{B}_3)$, we deduce that

$$\int_E |(\mathcal{N}_f\psi)(x,v)|\, dxdv \leq \int_E h(x,v)\, dxdv + k\int_E |\psi(x,v)|\, dxdv.$$

Hence,

$$\lim_{|E|\to 0} \int_E |(\mathcal{N}_f\psi)(x,v)|\, dxdv = 0,$$

uniformly for ψ in Ω, since

$$\lim_{|E|\to 0} \int_E h(x,v)\, dxdv = 0.$$

Now, by using Theorem 1.3.7, we may conclude that the family $(\mathcal{N}_f\psi_{n_p})$ is weakly compact in X. Thus, $(\mathcal{N}_f\psi_{n_p})_p$ is weakly convergent and $\mathcal{N}_f(\mathcal{O})$ is weakly compact. Finally, the use of Theorem 2.1.1 and Corollary 2.1.2 for the bounded closed convex subset B_r, shows that $\mathcal{F}(\lambda)$ has, at least, one fixed point in $B_r \cap \mathcal{D}(T_H)$. Q.E.D.

5.1.3 Positive solutions of the boundary value problem

In the following, we will focus our attention on the existence of positive solutions to the boundary value problem (5.13). Let us notice that our functional spaces X, D^i, and D^0 are Banach lattice spaces. Their positive cônes will be denoted, respectively, by X^+, D^{i+}, and D^{0+}.

Proposition 5.1.3 *Let $(\mathcal{B}_1)$–$(\mathcal{B}_3)$ be satisfied. If the operators H, F, and $\mathcal{N}_f$ are positive, then for each $r > 0$, the boundary value problem (5.13) has, at least, one solution in B_r^+.*

Proof. Let λ be a real such that $\lambda < \min(0, \underline{\sigma})$. Using the fact that the operators $M_\lambda, B_\lambda, G_\lambda$, C_λ, and F are positive and Eq. (5.8), we deduce that $(T_H - \lambda)^{-1}F$ is positive. Now, the remaining part of the proof is similar to that of Theorem 5.1.1; it suffices to replace the set $\mathcal{N} := \overline{co}(\mathcal{F}(\lambda)(B_r))$ by $\mathcal{N}^+ := \overline{co}(\mathcal{F}(\lambda)(B_r^+))$. Q.E.D.

Proposition 5.1.4 *Let $(\mathcal{B}_1)$–$(\mathcal{B}_3)$ be satisfied and assume that H and F are positive. If there exist $\tau > 0$ and $0 \neq \psi_0 \in B_r^+$ such that :*

(i) $\psi_0 \notin N(F)$ where $N(F)$ is the null space of F,

(ii) $(\mathcal{N}_f\psi)(x,v) \geq \tau\psi_0(x,v)$ for all $\psi \in B_r^+$.

Then, there exists $\lambda_1 < 0$ such that, for any $\lambda < \lambda_1$, there is $\eta > 0$ such that the boundary value problem

$$\begin{cases} v_3\dfrac{\partial \psi}{\partial x}(x,v) + \sigma(x,v)\psi(x,v) - \lambda\psi(x,v) = \eta \displaystyle\int_K \xi(x,v,v')f(x,v',\psi(x,v'))\,dv', \\ \text{for all } (x,v) \in D, \\ \psi^i = H(\psi^0) \end{cases}$$

has, at least, one solution $\psi^* \in B_r^+$ satisfying $\|\psi^*\| = r$.

Proof. Proceeding as in the proof of Theorem 5.1.1 and Proposition 5.1.2, there is a constant $\lambda_1 < 0$, such that for all $\lambda \leq \lambda_1$, the operator $\mathcal{F}(\lambda)$ maps B_r^+ into itself. We first claim that

$$\inf\{\|\mathcal{F}(\lambda)\psi\|, \psi \in B_r^+\} > 0.$$

Indeed, since $\mathcal{N}_f$ satisfies the assumption (ii), it follows from (5.9) and the positivity of $\sum_{n\geq 0} B_\lambda H(M_\lambda H)^n G_\lambda$, that

$$\mathcal{F}(\lambda)(\psi) \geq \tau(C_\lambda F\psi_0) \text{ for all } \psi \in B_r^+.$$

Using assumption (i), one sees that $F\psi_0 \geq 0$ and $F\psi_0 \neq 0$. Since $\lambda \in \mathbb{R}$, then it follows, from the positivity of C_λ and the fact that it is the resolvent of the linear operator T_0 (i.e., $H = 0$), that

$$\mathcal{F}(\lambda)\psi \geq \tau(C_\lambda F\psi_0) \geq 0 \text{ and } C_\lambda F\psi_0 \neq 0.$$

Hence,

$$\|\mathcal{F}(\lambda)(\psi)\| \geq \tau\|C_\lambda F\psi_0\| \text{ for all } \psi \in B_r^+.$$

This proves the claim. Consequently, for each $\lambda \leq \lambda_1$, we define the operator $\mathcal{G}(\lambda)$ on B_r^+ by:

$$\mathcal{G}(\lambda)(\psi) = r\frac{\mathcal{F}(\lambda)(\psi)}{\|\mathcal{F}(\lambda)(\psi)\|} \text{ for all } \psi \in B_r^+.$$

Next, let us prove that $\mathcal{G}(\lambda)$ is a weakly compact operator on B_r^+. Indeed, set $m = \inf\{\|\mathcal{F}(\lambda)\psi\|, \psi \in B_r^+\} > 0$. Then,

$$0 \leq \mathcal{G}(\lambda)(\psi) \leq \frac{r}{m}\mathcal{F}(\lambda)(\psi) \text{ for all } \psi \in B_r^+.$$

Since B_r^+ is a bounded convex subset of X^+, then $\mathcal{N}_f(B_r^+)$ is a bounded subset of X^+ and therefore, by using Lemma 5.1.2, we deduce that $\mathcal{F}(\lambda)(B_r^+)$ is weakly compact. Moreover, by using Theorem 1.3.7, we may conclude that

$$\lim_{|E|\to 0} \int_E (\mathcal{F}(\lambda)\psi)(x,v)\, dxdv = 0,$$

uniformly for ψ in B_r^+. Now, using the fact that

$$0 \leq \mathcal{G}(\lambda)(\psi) \leq \frac{r}{m}\mathcal{F}(\lambda)(\psi) \text{ for all } \psi \in B_r^+,$$

we find that

$$0 \leq \int_E (\mathcal{G}(\lambda)\psi)(x,v)\, dxdv \leq \frac{r}{m}\int_E (\mathcal{F}(\lambda)\psi(x,v))\, dxdv, \text{ for all } \psi \in B_r^+.$$

Thus,

$$\lim_{|E|\to 0} \int_E (\mathcal{G}(\lambda)\psi)(x,v)\, dxdv = 0,$$

uniformly for ψ in B_r^+, and $\mathcal{G}(\lambda)(B_r^+)$ is weakly compact. Finally, the use of Theorem 2.1.1 and Remark 2.1.2 for the bounded closed convex set B_r^+ shows that $\mathcal{G}(\lambda)$ has, at least, a fixed point ψ^* in B_r^* satisfying $\|\psi^*\| = r$. Setting $\eta = \frac{r}{\|\mathcal{F}(\lambda)(\psi^*)\|}$, we get

$$(T_H - \lambda)^{-1} F\mathcal{N}_f(\psi^*) = \eta^{-1}\psi^*.$$

Thus, $\psi^* \in \mathcal{D}(T_H) \cap B_r^+$, and

$$v_3 \frac{\partial \psi^*}{\partial x}(x,v) + \sigma(x,v)\psi^*(x,v) - \lambda\psi^*(x,v) = \eta \int_K \xi(x,v,v') f(x,v',\psi^*(x,v')\, dv'.$$

This completes the proof. Q.E.D.

5.1.4 Existence of solutions for a general nonlinear boundary value problem

Our main interest here is dealing with the existence of solutions for the more general nonlinear boundary value problem (5.1)–(5.2). We need the following hypothesis

$$(\mathcal{B}_4) \begin{cases} H \text{ is a bounded linear operator from } X^0 \text{ into } X^i \\ \text{and for each } r > 0, \mathcal{V}(.,.,.) \text{ satisfies:} \\ |\mathcal{V}(x,v,\psi_1(x,v)) - \mathcal{V}(x,v,\psi_2(x,v))| \leq |\rho(x,v)||\psi_1 - \psi_2| \text{ for every} \\ (\psi_1,\psi_2) \in B_r, \text{ where } \rho \in L_\infty(D) \text{ and } \mathcal{N}_{\mathcal{V}} \text{ acts from } B_r \text{ into } B_r. \end{cases}$$

Let us define the free streaming operator $\widehat{T}_H$ by:

$$\begin{cases} \mathcal{D}(\widehat{T}_H) \ni \psi \longrightarrow \widehat{T}_H\psi(x,v) = v_3 \dfrac{\partial \psi}{\partial x}(x,v), \\ \\ \mathcal{D}(\widehat{T}_H) = \left\{ \psi \in X, v_3 \dfrac{\partial \psi}{\partial x} \in X, \ \psi^i \in X^i, \ \psi^0 \in X^0 \text{ and } \psi^i = H(\psi^0) \right\}, \end{cases}$$

where $\psi_{|D^i} := \psi^i$ and $\psi_{|D^0} := \psi^0$. Since H is linear, $\widehat{T}_H$ is a closed densely defined linear operator. Moreover, easy calculations can show that the resolvent set of $\varrho(\widehat{T}_H)$ contains the half plane $\{Re\lambda < \lambda_0\}$, λ_0 is calculated by taking $\sigma = 0$ denoted by λ_{00} and defined by:

$$\lambda_{00} := \begin{cases} 0, & \text{if } \|H\| \leq 1 \\ \\ -\log(\|H\|), & \text{if } \|H\| > 1. \end{cases}$$

For any λ belonging to this half plane, we have

$$(\widehat{T}_H - \lambda)^{-1} = \sum_{n \geq 0} B^0_\lambda H (M^0_\lambda H)^n G^0_\lambda + C^0_\lambda,$$

where $B^0_\lambda, M^0_\lambda, G^0_\lambda$, and C^0_λ are the bounded linear operators derived from $B_\lambda, M_\lambda, G_\lambda$, and C_λ by taking $\sigma = 0$. Their norms are bounded above, respectively, by $\frac{-1}{Re\lambda}$, $e^{Re\lambda}$, 1, and $\frac{-1}{Re\lambda}$. These observations lead to the estimate

$$\|(\widehat{T}_H - \lambda)^{-1}\| \leq \frac{-1}{Re\lambda} \left\{ \frac{\|H\|}{1 - e^{Re\lambda}\|H\|} + 1 \right\} \tag{5.15}$$

for any λ in the half plane $Re\lambda < \lambda_{00}$. Let $\varepsilon < 0$ and let λ be such that $Re\lambda < \min(\varepsilon, \lambda_{00})$. Since $\lambda \in \varrho(\widehat{T}_H)$, we can consider the operators $\mathcal{F}(\lambda)\psi = (\widehat{T}_H - \lambda)^{-1} F \mathcal{N}_f$ and $\mathcal{H}(\lambda) = (\widehat{T}_H - \lambda)^{-1} \mathcal{N}_{-\mathcal{V}}$. Now, let us check that, for a suitable λ, the operator $\mathcal{H}(\lambda)$ is a contraction mapping on B_r. Indeed, let $\psi_1, \psi_2 \in B_r$. We have

$$\begin{aligned} \|\mathcal{H}(\lambda)\psi_1 - \mathcal{H}(\lambda)\psi_2\| &= \|(\widehat{T}_H - \lambda)^{-1}\mathcal{N}_{-\mathcal{V}}(\psi_1) - (\widehat{T}_H - \lambda)^{-1}\mathcal{N}_{-\mathcal{V}}(\psi_2)\| \\ &\leq -\frac{1 + \|H\|(1 - \|H\|e^{\varepsilon})^{-1}}{Re\lambda} \|\mathcal{N}_{-\mathcal{V}}(\psi_1) - \mathcal{N}_{-\mathcal{V}}(\psi_2)\|. \end{aligned}$$

Moreover, according to $(\mathcal{B}_4)$, we have

$$\begin{aligned} \|\mathcal{H}(\lambda)\psi_1 - \mathcal{H}(\lambda)\psi_2\| &\leq \|\rho\|_\infty \frac{1 + \|H\|(1 - \|H\|e^{\varepsilon})^{-1}}{Re\lambda} \|\psi_1 - \psi_2\| \\ &\leq \hat{Q}(Re\lambda)\|\psi_1 - \psi_2\|, \end{aligned} \tag{5.16}$$

where $\hat{Q}(t) = \|\rho\|_\infty \frac{1+\|H\|(1-\|H\|e^\varepsilon)^{-1}}{t}$. Therefore, the function $\hat{Q}$ has the same properties as Q defined in the proof of Theorem 5.1.1. Thus, there is $\lambda_2 < 0$ such that, for $Re\lambda < \lambda_2$, we have

$$\hat{Q}(Re\lambda) < 1. \tag{5.17}$$

Now, we need the following hypothesis.

$(\mathcal{B}_5)$ For λ such that $Re\lambda < \lambda_2, (I-\mathcal{H}(\lambda))^{-1}\mathcal{F}(\lambda)$ is a weakly compact operator.

Theorem 5.1.2 *Let $(\mathcal{B}_1)$–$(\mathcal{B}_5)$ be satisfied. Then, there exists $\lambda^* < 0$ such that, for each λ satisfying $Re\lambda < \lambda^*$, the boundary value problem (5.1)–(5.2) has, at least, one solution in B_r.*

Proof. Let $\varepsilon < 0$ and let λ be such that $Re\lambda < \min(\varepsilon, \lambda_{00})$. Since $\lambda \in \varrho(\widehat{T}_H)$, the problem (5.1)–(5.2) may be written in the form

$$\begin{cases} \psi = \mathcal{F}(\lambda)\psi + \mathcal{H}(\lambda)\psi, \\ \psi^i = H(\psi^0). \end{cases}$$

We will show that, for a suitable λ, $\mathcal{F}(\lambda)(B_r) + \mathcal{H}(\lambda)(B_r) \subset B_r$. Indeed, Let φ, $\psi \in B_r$. According to $(\mathcal{B}_4)$ and Inequation (5.15), we have

$$\begin{aligned}\|\mathcal{F}(\lambda)\varphi + \mathcal{H}(\lambda)\psi\| &\leq \frac{-1}{Re\lambda}\Theta(\|H\|, \varepsilon)\left\{r\left(\|\rho\|_\infty + k\|F\|\right) + \|\mathcal{N}_{-\mathcal{V}}(0)\| + \|F\|\|h\|\right\} \\ &:= Q^*(Re\lambda),\end{aligned}$$

where

$$\Theta(\|H\|, \varepsilon) := \left\{\frac{\|H\|}{1 - e^\varepsilon\|H\|} + 1\right\}.$$

Therefore, the function Q^* has the same properties as Q defined in the proof of Theorem 5.1.1. Thus, there is $\lambda_3 < 0$ such that, for $Re\lambda < \lambda_3$, $\mathcal{F}(\lambda)\varphi + \mathcal{H}(\lambda)\psi \in B_r$. Let $\lambda^* := \min(\varepsilon, \lambda_{00}, \lambda_2, \lambda_3)$. As a summary, the above steps show that, for any λ satisfying $Re\lambda < \lambda^*$, the operators $\mathcal{F}(\lambda)$ and $\mathcal{H}(\lambda)$ satisfy the hypotheses of Theorem 2.1.2 on the nonempty bounded, closed and convex subset B_r. Hence, the problem (5.1)–(5.2) has a solution in B_r. Q.E.D.

For our subsequent analysis, we need the hypothesis:

$(\mathcal{B}_6)$ $r(x, v, v', \psi(x, v')) = \kappa(x, v, v')f(x, v', L(\psi)(x, v'))$,

where

$$L : L_1([0,1] \times K) \longrightarrow L_\infty([0,1] \times K)$$

is a continuous linear map and f is a measurable function defined by:

$$\begin{cases} f : [0,1] \times K \times \mathbb{C} \longrightarrow \mathbb{C} \\ \qquad\qquad (x, v, u) \longrightarrow f(x, v, u). \end{cases}$$

The function $\kappa(.,.,.)$ is a measurable function from $[0,1] \times K \times K$ into $\mathbb{R}$. It defines a continuous linear operator F by:

$$\begin{cases} F : X \longrightarrow X \\ \qquad \psi \longrightarrow F\ \psi(x,v) = \displaystyle\int_K \kappa(x,v,v')\psi(x,v')\ dv'. \end{cases} \tag{5.18}$$

Note that

$$dx \otimes d\mu - \text{ess} \sup_{(x,v') \in [0,1] \times K} \int_K |\kappa(x,v,v')| dv' = \|F\| < \infty.$$

Definition 5.1.1 *Let F be the collision operator defined by Eq.(5.18). Then, F is said to be regular if $\{\kappa(x,.,v')$ such that $(x,v') \in [0,1] \times K\}$ is a relatively weakly compact subset of $L_1(K,dx)$.*

We assume that:

$(\mathcal{B}_7)$ f is a weakly Carathéodory map satisfying

$$|f(x,v,u)| \leq A(x,v)h(|u|),$$

where $A \in L_1([0,1] \times [a,b], d\mu dv)$ and $h \in L_\infty^{loc}(\mathbb{R}^+)$ is a nondecreasing function.

The interest of the operators in the form that satisfies $(\mathcal{B}_7)$ lies in the following lemma which can be found in [23, Lemma 7.11].

Lemma 5.1.3 *Let $L : L_1([0,1] \times K, dxdv) \longrightarrow L_\infty([0,1] \times K, dxdv)$ be a continuous linear map and let $f : [0,1] \times K \times \mathbb{C} \longrightarrow \mathbb{C}$ be a map satisfying the hypothesis $(\mathcal{B}_7)$. Then, the map*

$$\mathcal{N}_f \circ L : L_1([0,1] \times K, dxdv) \longrightarrow L_1([0,1] \times K, dxdv)$$

is weakly sequentially continuous.

Proof. For more simplicity, we will restrict ourselves to finite measure spaces. We will use Dunford's theorem, (see Theorem 1.3.8). Let $u_n \rightharpoonup u$ in $L_1([0,1] \times K, dxdv)$. According to Eberlein–Šmulian's theorem, the set $K = \{u, u_n\}_{n=1}^{\infty}$

is weakly compact. Let us show that $\mathcal{N}_f \circ L(K)$ is relatively weakly compact in $L_1([0,1] \times K, dxdv)$. Clearly, $\mathcal{N}_f \circ L(K)$ is bounded. Hence,

$$\|\mathcal{N}_f \circ L(v)\|_{L_1([0,1]\times K, dxdv)} \leq \|A\|_{L_1} \mathcal{N}_f \circ L(\|L\|.\|v\|_{L_1([0,1]\times K, dxdv)}).$$

The last inequality also shows that $\mathcal{N}_f \circ L(K)$ is uniformly integrable. Since $[0,1] \times K$ is reflexive, we get item (iii) of Dunford's theorem (Theorem 1.3.8) for free. Hence, $\mathcal{N}_f \circ L(K)$ is relatively weakly compact in $L_1([0,1] \times K, dxdv)$. Q.E.D.

The following hypothesis will play a crucial role:

$$(\mathcal{B}_8) \begin{cases} \mathcal{N}_{-\mathcal{V}} \text{ is weakly sequentially continuous and acts from } B_r \text{ into } B_r, \\ |\mathcal{V}(x,v,\psi_1(x,v)) - \mathcal{V}(x,v,\psi_2(x,v))| \leq |\rho(x,v)|\ |\psi_1 - \psi_2|, \text{ for every} \\ (\psi_1, \psi_2) \in B_r, \text{ where } r > 0, B_r := \{\psi \in X \text{ such that } \|\psi\| \leq r\}, \\ \text{and } \rho \in L_\infty(D). \end{cases}$$

Let λ be a complex number such that $Re\lambda < \lambda_{00}$. Then, according to Proposition 5.1.1, the mapping $T_H - \lambda$ is invertible and therefore, the problem (5.1)–(5.2) is equivalent to the following fixed point problem:

$$\begin{cases} \psi = \mathcal{F}(\lambda)\psi + \mathcal{H}(\lambda)\psi \\ \psi \in \mathcal{D}(T_H), \ Re\lambda < \lambda_{00}, \end{cases}$$

where

$$\begin{cases} \mathcal{F}(\lambda) := (T_H - \lambda)^{-1} F \mathcal{N}_f L \\ \mathcal{H}(\lambda) := (T_H - \lambda)^{-1} \mathcal{N}_{-\mathcal{V}}. \end{cases}$$

Now, we are ready to state the main result of this section.

Theorem 5.1.3 *Assume that* $(\mathcal{B}_6)$–$(\mathcal{B}_8)$ *hold and that* F *is a regular operator on* X. *Let* U_r *be a weakly open subset of* B_r *with* $0 \in U_r$. *In addition, suppose that*

$$[\text{for any solution } \psi \in X \text{ to } \psi = \alpha\mathcal{F}(\lambda)\psi + \alpha\mathcal{H}(\lambda)\left(\frac{\psi}{\alpha}\right) \text{ a.e., } 0 < \alpha < 1,$$

$$\text{we have } \psi \notin \partial^w_{B_r} U_r (\text{the weak boundary of } U_r \text{ in } B_r)] \text{ holds.}$$

Then, for a small enough $Re\lambda$, *the problem* (5.1)–(5.2) *has a solution in* $\overline{U_r^w}$.

Proof. We first check that, for a suitable λ, $\mathcal{F}(\lambda)$ and $\mathcal{H}(\lambda)$ are weakly sequentially continuous. Indeed, we have $\mathcal{N}_{-\mathcal{V}}$ is weakly sequentially continuous and for $Re\lambda < \lambda_{00}$, the linear operator $(T_H - \lambda)^{-1}$ is bounded. So, $\mathcal{H}(\lambda) := (T_H - \lambda)^{-1}\mathcal{N}_{-\mathcal{V}}$ is weakly sequentially continuous, for $Re\lambda < \lambda_{00}$. Moreover, we have $\mathcal{F}(\lambda) := (T_H - \lambda)^{-1}F\mathcal{N}_f L$ is weakly sequentially continuous for $Re\lambda < \lambda_{00}$. Second, by using the hypothesis $(\mathcal{B}_7)$, we show that $\mathcal{N}_f L(\overline{U_r^w})$ is a bounded subset of X. So, from Lemma 5.1.2, we deduce that

$$\mathcal{F}(\lambda)(\overline{U_r^w}) := (\lambda - T_H)^{-1}B\mathcal{N}_f L(\overline{U_r^w})$$

is relatively weakly compact. Furthermore, according to (5.16) and (5.17), we infer that $\mathcal{H}(\lambda)$ is a contraction on B_r, for $Re\lambda < \min(\lambda_{00}, \lambda_1)$, where $\lambda_1 < 0$. Now, we may show that for a suitable λ, we have $\mathcal{F}(\lambda)(\overline{U_r^w}) + \mathcal{H}(\lambda)(B_r) \subset B_r$. To do so, let $\varphi \in \overline{U_r^w}$ and $\psi \in B_r$. Then, we have

$$\begin{aligned} \|\mathcal{F}(\lambda)(\varphi) + \mathcal{H}(\lambda)(\psi)\| &\leq \|(T_H - \lambda)^{-1}F\mathcal{N}_f L\varphi\| + \|(T_H - \lambda)^{-1}\mathcal{N}_{-\mathcal{V}}\psi\| \\ &\leq \|(T_H - \lambda)^{-1}\| \left(\|F\| \, \|\mathcal{N}_f L\varphi\| + \|\mathcal{N}_{-\mathcal{V}}\psi\|\right) \\ &\leq \|(T_H - \lambda)^{-1}\| \left(\|F\| \, \|A\| \, \|h\|_\infty + M(r)\right), \quad (5.19) \end{aligned}$$

where $M(r)$ denotes the upper bound of $\mathcal{N}_{-\mathcal{V}}$ on B_r. Moreover, we have

$$\|(T_H - \lambda)^{-1}\| \leq \frac{-1}{Re\lambda}\left(1 + \frac{\|H\|}{1 - e^{Re\lambda\|H\|}}\right),$$

for $Re\lambda < \lambda_{00}$. So, for $Re\lambda < \min(\lambda_{00}, \lambda_1, \varepsilon)$, $\varepsilon < 0$, Inequality (5.19) implies that

$$\begin{aligned} \|\mathcal{F}(\lambda)(\varphi) + \mathcal{H}(\lambda)(\psi)\| &\leq \frac{-1}{Re\lambda}\left(1 + \frac{\|H\|}{1 - e^{\varepsilon\|H\|}}\right)\left(\|F\| \, \|A\| \, \|h\|_\infty + M(r)\right) \\ &= \mathcal{G}(Re\lambda), \end{aligned}$$

where

$$\mathcal{G}(t) := \frac{-1}{t}\left(1 + \frac{\|H\|}{1 - e^{\varepsilon\|H\|}}\right)\left(\|F\| \, \|A\| \, \|h\|_\infty + M(r)\right).$$

Clearly, $\mathcal{G}$ is continuous, strictly increasing in t, $t < 0$ and satisfies $\lim_{t\to-\infty}\mathcal{G}(t) = 0$. Hence, there exists $\lambda_2 < \min(\lambda_{00}, \lambda_1, \varepsilon)$ such that for $Re\lambda < \lambda_2$, we have $\mathcal{F}(\lambda)(\varphi) + \mathcal{H}(\lambda)(\psi) \in B_r$. Consequently, for $Re\lambda < \lambda_2$, the mappings $\mathcal{F}(\lambda)$ and $\mathcal{H}(\lambda)$ satisfy the assumptions of Corollary 2.5.1 on the nonempty bounded, closed, and convex subset B_r. Hence, the problem (5.1)–(5.2) has a solution in $\overline{U_r^w}$ for all λ such that $Re\lambda < \lambda_2$. Q.E.D.

5.2 Transport Equations Arising in Growing Cell Population

In this section, we study the solution of the stationary nonlinear model arising in the theory of growing cell population:

$$v\frac{\partial \psi}{\partial \mu}(\mu, v) + \lambda\psi(\mu, v) + \sigma(\mu, v, \psi(\mu, v)) = \int_a^b r(\mu, v, v', \psi(\mu, v'))dv' \quad (5.20)$$

where $\mu \in [0, 1], v, v' \in [a, b]$ with $0 \leq a < b < \infty$, $\sigma(., ., .)$ and $r(., ., ., .)$ are nonlinear functions of ψ and λ is a complex number. This equation describes the number density $\psi(\mu, v)$ of cell population as a function of the degree of maturation μ and the maturation velocity v. The degree of maturation is defined so that $\mu = 0$ at birth and $\mu = 1$ at the death of a cell. The boundary conditions are modeled by:

$$\psi_{|\Gamma_0} = K(\psi_{|\Gamma_1}), \quad (5.21)$$

where $\Gamma_0 = \{0\} \times [a, b]$ and $\Gamma_1 = \{1\} \times [a, b]$. $\psi_{|\Gamma_0}$ (resp. $\psi_{|\Gamma_1}$) denotes the restriction of ψ to Γ_0 (resp. Γ_1) whereas K is a nonlinear operator from a suitable function space on Γ_1 to a similar one on Γ_0. In [142], M. Rotenberg studied essentially the Fokker–Plank approximation of Eq. (5.20) for which he obtained numerical solutions. Using an eigenfunction expansion technique, C. Van der Mee and P. Zweifel [152] obtained analytical solutions for a variety of linear boundary conditions. Using J. L. Lebowitz and S. I. Rubinow's boundary conditions (see [121] or [142]), M. Boulanouar and L. Leboucher [38] proved that the associated Cauchy problem to the Rotenberg model is governed by a positive C^0-semigroup and they gave sufficient conditions guaranteeing its irreducibility. Recently, K. Latrach and A. Jeribi [117, 118] obtained several existence results for the boundary value problem (5.20)–(5.21) in L_p spaces with $1 < p < \infty$. The analysis which started in [118] was based essentially on compactness results established only for $1 < p < \infty$. This analysis used the Schauder and Krasnosel'skii fixed point theorems. Due to the lack of compactness in L_1 spaces, this approach failed in the L_1 context (which represents the convenient and natural setting of the problem) and therefore, the solvability of the problem (5.20)–(5.21) remained open in the L_1 framework [118].

In the following, we will present some existence results of the stationary model (5.20)–(5.21) in L_1 spaces. The main points in this framework are the nonlinearity of the boundary condition K and the nonlinear dependence of the function $r(.,.,.,.)$ on ψ. More specifically, we suppose that

$$r(\mu, v, v', \psi(\mu, v')) = \kappa(\mu, v, v')f(\mu, v', L(\psi)(\mu, v')),$$

where

$$L \ : \ L_1([0,1] \times [a,b]) \longrightarrow L^\infty([0,1] \times [a,b])$$

represents a continuous map and f is a measurable weak Carathéodory function defined by:

$$\begin{cases} f : [0,1] \times [a,b] \times \mathbb{C} \longrightarrow \mathbb{C} \\ \qquad\qquad\qquad (\mu, v, u) \longrightarrow f(\mu, v, u) \end{cases}$$

and satisfying the condition

$$|f(\mu, v, u)| \leq A(\mu, v)h(|u|),$$

where $A \in L_1([0,1] \times [a,b])$ and $h \in L^\infty(\mathbb{R}^+)$. The function $\kappa(.,.,.)$ is a measurable function from $[0,1] \times [a,b] \times [a,b]$ into $\mathbb{R}$, satisfying the condition,

$$\{\kappa(\mu, ., v') \text{ such that } (\mu, v') \in [0,1] \times [a,b]\}$$

is a relatively weakly compact subset of $L_1([a,b], dv)$.

5.2.1 A particular case

Let us consider the following problem:

$$v\frac{\partial \psi}{\partial \mu}(\mu, v) + \sigma(\mu, v)\psi(\mu, v) + \lambda\ \psi(\mu, v) = \int_a^b r(\mu, v, v', \psi(\mu, v'))dv' \qquad (5.22)$$

with the boundary conditions

$$\psi_{|\Gamma_0} = K\left(\psi_{|\Gamma_1}\right), \qquad (5.23)$$

where $\sigma(.,.) \in L^\infty([0,1] \times [a,b])$, λ is a complex number, $\Gamma_0 = \{0\} \times [a,b]$ and $\Gamma_1 = \{1\} \times [a,b]$. $\psi_{|\Gamma_0}$ (resp. $\psi_{|\Gamma_1}$) denotes the restriction of ψ to Γ_0 (resp. Γ_1) whereas K is a nonlinear operator from a suitable function space on Γ_1 to a similar one on Γ_0. Let

$$X := L_1([0,1] \times [a,b]; d\mu dv),$$

where $0 \leq a < b < \infty$. We denote by X^0 and X^1 the following boundary spaces

$$X^0 := L_1(\{0\} \times [a,b]; vdv),$$

$$X^1 := L_1(\{1\} \times [a,b]; vdv)$$

endowed with their natural norms. Let W be the space defined by:

$$W = \left\{ \psi \in X \text{ such that } v\frac{\partial \psi}{\partial \mu} \in X \right\}.$$

It is well known (see [52], [53] or [63]) that any ψ in W has traces on the spatial boundaries $\{0\}$ and $\{1\}$ which belong respectively to the spaces X^0 and X^1. We define the free streaming operator S_K by:

$$\begin{cases} S_K : \mathcal{D}(S_K) \subset X \longrightarrow X \\ \qquad\qquad \psi \longrightarrow S_K\psi(\mu, v) = -v\dfrac{\partial \psi}{\partial \mu}(\mu, v) - \sigma(\mu, v)\psi(\mu, v) \\ \mathcal{D}(S_K) = \{\psi \in W \text{ such that } \psi^0 = K(\psi^1)\}, \end{cases}$$

where $\psi^0 = \psi_{|\Gamma_0}$, $\psi^1 = \psi_{|\Gamma_1}$ and K is the following nonlinear boundary operator

$$\begin{cases} K : X^1 \longrightarrow X^0 \\ \qquad u \longrightarrow Ku \end{cases}$$

satisfying the following conditions:

$(\mathcal{B}_9)$ there exists $\alpha > 0$ such that

$$\|Kf_1 - Kf_2\| \leq \alpha\|f_1 - f_2\| \qquad (f_1,\ f_2 \in X^1).$$

As immediate consequences of $(\mathcal{B}_9)$, we have the continuity of the operator K from X^1 into X^0 and

$$\|Kf\| \leq \alpha\|f\| + \|K(0)\| \qquad \forall\ f \in X^1. \tag{5.24}$$

Let us consider the equation

$$(\lambda - S_K)\psi = g. \tag{5.25}$$

Our objective is to determine a solution $\psi \in \mathcal{D}(S_K)$, where g is given in X and $\lambda \in \mathbb{C}$. Let $\underline{\sigma}$ be the real defined by:

$$\underline{\sigma} := \text{ess-}\inf\{\sigma(\mu, v), (\mu, v) \in [0,1] \times [a,b]\}.$$

For $Re\lambda > -\underline{\sigma}$, the solution of Eq. (5.25) is formally given by:

$$\psi(\mu, v) = \psi(0, v)\, e^{-\frac{1}{v}\int_0^{\mu}(\lambda+\sigma(\mu',v))d\mu'} + \frac{1}{v}\int_0^{\mu} e^{-\frac{1}{v}\int_{\mu'}^{\mu}(\lambda+\sigma(\tau,v))d\tau}\varphi(\mu', v)\, d\mu'. \tag{5.26}$$

Accordingly, for $\mu = 1$, we get

$$\psi(1, v) = \psi(0, v)\, e^{-\frac{1}{v}\int_0^{1}(\lambda+\sigma(\mu',v))d\mu'} + \frac{1}{v}\int_0^{1} e^{-\frac{1}{v}\int_{\mu'}^{1}(\lambda+\sigma(\tau,v))d\tau}\varphi(\mu', v)\, d\mu'. \tag{5.27}$$

Let the following operators P_λ, Q_λ, Π_λ, and R_λ be defined by:

$$\begin{cases} P_\lambda : X^0 \longrightarrow X^1 \\ \qquad u \longrightarrow (P_\lambda u)(1, v) := u(0, v)\, e^{-\frac{1}{v}\int_0^{1}(\lambda+\sigma(\mu',v))d\mu'}; \end{cases}$$

$$\begin{cases} Q_\lambda : X^0 \longrightarrow X \\ \qquad u \longrightarrow (Q_\lambda u)(\mu, v) := u(0, v)\, e^{-\frac{1}{v}\int_0^{\mu}(\lambda+\sigma(\mu',v))d\mu'}; \end{cases}$$

$$\begin{cases} \Pi_\lambda : X \longrightarrow X^1 \\ \qquad u \longrightarrow (\Pi_\lambda u)(1, v) := \dfrac{1}{v}\displaystyle\int_0^{1} e^{-\frac{1}{v}\int_{\mu'}^{1}(\lambda+\sigma(\tau,v))d\tau} u(\mu', v)\, d\mu'; \end{cases}$$

and finally,

$$\begin{cases} R_\lambda : X \longrightarrow X \\ \qquad u \longrightarrow (R_\lambda u)(\mu, v) := \dfrac{1}{v}\displaystyle\int_0^{\mu} e^{-\frac{1}{v}\int_{\mu'}^{\mu}(\lambda+\sigma(\tau,v))d\tau} u(\mu', v)\, d\mu'. \end{cases}$$

Clearly, for λ satisfying $Re\lambda > -\underline{\sigma}$, the operators P_λ, Q_λ, Π_λ, and R_λ are bounded. It is not difficult to check that

$$\|P_\lambda\| \leq e^{-\frac{1}{b}(Re\lambda+\underline{\sigma})}, \tag{5.28}$$

and

$$\|Q_\lambda\| \leq (Re\lambda + \underline{\sigma})^{-1}. \tag{5.29}$$

Moreover, by making some simple calculations, we may show that

$$\|\Pi_\lambda\| \leq 1, \tag{5.30}$$

and

$$\|R_\lambda\| \leq (Re\lambda + \underline{\sigma})^{-1}. \tag{5.31}$$

Thus, Eq. (5.27) may be abstractly written as follows

$$\psi^1 = P_\lambda \psi^0 + \Pi_\lambda g. \tag{5.32}$$

Moreover, ψ must satisfy the boundary conditions (5.21). Hence, we obtain

$$\psi^1 = P_\lambda K \psi^1 + \Pi_\lambda g.$$

Let us notice that the operator $P_\lambda K$ is defined from X^1 into X^1. Let $\varphi_1,\ \varphi_2 \in X^1$. By using $(\mathcal{B}_9)$ and (5.28), we deduce that

$$\|P_\lambda K\varphi_1 - P_\lambda K\varphi_2\| \leq \alpha\, e^{-\frac{Re\lambda + \underline{\sigma}}{b}} \|\varphi_1 - \varphi_2\| \quad \forall \varphi_1,\ \varphi_2 \in X^1. \tag{5.33}$$

Now, let us consider the equation

$$u = P_\lambda K u + f, \quad f \in X^1, \tag{5.34}$$

where u is the unknown function, and let us define the operator $A_{(\lambda,f)}$ on X^1 by:

$$\begin{cases} A_{(\lambda,f)} : X^1 \longrightarrow X^1, \\ \qquad\qquad u \longrightarrow (A_{(\lambda,f)}u)(1,v) := P_\lambda K u + f. \end{cases}$$

From the estimate (5.33), it follows that

$$\|A_{(\lambda,f)}\varphi_1 - A_{(\lambda,f)}\varphi_2\| = \|P_\lambda K\varphi_1 - P_\lambda K\varphi_2\| \leq \alpha\, e^{-\frac{Re\lambda + \underline{\sigma}}{b}} \|\varphi_1 - \varphi_2\|. \tag{5.35}$$

Consequently, for $Re\lambda > -\underline{\sigma} + b\log(\alpha)$, the operator $A_{(\lambda,f)}$ is a contraction mapping and therefore, Eq. (5.34) has a unique solution

$$u_{(\lambda,f)} = u.$$

Now, let W_λ be the nonlinear operator defined by:

$$\begin{cases} W_\lambda \ :\ X^1 \longrightarrow X^1 \\ \qquad\qquad f \longrightarrow u_{(\lambda,f)} = u, \end{cases}$$

where u is the solution of Eq. (5.34). Arguing as in the proof of Lemma 2.1 and Proposition 2.1 in [118], we deduce the following result :

Lemma 5.2.1 *Assume that $(\mathcal{B}_9)$ holds. Then,*

(i) for every λ satisfying $Re\lambda > -\underline{\sigma} + b\log(\alpha)$, the operator W_λ is continuous and maps bounded sets into bounded ones and satisfying the following estimate

$$\|W_\lambda f_1 - W_\lambda f_2\| \leq (1 - \alpha e^{-(\frac{Re\lambda + \underline{\sigma}}{b})})^{-1} \|f_1 - f_2\| \qquad (f_1, f_2 \in X^1). \tag{5.36}$$

(ii) If $Re\lambda > \max(-\underline{\sigma}, -\underline{\sigma} + b\log(\alpha))$, *then the operator* $(\lambda - S_K)$ *is invertible and* $(\lambda - S_K)^{-1}$ *is given by:*

$$(\lambda - S_K)^{-1} = Q_\lambda K W_\lambda \Pi_\lambda + R_\lambda. \tag{5.37}$$

Moreover, $(\lambda - S_K)^{-1}$ *is continuous on* X *and maps bounded sets into bounded ones.*

Proof. (i) According to the definition of W_λ, Eq. (5.34) may be written in the form

$$W_\lambda f = P_\lambda K(W_\lambda f) + f$$

and therefore,

$$\begin{aligned}\|W_\lambda f_1 - W_\lambda f_2\| &\leq \|P_\lambda K(W_\lambda f_1) - P_\lambda K(W_\lambda f_2)\| + \|f_1 - f_2\| \\ &\leq \alpha\, e^{-\frac{Re\lambda + \underline{\sigma}}{b}} \|W_\lambda f_1 - W_\lambda f_2\| + \|f_1 - f_2\|\end{aligned}$$

for any $f_1,\ f_2 \in X^1$. This leads to the following estimate

$$\|W_\lambda f_1 - W_\lambda f_2\| \leq \left(1 - \alpha\, e^{-\frac{Re\lambda + \underline{\sigma}}{b}}\right)^{-1} \|f_1 - f_2\|$$

which proves the continuity of W_λ. The second part of the lemma follows from the estimate

$$\|W_\lambda f\| \leq \left(1 - \alpha\, e^{-\frac{Re\lambda - \underline{\sigma}}{b}}\right)^{-1} \|f\| + \|W_\lambda(0)\|.$$

This completes the proof of (i).

(ii) Since $Re\lambda > \max(-\underline{\sigma}, -\underline{\sigma} + b\log(\alpha))$, the solution of the problem (5.32) is given by:

$$\psi^1 = W_\lambda \Pi_\lambda g. \tag{5.38}$$

Moreover, Eq. (5.26) may be written as

$$\psi = Q_\lambda K \psi^1 + R_\lambda g.$$

Substituting Eq. (5.38) into the above equation, we obtain

$$\psi = Q_\lambda K W_\lambda \Pi_\lambda g + R_\lambda g$$

from which we can infer that $(\lambda - S_K)$ is invertible and

$$(\lambda - S_K)^{-1} = Q_\lambda K W_\lambda \Pi_\lambda + R_\lambda.$$

The second assertion follows from the boundedness of the linear operators Q_λ, Π_λ, R_λ and from (i). Q.E.D.

5.2.2 Regular collision and weak compactness results

For our subsequent analysis, we need the following hypothesis

$(\mathcal{B}_{10})$ $$r(\mu, v, v', \psi(\mu, v')) = \kappa(\mu, v, v')f(\mu, v', L(\psi)(\mu, v')),$$

where

$$L \; : \; L_1([0,1] \times [a,b]) \longrightarrow L^\infty([0,1] \times [a,b])$$

is a continuous linear map and f is a measurable function defined by:

$$\begin{cases} f : [0,1] \times [a,b] \times \mathbb{C} \longrightarrow \mathbb{C} \\ \qquad\qquad (\mu, v, u) \longrightarrow f(\mu, v, u). \end{cases}$$

The function $\kappa(.,.,.)$ is a measurable function from $[0,1] \times [a,b] \times [a,b]$ into $\mathbb{R}$. It defines a continuous linear operator B by:

$$\begin{cases} B \; : X \longrightarrow X \\ \qquad \varphi \longrightarrow B\varphi(\mu, v) = \displaystyle\int_a^b \kappa(\mu, v, v')\varphi(\mu, v')\, dv'. \end{cases} \tag{5.39}$$

Note that

$$d\mu \otimes dv - \operatorname*{ess-sup}_{(\mu,v)\in[0,1]\times[a,b]} \int_a^b |\kappa(\mu, v, v')| \, dv' = \|B\| < \infty.$$

Definition 5.2.1 *Let B be the operator defined by (5.39). Then, B is said to be a regular operator if $\{\kappa(\mu, ., v')$ such that $(\mu, v') \in [0,1] \times [a,b]\}$ is a relatively weakly compact subset of $L_1([a,b], d\mu)$.*

Remark 5.2.1 *Definition 5.2.1 asserts that, for every $\mu \in [0,1]$,*

$$f \in L_1([a,b]) \longrightarrow \int_a^b \kappa(\mu, v, v')f(v')\, dv' \in L_1([a,b])$$

is a weakly compact operator and this weak compactness holds collectively in $\mu \in [0,1]$.

Let the operator B be defined by (5.39) and let $\kappa^+(.,.,.)$ (resp. $\kappa^-(.,.,.)$) denote the positive part (resp. the negative part) of $\kappa(.,.,.)$:

$$\kappa(\mu, v, v') = \kappa^+(\mu, v, v') - \kappa^-(\mu, v, v') \qquad (\mu, v, v') \in [0,1] \times [a,b] \times [a,b].$$

We define the following non-negative operators :

$$B^\pm \; : \; \psi \longrightarrow B^\pm\psi(\mu, v) := \int_a^b \kappa^\pm(\mu, v, v')\psi(\mu, v')\, dv'.$$

Clearly,

$$B = B^+ - B^-.$$

Now, let $|B|$ denote the following nonnegative operator:

$$|B| := B^+ + B^-$$

i.e.,

$$|B|\,\varphi(\mu, v) = \int_a^b |\kappa|\,(\mu, v, v')\varphi(\mu, v')\,dv', \quad \varphi \in X.$$

Remark 5.2.2 *Thanks to Dunford–Pettis criterion, the following assertions are equivalent:*

(i) B *is a regular collision operator.*

(ii) $|B|$ *is a regular collision operator.*

(iii) B^+ *and* B^- *are regular collision operators.*

Let Ω be a bounded smooth open subset of $\mathbb{R}^N$ $(N \geq 1)$ and let $d\mu$ and $d\nu$ be two positive radon measures on $\mathbb{R}^N$ with a common support V. Let

$$K \in \mathcal{L}(L^1(\Omega \times V, dxd\mu(v)), L^1(\Omega \times V, dxd\nu(v)))$$

be defined by:

$$\begin{cases} K : L^1(\Omega \times V, dxd\mu(v)) \longrightarrow L^1(\Omega \times V, dxd\nu(v)) \\ \qquad \psi \longrightarrow \displaystyle\int_V \kappa(x, v, v')\psi(x, v')\,d\mu(v'), \end{cases} \tag{5.40}$$

where the kernel $\kappa(., ., .)$ is measurable. Note that

$$dx \otimes d\mu - \text{ess} \sup_{(x,v') \in \Omega \times V} \int_V |\kappa(x, v, v')| d\nu(v) = \|K\| < \infty. \tag{5.41}$$

The class of regular operators satisfies the following approximate property:

Theorem 5.2.1 *[126] Let* $B \in \mathcal{L}\big(L_1([0,1] \times [a,b], d\mu dv)\big)$*, be a regular and nonnegative operator. Then, there exists* $(B_m)_m \subset \mathcal{L}\big(L_1([0,1] \times [a,b], d\mu dv)\big)$ *such that:*

(i) $0 \leq B_m \leq B$ *for any* $m \in \mathbb{N}$.

(ii) For any $m \in \mathbb{N}$, B_m *is dominated by a rank-one operator in* $\mathcal{L}\big(L_1([a,b], dv)\big)$.

(iii) $\lim_{m \to +\infty} \|B - B_m\| = 0$.

Proof. According to Definition 5.2.1, the operator K, defined in (5.40) and satisfying (5.41) and $\{\kappa(x,.,v') \text{ such that } (x,v') \in \Omega \times V\}$, is a relatively weak compact subset of $L^1(V, d\nu)$. According to Takac's version of Dunford–Pettis criterion, we have

$$\lim_{m\to\infty} \sup_{(x,v')\in\Omega\times V} \int_{S_m(x,v')} |\kappa(x,v,v')| d\nu(v) = 0, \tag{5.42}$$

where

$$S_m(x,v') := \{v \in V : |v| \geq m\} \cup \{v \in V : \kappa(x,v,v') \geq m\}, \quad (x,v') \in \Omega \times V.$$

For any $m \in \mathbb{N}$ and $\varphi \in L^1(\Omega \times V, dx d\mu(v))$, let us define

$$K_m(\varphi) = \int_V \kappa_m(x,v,v')\varphi(x,v')\, d\mu(v') \in L^1(\Omega \times V, dx d\nu(v)),$$

with

$$\kappa_m(x,v,v') = \inf\{\kappa(x,v,v') \geq m\chi_{B_m}(v)\} \quad (x,v,v') \in \Omega \times V \times V,$$

where $\chi_{B_m}(.)$ denotes the characteristic function of the set

$$\{v \in V \text{ such that } \|v\| \leq m\}.$$

Clearly,

$$0 \leq K_m \leq K.$$

Moreover, we can easily check that

$$\|K - K_m\| \leq dx \otimes d\mu - \text{ess} \sup_{(x,v')\in\Omega\times V} \int_V |\kappa(x,v,v') - \kappa_m(x,v,v')| d\nu(v).$$

Besides, for any $(x,v') \in \Omega \times V$, the construction of $\kappa_m(x,.,v')$ implies that

$$\begin{aligned} \int_V \Delta_m(x,v,v') d\nu(v) &= \int_{\{\kappa(x,v,v')\geq m\chi_{B_m}(v)\}} \Delta_m(x,v,v') d\nu(v) \\ &\leq \int_{\{\kappa(x,v,v')\geq m\chi_{B_m}(v)\}} \kappa(x,v,v') d\nu(v), \end{aligned}$$

where $\Delta_m(x,v,v') := |\kappa(x,v,v') - \kappa_m(x,v,v')|$. Then, according to Eq. (5.42), we have

$$\lim_{m\to\infty} \|K - K_m\| = 0.$$

Finally, it is easy to check that, for any $\varphi \in L^1(\Omega \times V, dxd\mu(v))$, $\varphi \geq 0$, and

$$K_m\varphi(x,v) \leq m\chi_{B_m}(v)\int_V \varphi(x,v')d\mu(v'),$$

which proves the second assertion and achieves the proof. Q.E.D.

Remark 5.2.3 *Let us clarify the point (ii) of Theorem 5.2.1. This asserts that, for any $m \in \mathbb{N}$, there exists a nonnegative $f_m \in L_1([a,b],dv)$ such that, for any $\varphi \in L_1([0,1]\times[a,b], d\mu dv)$, $\varphi \geq 0$, we have*

$$B_m\varphi(\mu,v) \leq f_m(v)\int_a^b \varphi(\mu,v')\,dv'.$$

Proposition 5.2.1 *[82] Let (Ω,Σ,μ) be a σ-finite, positive measure space and let S and T be two bounded linear operators on $L_1(\Omega,d\mu)$. Then, the following assertions hold:*

(i) The set of all weakly compact operators is a norm-closed subset of $\mathcal{L}(L_1(\Omega,d\mu))$.

(ii) If T is weakly compact on $L_1(\Omega,d\mu)$ and satisfying $0 \leq S \leq T$, then S is also weakly compact.

Lemma 5.2.2 *If B is a regular operator, then $\Pi_\lambda B$ and $R_\lambda B$ are weakly compact on X, for $Re\lambda > \max(-\underline{\sigma}, -\underline{\sigma} + b\log(\alpha))$.*

Proof. Thanks to Remark 5.2.2, we can suppose that B is nonnegative. According to (5.30), we have

$$\|\Pi_\lambda B\| \leq \|B\| \qquad \text{for all } Re\lambda > \max(-\underline{\sigma}, -\underline{\sigma} + b\log(\alpha)).$$

Thus, $\Pi_\lambda B$ depends continuously on B, uniformly on $\{Re\lambda > \max(-\underline{\sigma}, -\underline{\sigma} + b\log(\alpha))\}$. According to Theorem 5.2.1 and Proposition 5.2.1 (*i*), it is sufficient to prove the result when B is dominated by a rank-one operator in $\mathcal{L}\big(L_1([a,b],dv)\big)$. Moreover, by using both Remark 5.2.3 and Proposition 5.2.1 (*ii*), we may assume that B itself is a rank-one collision operator in $\mathcal{L}\big(L_1([a,b],dv)\big)$. This asserts that B has a kernel

$$\kappa(v,v') = \kappa_1(v)\kappa_2(v'); \quad \kappa_1 \in L_1([a,b],dv), \kappa_2 \in L_\infty([a,b],dv).$$

Let us consider $\varphi \in X$. Then, we have

$$\begin{aligned}(\Pi_\lambda B\varphi)(v) &= \frac{1}{v}\int_0^1 e^{-\frac{1}{v}\int_{\mu'}^1(\lambda+\sigma(\tau,v))d\tau} B\varphi(\mu',v)\,d\mu' \\ &= \frac{1}{v}\int_0^1\int_a^b e^{-\frac{1}{v}\int_{\mu'}^1(\lambda+\sigma(\tau,v))d\tau}\,\kappa_1(v)\kappa_2(v')\varphi(\mu,v')\,d\mu dv',\ a<v<b \\ &= J_\lambda U_\lambda\varphi,\end{aligned}$$

where U_λ and J_λ denote the following bounded operators

$$\begin{cases} U_\lambda : X \longrightarrow L_1([0,1];\,d\mu) \\ \qquad \varphi \longrightarrow \displaystyle\int_a^b \kappa_2(v)\,\varphi(\mu,v)\,dv, \end{cases}$$

and

$$\begin{cases} J_\lambda : L_1([0,1];\,d\mu) \longrightarrow X^1 \\ \qquad \psi \longrightarrow \displaystyle\frac{1}{v}\int_0^1 e^{-\frac{1}{v}\int_{\mu'}^1(\lambda+\sigma(\tau,v))d\tau}\,\kappa_1(v)\,\psi(\mu)\,d\mu. \end{cases}$$

Now, it is sufficient to show that J_λ is weakly compact. To do so, let $\mathcal{O}$ be a bounded set of $L_1([0,1];\,d\mu)$, and let $\psi \in \mathcal{O}$. We have

$$\int_E |J_\lambda\psi(v)|\,v\,dv \leq \|\psi\|\int_E |\kappa_1(v)|\,dv,$$

for all measurable subsets E of $[a,b]$. Next, by applying Theorem 1.3.7, we deduce that the set $J_\lambda(\mathcal{O})$ is weakly compact, since $\lim_{|E|\to 0}\int_E |\kappa_1(v)|\,dv = 0$, ($\kappa_1 \in L_1([a,b];\,dv)$) where $|E|$ represents the measure of E. A similar reasoning allows us to reach the same result for the operator $R_\lambda B$. Q.E.D.

Theorem 5.2.2 *If B is a regular operator, then $(\lambda - S_K)^{-1}B$ is weakly compact on X, for $Re\lambda > \max(-\underline{\sigma}, -\underline{\sigma} + b\log(\alpha))$.*

Proof. According to Eq. (5.37), we have

$$(\lambda - S_K)^{-1}B = Q_\lambda K W_\lambda \Pi_\lambda B + R_\lambda B.$$

By using Lemma 5.2.1, we deduce that $R_\lambda B$ is weakly compact. We claim that $Q_\lambda K W_\lambda \Pi_\lambda B$ is weakly compact. Since Q_λ is linear, it suffices to show that $K W_\lambda \Pi_\lambda B$ is weakly compact. To do so, let $\mathcal{O}$ be a bounded subset of X. By using both hypothesis $(\mathcal{B}_9)$ and Inequality (5.36), we show that

$$\int_E |KW_\lambda f(v)|\,v\,dv \leq \int_E |K(0)(v)| + \alpha\,|W_\lambda(0)(v)|\,v\,dv + \alpha\zeta(\lambda)\int_E |f(v)|\,v\,dv$$

for all measurable subsets E of $[a,b]$ and $f \in \Pi_\lambda B(\mathcal{O})$, where

$$\zeta(\lambda) = \left(1 - \alpha e^{-(\frac{Re\lambda + \underline{\sigma}}{b})}\right)^{-1}.$$

Now, by applying Theorem 1.3.7 again and the fact that $\Pi_\lambda B(\mathcal{O})$ is weakly compact, we conclude that $KW_\lambda \Pi_\lambda B$ is weakly compact. This proves the claim and completes the proof. Q.E.D.

Throughout this section, our main interest is dealing with the existence results for the boundary value problem (5.22)–(5.23). For this purpose, we use the notations and the preliminary results presented in the above section.

Now, we recall some facts concerning the superposition operators required below. A function $f : [0,1] \times [a,b] \times \mathbb{C} \longrightarrow \mathbb{C}$ is a weak Carathéodory map, if the following conditions are satisfied:

$$\begin{cases} (\mu, v) \longrightarrow f(\mu, v, u) & \text{is measurable on } [0,1] \times [a,b] \text{ for all } u \in \mathbb{C} \\ u \longrightarrow f(\mu, v, u) & \text{is sequentially weakly continuous on } \mathbb{C} \text{ a.e. } (\mu, v) \end{cases}$$

Notice that, if f is a weak Carathéodory map, then we can define the operator $\mathcal{N}_f$ on the set of functions $\psi : [0,1] \times [a,b] \longrightarrow \mathbb{C}$ by:

$$(\mathcal{N}_f \psi)(\mu, v) = f(\mu, v, \psi(\mu, v))$$

for every $(\mu, v) \in [0,1] \times [a,b]$. Let us assume that

$(\mathcal{B}_{11})$ f is a weak Carathéodory map satisfying

$$|f(\mu, v, u)| \leq A(\mu, v) h(|u|),$$

where $A \in L_1([0,1] \times [a,b], d\mu dv)$ and $h \in L^\infty(\mathbb{R}^+)$.

Theorem 5.2.3 *Assume that* $(\mathcal{B}_9)$, $(\mathcal{B}_{10})$, *and* $(\mathcal{B}_{11})$ *hold. If* B *is a regular collision operator on* X, *then for each* $r > 0$, *there is* $\lambda_r > 0$ *such that, for each* λ *satisfying* $Re(\lambda) > \lambda_r$, *the problem* (5.22)–(5.23) *has, at least, one solution in* B_r.

Proof. Let λ be a complex number such that $Re\lambda > \max(-\underline{\sigma}, -\underline{\sigma} + b\log(\alpha))$. Then, according to Lemma 5.2.1 (ii), $\lambda - S_K$ is invertible and therefore, the problem (5.22)–(5.23) may be transformed into

$$\psi = \mathcal{F}(\lambda)\psi, \qquad \psi^0 = K\psi^1,$$

where

$$\mathcal{F}(\lambda) = (\lambda - S_K)^{-1} B \mathcal{N}_f L.$$

Let $r > 0$. We first check that, for a suitable λ, $\mathcal{F}(\lambda)$ is weakly sequentially continuous, leaves B_r invariant, and $\mathcal{F}(\lambda)(B_r)$ is relatively weakly compact. By using Lemma 5.1.3, we show that $\mathcal{N}_f L$ is weakly sequentially continuous. By using both Lemma 5.2.2 and Proposition 1.3.8, we deduce that $\Pi_\lambda B$ and $R_\lambda B$ are strongly continuous. The continuity of the operator $Q_\lambda K W_\lambda$ ensures that $(\lambda - S_K)^{-1} B$ is strongly continuous. So, $\mathcal{F}(\lambda)$ is weakly sequentially continuous. Moreover, since f satisfies $(\mathcal{B}_{11})$, then for all $\psi \in X$, we have:

$$|\mathcal{N}_f L(\psi)(\mu, v)| \leq A(\mu, v) h(|L(\psi)(\mu, v)|).$$

Let $\psi \in B_r$. From (5.24), (5.29)–(5.31), (5.36), and (5.37), it follows that

$$\begin{aligned} \|\mathcal{F}(\lambda)(\psi)\| &\leq \|(Q_\lambda K W_\lambda \Pi_\lambda B \mathcal{N}_f L)(\psi)\| + \|(R_\lambda B \mathcal{N}_f L)(\psi)\| \\ &\leq \frac{\|K(0)\| + \alpha \|W_\lambda(0)\|}{\underline{\sigma} + Re\lambda} + \frac{\|B\|\|A\|\|h\|_\infty (1 + \alpha(1 - \alpha e^{-(\frac{\underline{\sigma} + Re\lambda}{b})})^{-1})}{\underline{\sigma} + Re\lambda}. \end{aligned}$$

Let $\varepsilon > \max(-\underline{\sigma}, -\underline{\sigma} + b\log(\alpha))$. For $Re\lambda > \varepsilon$, we have

$$(1 - \alpha e^{-(\frac{\underline{\sigma} + Re\lambda}{b})})^{-1} \leq (1 - \alpha e^{-(\frac{\underline{\sigma} + \varepsilon}{b})})^{-1}$$

and therefore,

$$\begin{aligned} \|\mathcal{F}(\lambda)(\psi)\| &\leq \frac{\|K(0)\| + \alpha \|W_\lambda(0)\|}{\underline{\sigma} + Re\lambda} + \frac{\|B\|\|A\|\|h\|_\infty (1 + \alpha(1 - \alpha e^{-(\frac{\underline{\sigma} + \varepsilon}{b})})^{-1})}{\underline{\sigma} + Re\lambda} \\ &= Q(Re\lambda), \end{aligned}$$

where

$$Q(t) = \frac{\|K(0)\| + \alpha \|W_\lambda(0)\|}{\underline{\sigma} + t} + \frac{\|B\|\|A\|\|h\|_\infty (1 + \alpha(1 - \alpha e^{-(\frac{\underline{\sigma} + \varepsilon}{b})})^{-1})}{\underline{\sigma} + t}.$$

Clearly, $Q(.)$ is continuous, strictly decreasing in t, $t > 0$ and satisfies $\lim_{t \to +\infty} Q(t) = 0$. Hence, there exists λ_0, such that $Q(\lambda_0) \leq r$ and therefore, $\|\mathcal{F}(\lambda)\psi\| \leq r$ for $Re\lambda \geq \lambda_0$. This shows that $\mathcal{F}_\lambda$ leaves B_r invariant. Since $\mathcal{N}_f L(B_r)$ is a bounded subset of X, it follows from the weak compactness of $(\lambda - S_K)^{-1} B$ (see Theorem 5.2.2), that

$$\mathcal{F}(\lambda)(B_r) = (\lambda - S_K)^{-1} B \mathcal{N}_f L(B_r)$$

is relatively weakly compact. Finally, the use of Theorem 2.2.1 shows that $\mathcal{F}(\lambda)$ has, at least, one fixed point in B_r. Q.E.D.

Let us discuss the existence of positive solutions to the boundary value problem. For this purpose, we make the hypothesis:

$$(\mathcal{B}_{12}) \qquad K[(X^1)^+)] \subset (X^0)^+,$$

where $(X^1)^+$ (resp. $(X^0)^+$) denotes the positive cone of the space X^1 (resp. X^0). Let $r > 0$. We define the set B_r^+ by $B_r^+ := B_r \cap X^+$.

Theorem 5.2.4 *Assume that* $(\mathcal{B}_9)$, $(\mathcal{B}_{10})$, $(\mathcal{B}_{11})$, *and* $(\mathcal{B}_{12})$ *hold. If* B *is a regular positive operator and* $\mathcal{N}_f L(X^+) \subset X^+$, *then for each* $r > 0$, *there is* $\lambda_r > 0$ *such that for all* $\lambda > \lambda_r$, *the problem* (5.22)–(5.23) *has, at least, one solution in* B_r^+.

Proof. Let λ be real such that $\lambda > \max(-\underline{\sigma}, -\underline{\sigma} + b\log(\alpha))$. In order to prove that the operator $(\lambda - S_K)^{-1}B$ carries elements of X^+ onto elements of X^+, it suffices to establish that W_λ is positive, i.e., $W_\lambda((X^1)^+) \subset (X^1)^+$. For this purpose, let $f \in (X^1)^+$ and consider the sequence in $(X^1)^+$ defined by:

$$\begin{cases} u_0 = 0 \\ u_{n+1} = A_{(\lambda,f)}(u_n). \end{cases}$$

Since $f \in (X^1)^+$, by using the positivity of $P_\lambda K$, one sees by induction, that $u_n \in (X^1)^+$. Moreover, since the operator $A_{(\lambda,f)}$ is a contraction mapping, it follows from Banach's fixed point theorem, that $u_n \to W_\lambda(f)$. We claim that $W_\lambda(f)$ is positive. To this end, from hypothesis $(\mathcal{B}_9)$ and (5.35), we have

$$|u_{n+p}(v) - u_n(v)| \leq \left[\alpha e^{-(\frac{\underline{\sigma}+Re\lambda}{b})}\right]^n \frac{1 - [\alpha e^{-(\frac{\underline{\sigma}+Re\lambda}{b})}]^p}{1 - \alpha e^{-(\frac{\underline{\sigma}+Re\lambda}{b})}} \,|u_1(v)|\,.$$

So, $u_n(v)$ converges to $u(v)$ almost everywhere, $v \in [a,b]$ and we have $u(v) \geq 0$. Now, as u_n converges to $W_\lambda(f)$ on X^1 and $u_n(v)$ converges to $u(v)$ *a.e.* $v \in [a,b]$, we get from [148, Chapter 2, Lemma 3.9], $W_\lambda(f) = u$ which is positive. Now, the result follows from Theorem 2.2.2. Q.E.D.

Theorem 5.2.5 *Let the hypotheses* $(\mathcal{B}_9)$, $(\mathcal{B}_{10})$, $(\mathcal{B}_{11})$ *and* $(\mathcal{B}_{12})$ *be satisfied and suppose that* B *is a regular positive operator and that there is* $c > 0$ *and* $0 \neq \psi_0 \in B_r^+$ *such that*

(i) $\psi_0 \notin N(B)$, *where* $N(B)$ *denotes the kernel of* B,

(ii) $(\mathcal{N}_f L\psi)(\mu, v) \geq c\, \psi_0(\mu, v)$ *for all* $\psi \in B_r^+$.

Then, for each $\lambda > \max(-\underline{\sigma}, -\underline{\sigma} + b\log(\alpha))$, *there is* $\eta > 0$ *such that the problem*

$$\begin{cases} v\dfrac{\partial \psi}{\partial \mu}(\mu, v) + \sigma(\mu, v)\psi(\mu, v) + \lambda\psi\ (\mu, v) = \eta \displaystyle\int_a^b r(\mu, v, v', \psi(\mu, v'))dv', \\ \qquad\qquad\qquad\qquad \psi^0 = K(\psi^1) \end{cases}$$

has, at least, one solution ψ^* *in* B_r^+ *satisfying* $\|\psi^*\| = r$.

Proof. Arguing as in the proof of Theorem 5.2.4, we show that the operator

$$\mathcal{F}(\lambda) := (\lambda - S_K)^{-1} B\mathcal{N}_f L$$

is weakly sequentially continuous on X. Besides, $\mathcal{F}(\lambda)$ leaves B_r^+ invariant. Then, it follows from (5.37) the estimate

$$(\lambda - S_K)^{-1} B\mathcal{N}_f L\psi \geq R_\lambda B\mathcal{N}_f L\psi \quad \text{for all} \quad \psi \in B_r^+.$$

Moreover, the hypothesis (ii) implies that

$$(R_\lambda B\mathcal{N}_f L\psi)(\mu, v) \geq c(R_\lambda B\psi_0)(\mu, v) \quad \text{for all} \quad \psi \in B_r^+.$$

Since $\psi_0 \notin N(B)$, and $\psi_0 \geq 0$, we obtain $B\psi_0 \geq 0$ and $B\psi_0 \neq 0$. By using the positivity of R_λ and the fact that R_λ is the resolvent of the operator S_0 (i.e., $K = 0$), we deduce that

$$(R_\lambda B\mathcal{N}_f L\psi)(\mu, v) \geq c(R_\lambda B\psi_0)(\mu, v) \geq 0 \quad \text{and} \quad R_\lambda B\psi_0 \neq 0.$$

Accordingly,

$$\|R_\lambda B\mathcal{N}_f L\psi\| \geq c\|R_\lambda B\psi_0\| \neq 0 \quad \text{for all} \quad \psi \in B_r^+.$$

Therefore,

$$\inf\{\|\mathcal{F}(\lambda)\psi\|, \quad \psi \in B_r^+\} > 0. \tag{5.43}$$

Let us define the operator $\Lambda(\lambda)$ by:

$$\Lambda(\lambda)\psi = \frac{r\mathcal{F}(\lambda)\psi}{\|\mathcal{F}(\lambda)\psi\|} \quad \text{for all} \quad \psi \in B_r^+.$$

By using (5.43) and the fact that $\mathcal{F}(\lambda)$ is weakly sequentially continuous, we infer that $\Lambda(\lambda)$ is weakly sequentially continuous on X. In fact, let a sequence $(f_n)_n$ such that $f_n \rightharpoonup f$ on X. We have, for $l \in X'$, where X' denotes the dual of X,

$$l(\Lambda(\lambda)f_n) = \frac{r}{\|\mathcal{F}(\lambda)f_n\|} l(\mathcal{F}(\lambda)f_n).$$

Since $\mathcal{F}(\lambda)$ is weakly sequentially continuous, we deduce that $l(\mathcal{F}(\lambda)f_n)$ converges to $l(\mathcal{F}(\lambda)f)$. Moreover,

$$\mathcal{F}(\lambda)f_n = Q_\lambda K W_\lambda \Pi_\lambda B \mathcal{N}_f L f_n + R_\lambda B \mathcal{N}_f L f_n.$$

Since $\mathcal{N}_f L f_n \rightharpoonup \mathcal{N}_f L f$, then using both the weak compactness of $\Pi_\lambda B$ and $R_\lambda B$ (see Lemma 5.2.2) and Proposition 1.3.8, we deduce that $\Pi_\lambda B \mathcal{N}_f L f_n$ converges to $\Pi_\lambda B \mathcal{N}_f L f$ on X. By using the continuity of the operator $Q_\lambda K W_\lambda$, we infer that $\mathcal{F}(\lambda)f_n$ converges on X to $\mathcal{F}(\lambda)f$. Hence, $l(\Lambda(\lambda)f_n)$ converges to $\frac{r}{\|\mathcal{F}(\lambda)f\|} l(\mathcal{F}(\lambda)f)$. So, $\Lambda(\lambda)f_n \rightharpoonup \Lambda(\lambda)f$ and $\Lambda(\lambda)$ is weakly sequentially continuous on B_r^+ and takes B_r^+ into itself. We claim that $\Lambda(\lambda)(B_r^+)$ is relatively weakly compact. Indeed, set

$$m := \inf\{\|\mathcal{F}(\lambda)\psi\|, \quad \psi \in B_r^+\} > 0.$$

Then,

$$0 \leq \Lambda(\lambda)\psi \leq \frac{r}{m}\mathcal{F}(\lambda)\psi \text{ for all } \psi \in B_r^+.$$

Since B_r^+ is a bounded convex subset of X^+, then $\mathcal{N}_f L(B_r^+)$ is a bounded subset of X^+ and therefore, by using Theorem 5.2.2, we show that $\mathcal{F}(\lambda)(B_r^+)$ is weakly compact. Moreover, by using Theorem 1.3.7, we may conclude that

$$\lim_{|E|\to 0} \int_E (\mathcal{F}(\lambda)\psi)(\mu, v)\, d\mu dv = 0,$$

uniformly for ψ in B_r^+. Now, by using the fact that

$$0 \leq \Lambda(\lambda)\psi \leq \frac{r}{m}\mathcal{F}(\lambda)\psi \text{ for all } \psi \in B_r^+,$$

we find that

$$0 \leq \int_E (\Lambda(\lambda)\psi)(\mu, v)\, d\mu dv \leq \frac{r}{m}\int_E (\mathcal{F}(\lambda)\psi)(\mu, v)\, d\mu dv, \text{ for all } \psi \in B_r^+.$$

Hence,

$$\lim_{|E|\to 0} \int_E (\mathcal{F}(\lambda)\psi)(\mu, v)\, d\mu dv = 0,$$

uniformly for ψ in B_r^+, and $\Lambda(\lambda)(B_r^+)$ is weakly compact. Finally, the use of Theorem 2.2.1 shows that $\Lambda(\lambda)$ has, at least, a fixed point ψ^* in B_r^+ satisfying $\|\psi^*\| = r$. Setting $\eta = \frac{r}{\|\mathcal{F}(\lambda)\psi^*\|}$, we obtain

$$(\lambda - S_K)^{-1} B \mathcal{N}_f L(\psi^*) = \eta^{-1}\psi^*.$$

Consequently,

$$\psi^* \in \mathcal{D}(S_K) \cap B_r,$$

and

$$v\frac{\partial\psi^*}{\partial\mu}(\mu,v)+\sigma(\mu,v)\psi^*(\mu,v)+\lambda\psi^*(\mu,v)$$
$$=\eta\int_a^b\kappa(\mu,v,v')f(\mu,v',L(\psi^*)(\mu,v'))dv'$$

which achieves the proof. Q.E.D.

5.2.3 The general case

In the following, we are concerned with the existence of solutions for the more general nonlinear boundary value problem

$$\begin{cases} v\frac{\partial\psi}{\partial\mu}+\sigma(\mu,v,\psi(\mu,v))+\lambda\psi(\mu,v)=\int_a^b r(\mu,v,v',\psi(\mu,v'))dv' \\ \\ \psi^0=K(\psi^1),\ \lambda\in\mathbb{C}. \end{cases} \tag{5.44}$$

When dealing with this problem, some technical difficulties arise. So, we need the following assumptions:

$(\mathcal{B}_{13})$ $K\in\mathcal{L}(X^1,X^0)$ and for some $r>0$, we have

$$|\sigma(\mu,v,\psi_1)-\sigma(\mu,v,\psi_2)|\ \le\ |\rho(\mu,v)|\,|\psi_1-\psi_2| \qquad (\psi_1,\psi_2\in\ X),$$

where $\mathcal{L}(X^1,X^0)$ denotes the set of all bounded linear operators from X^1 into X^0, $\rho(.,.)\in L^\infty([0,1]\times[a,b],d\mu dv)$, and $\mathcal{N}_\sigma$ acts from B_r into B_r. Let us define the free streaming operator $\widetilde{S}_K$ by:

$$\begin{cases} \widetilde{S}_K:\mathcal{D}(\widetilde{S}_K)\subset X\longrightarrow X \\ \qquad\qquad \psi\longrightarrow\widetilde{S}_K\psi(\mu,v)=-v\frac{\partial\psi}{\partial\mu}(\mu,v) \\ \mathcal{D}(\widetilde{S}_K)=\left\{\psi\in W \text{ such that } \psi^0=K(\psi^1)\right\}. \end{cases}$$

Theorem 5.2.6 *Let $r>0$. If $(\mathcal{B}_{10})$, $(\mathcal{B}_{11})$, and $(\mathcal{B}_{13})$ are satisfied, and if B is a regular collision operator on X, then there exists $\lambda_0>0$ such that, for all λ satisfying $Re\lambda>\lambda_0$, the problem (5.44) has, at least, one solution in B_r.*

Proof. Since K is linear (according to $(\mathcal{B}_{13})$), the operator $\widetilde{S}_K$ is linear too and by using Lemma 5.2.1, we infer that

$$\{\lambda\in\mathbb{C}\ \text{ such that }\ Re\lambda>\max(0,b\log(\|K\|))\}\subset\varrho(\widetilde{S}_K),$$

where $\varrho(\widetilde{S}_K)$ denotes the resolvent set of $\widetilde{S}_K$. Let λ be such that $Re\lambda > \max(0, b\log(\|K\|))$. Then, by using the linearity of the operator $(\lambda - \widetilde{S}_K)^{-1}$, the problem (5.44) may be written in the following form:

$$\begin{cases} \psi = (\lambda - \widetilde{S}_K)^{-1}\mathcal{N}_\sigma\psi + (\lambda - \widetilde{S}_K)^{-1}B\mathcal{N}_f L\psi \\ \quad = \mathcal{H}(\lambda)\psi + \widetilde{\mathcal{F}}(\lambda)\psi \\ \psi \in \mathcal{D}(\widetilde{S}_K), Re\lambda > \max(0, b\log(\|K\|)), \end{cases}$$

where

$$\mathcal{H}(\lambda) := (\lambda - \widetilde{S}_K)^{-1}\mathcal{N}_\sigma$$

and

$$\widetilde{\mathcal{F}}(\lambda) := (\lambda - \widetilde{S}_K)^{-1}B\mathcal{N}_f L.$$

Let us check that, for a suitable λ, the operator $\mathcal{H}(\lambda)$ is a contraction mapping. Indeed, let ψ_1, ψ_2 in X. Then, we have

$$\begin{aligned} \|\mathcal{H}(\lambda)\psi_1 - \mathcal{H}(\lambda)\psi_2\| &= \|(\lambda - \widetilde{S}_K)^{-1}\mathcal{N}_\sigma(\psi_1) - (\lambda - \widetilde{S}_K)^{-1}\mathcal{N}_\sigma(\psi_2)\| \\ &\leq \|(\lambda - \widetilde{S}_K)^{-1}\|\|\mathcal{N}_\sigma(\psi_1) - \mathcal{N}_\sigma(\psi_2)\|. \end{aligned}$$

A simple calculation using the estimates (5.28)–(5.31) leads to

$$\|(\lambda - \widetilde{S}_K)^{-1}\| \leq \frac{1}{Re\lambda}\left\{1 + \frac{\|K\|}{1 - \|K\|e^{-\frac{Re\lambda}{b}}}\right\}. \tag{5.45}$$

Moreover, taking into account the assumption on $\sigma(.,.,.)$, we get

$$\|\mathcal{N}_\sigma(\psi_1) - \mathcal{N}_\sigma(\psi_2)\| \leq \|\rho\|_\infty\|\psi_1 - \psi_2\|.$$

Therefore, we have

$$\begin{aligned} \|\mathcal{H}(\lambda)\psi_1 - \mathcal{H}(\lambda)\psi_2\| &\leq \frac{\|\rho\|_\infty}{Re\lambda}\left\{1 + \frac{\|K\|}{1 - \|K\|\, e^{-\frac{Re\lambda}{b}}}\right\}\|\psi_1 - \psi_2\| \\ &= \Xi(Re\lambda)\,\|\psi_1 - \psi_2\|. \end{aligned}$$

We notice that Ξ is a continuous strictly decreasing function defined on $]0, \infty[$, and

$$\lim_{x\to\infty} \Xi(x) = 0.$$

So, there exists $\lambda_1 \in]\max(0, b\log(\|K\|)), \infty[$ such that $\Xi(\lambda_1) < 1$. Hence, for $Re\lambda \geq \lambda_1$, $\mathcal{H}(\lambda)$ is a contraction mapping. Now, let φ and ψ be two elements of B_r. Then, we have

$$\begin{aligned}\|\mathcal{H}(\lambda)\varphi + \widetilde{\mathcal{F}}(\lambda)\psi\| &\leq \|(\lambda - \widetilde{S}_K)^{-1}B\mathcal{N}_f L\psi\| + \|(\lambda - \widetilde{S}_K)^{-1}\mathcal{N}_\sigma\varphi\| \\ &\leq \|(\lambda - \widetilde{S}_K)^{-1}\| \;\{\|B\|\;\|\mathcal{N}_f L\psi\| + \|\mathcal{N}_\sigma\varphi\|\} \\ &\leq \|(\lambda - \widetilde{S}_K)^{-1}\| \;\{\|B\|\;\|A\|\|h\|_\infty + M(r)\},\end{aligned}$$

where $M(r)$ denotes the upper bound of $\mathcal{N}_\sigma$ on B_r. By using the estimate (5.45), we get

$$\begin{aligned}\|\mathcal{H}(\lambda)\varphi + \widetilde{\mathcal{F}}(\lambda)\psi\| &\leq \frac{1}{Re\lambda}\;[\|B\|\;\|A\|\|h\|_\infty + M(r)]\left[1 + \frac{\|K\|}{1 - \|K\|\;e^{-\frac{Re\lambda}{b}}}\right] \\ &= \mathcal{G}(Re\lambda),\end{aligned}$$

where $\mathcal{G}(.)$ has the same properties as $\Xi(.)$. Arguing as above, we show that there exists $\lambda_2 > \max\,(0, b\log(\|K\|))$ such that, for $Re\lambda > \lambda_2$, $\mathcal{G}(\lambda) < r$. Accordingly, for $Re\lambda \geq \lambda_2$,

$$\mathcal{H}(\lambda)\varphi + \widetilde{\mathcal{F}}(\lambda)\psi \;\in B_r$$

when $\varphi,\ \psi \in B_r$. Arguing as in the proof of Theorem 5.2.3, we show that the operator

$$(I - \mathcal{H}(\lambda))^{-1}\widetilde{\mathcal{F}}(\lambda)$$

is weakly sequentially continuous on X. In fact,

$$(I - \mathcal{H}(\lambda))^{-1}\widetilde{\mathcal{F}}(\lambda) = (I - \mathcal{H}(\lambda))^{-1}(\lambda - \widetilde{S}_K)^{-1}B\mathcal{N}_f L.$$

By using the weak compactness of $(\lambda - \widetilde{S}_K)^{-1}B$ and Proposition 1.3.8, we deduce that $(\lambda - \widetilde{S}_K)^{-1}B$ is strongly continuous. The continuity of the operator $(I - \mathcal{H}(\lambda))^{-1}$ and the sequentially weak continuity of $\mathcal{N}_f L$ imply the sequentially weak continuity of $(I - \mathcal{H}(\lambda))^{-1}\widetilde{\mathcal{F}}(\lambda)$. Finally, we claim that

$$(I - \mathcal{H}(\lambda))^{-1}\widetilde{\mathcal{F}}(\lambda)$$

is weakly compact on B_r. Indeed, let $\varphi \in B_r$. Then, we have

$$|\mathcal{N}_f L(\varphi)| \leq A(\mu, v)\, h(|L\varphi(\mu, v)|).$$

For all measurable subsets E of $[0, 1] \times [a, b]$, we have :

$$\int_E |\mathcal{N}_f L(\varphi)(\mu, v)|\; d\mu dv \leq \|h\|_\infty \int_E |A(\mu, v)|\; d\mu dv,$$

and we conclude that

$$\lim_{|E|\to 0} \int_E |\mathcal{N}_f L(\varphi)(\mu, v)|\; d\mu dv = 0,$$

uniformly for $\varphi \in B_r$. So, $\mathcal{N}_f L$ is weakly compact on B_r. Let $(x_n)_n \in B_r$. Since, $\mathcal{N}_f L$ is weakly compact, there exists a subsequence $(x_{\rho(n)})_n$ such that $\mathcal{N}_f L(x_{\rho(n)})$ is weakly convergent. Since $(\lambda - \widetilde{S}_K)^{-1}B$ is a linear weakly compact operator and since X is a Dunford–Pettis space, then Proposition 1.3.8 allows us to deduce that $(\lambda - \widetilde{S}_K)^{-1}B\mathcal{N}_f L(x_{\rho(n)})$ is convergent. The continuity of the operator $(I - \mathcal{H}(\lambda))^{-1}$ implies that $(I - \mathcal{H}(\lambda))^{-1}\widetilde{\mathcal{F}}(\lambda)(x_{\rho(n)})$ is convergent. Hence, $(I - \mathcal{H}(\lambda))^{-1}\widetilde{\mathcal{F}}(\lambda)$ is weakly compact and so, $(I - \mathcal{H}(\lambda))^{-1}\widetilde{\mathcal{F}}(\lambda)(B_r)$ is relatively weakly compact. Obviously, if $\lambda_0 = \max(\lambda_1, \lambda_2)$, then for all λ satisfying $Re\lambda \geq \lambda_0$, the operators $\mathcal{H}(\lambda)$ and $\widetilde{\mathcal{F}}(\lambda)$ satisfy the conditions of Theorem 2.2.3. Consequently, the problem (5.44) has a solution ψ in B_r, for all λ such that $Re\lambda \geq \lambda_0$. Q.E.D.

Chapter 6

Existence of Solutions for Nonlinear Integral Equations

We start this chapter by studying the existence of solutions for some variants of Hammerstein's integral equation. Next, we investigate the existence of solutions for several nonlinear functional integral and differential equations, in the Banach algebra $\mathcal{C}([0,T],X)$, where X is a Banach algebra satisfying the condition $(\mathcal{P})$. An application of Leray–Schauder type fixed point theorem under the weak topology is given.

6.1 Existence of Solutions for Hammerstein's Integral Equation

6.1.1 Hammerstein's integral equation

The objective of this section is to prove the existence of a solution for the following nonlinear integral equation

$$x(t) = f(t, x(t)) + \lambda \int_0^t g(s, x(s))ds, \quad x \in \mathcal{C}(J, X), \tag{6.1}$$

where $J = [0,T]$, $\lambda \in (\frac{1}{2}, 1)$, $(X, \|.\|)$ is a reflexive Banach space, $\mathcal{C}(J, X)$ is the Banach space of all continuous functions from J into X endowed with the sup-norm $\|.\|_\infty$, defined by $\|x\|_\infty = \sup \{\|x(t)\|;\ t \in J\}$, for each $x \in \mathcal{C}(J, X)$, and f and g satisfy some conditions. Let us suppose that the functions involved in Eq. (6.1) satisfy the following conditions:

$(\mathcal{J}_1)$ The mapping $f : J \times X \longrightarrow X$ is such that:

(i) f is a nonlinear contraction with respect to the second variable, i.e., there exists a continuous nondecreasing function $\varphi : \mathbb{R}_+ \longrightarrow \mathbb{R}_+$ satisfying

$\varphi(r) < r$, for $r > 0$ and

$$\|f(t,x) - f(t,y)\| \leq \varphi(\|x - y\|) \text{ for all } x, y \in X \text{ and } t \in J,$$

(*ii*) $\varphi(r) < (1 - \lambda)r$, for all $r > 0$, and

(*iii*) $f(t,0) = f(0,0)$, for all $t \in J$.

$(\mathcal{J}_2)$ The mapping $x \longrightarrow f(.,x(.))$ is weakly sequentially continuous on $\mathcal{C}(J,X)$.

$(\mathcal{J}_3)$ For every $t \in J$, the mapping $g_t = g(t,.) : X \longrightarrow X$ is weakly sequentially continuous.

$(\mathcal{J}_4)$ For all $x \in \mathcal{C}(J,X)$, $g(.,x(.))$ is Pettis integrable on J.

$(\mathcal{J}_5)$ There exist $\alpha \in L^1([0,T])$ and a nondecreasing continuous function ϕ from $[0,+\infty)$ to $(0,+\infty)$ such that $\|g(t,x)\| \leq \alpha(t)\phi(|\|x\| - \|f(0,0)\||)$ for a.e. $t \in [0,T]$ and all $x \in X$. Further, we assume that $\int_0^T \alpha(s)ds < \int_{\|f(0,0)\|}^{+\infty} \frac{dr}{\phi(r)}$.

The existence result for Eq. (6.1) is given by:

Theorem 6.1.1 *Assume that the assumptions* $(\mathcal{J}_1)$–$(\mathcal{J}_5)$ *hold. Then, Eq.* (6.1) *has, at least, one solution* $x \in \mathcal{C}(J,X)$.

Proof. We set

$$\beta(t) = \int_{\|f(0,0)\|}^{t} \frac{dr}{\phi(r)} \quad \text{and} \quad b(t) = \beta^{-1}\left(\int_0^t \alpha(s)ds\right).$$

Then,

$$\int_{\|f(0,0)\|}^{b(t)} \frac{dr}{\phi(r)} = \int_0^t \alpha(s)ds. \tag{6.2}$$

Now, we define the set Ω by:

$$\Omega = \{x \in \mathcal{C}(J,X) \text{ such that } \|x(t)\| \leq b(t) + \|f(0,0)\| \text{ for all } t \in J\}.$$

Clearly, Ω is a closed convex and bounded subset of $\mathcal{C}(J,X)$. Let us consider the nonlinear mappings A, $B : \mathcal{C}(J,X) \longrightarrow \mathcal{C}(J,X)$ defined as follows:

$$(Ax)(t) = f(0,0) + \lambda \int_0^t g(s,x(s))ds,$$

and

$$(Bx)(t) = f(t,x(t)) - f(0,0).$$

In the following, we will prove that A and B satisfy the assumptions of Theorem 2.3.6. For this purpose, we need four steps:

Step 1: Let us check that $A(\Omega) \subset \Omega$, $A(\Omega)$ is weakly equicontinuous and $A(\Omega)$ is relatively weakly compact. (i) Let $x \in \Omega$ be an arbitrary point. We will prove that $Ax \in \Omega$. Let $t \in J$. Without loss of generality, we may assume that $(Ax)(t) \neq 0$. According to Hahn–Banach's theorem, there exists $x_t \in X^*$ such that $\|x_t\| = 1$ and $\|(Ax)(t)\| = x_t((Ax)(t))$. By using assumption $(\mathcal{J}_5)$ and Eq. (6.2), we get

$$\begin{aligned} \|(Ax)(t)\| &= x_t\left(f(0,0) + \lambda\int_0^t g(s,x(s))ds\right) \\ &\leq \|f(0,0)\| + \lambda\int_0^t \alpha(s)\phi(\|x(s)\|)ds \\ &\leq \|f(0,0)\| + \lambda\int_0^t \alpha(s)\phi(b(s))ds \\ &= \|f(0,0)\| + \lambda(b(t) - \|f(0,0)\|) \\ &= \lambda b(t) + (1-\lambda)\|f(0,0)\| \qquad (6.3) \\ &< b(t) + \|f(0,0)\|. \end{aligned}$$

Hence, $Ax \in \Omega$.

(ii) Let $\varepsilon > 0$; $x \in \Omega$; $x^* \in X^*$; t, $t' \in J$ such that $t \leq t'$ and $t' - t \leq \varepsilon$. Let us show that $\|x^*\left((Ax)(t) - (Ax)(t')\right)\| \leq \varepsilon$. We have

$$\begin{aligned} \|(Ax)(t) - (Ax)(t')\| &= \lambda\left\|\int_0^t g(s,x(s))ds - \int_0^{t'} g(s,x(s))ds\right\| \\ &\leq \int_t^{t'} \|g(s,x(s))\|ds. \end{aligned}$$

By using assumption $(\mathcal{J}_5)$, we obtain

$$\|(Ax)(t) - (Ax)(t')\| \leq \int_t^{t'} \alpha(s)\phi(b(s))ds \leq |b(t) - b(t')|.$$

(iii) The reflexiveness of X implies that, for all $t \in J$, the subset

$$A(\Omega)(t) = \{(Ax)(t); \quad x \in \Omega\}$$

is relatively weakly compact. Since $A(\Omega)$ is weakly equicontinuous, then, according to Arzelà–Ascoli's theorem theorem (see Theorem 1.3.9), we deduce that $A(\Omega)$ is relatively weakly compact.

Step 2: Let us show that $A : \Omega \longrightarrow \Omega$ is weakly sequentially continuous.
To do so, let $(x_n)_n \subset \Omega$ such that $x_n \rightharpoonup x \in \Omega$. Taking into account both Dobrakov's theorem (see Theorem 1.4.1) and the fact that $(x_n)_n$ is bounded, we deduce that $x_n(t) \rightharpoonup x(t)$ in X, for all $t \in J$. From the assumption $(\mathcal{J}_3)$, we get $g(t, x_n(t)) \rightharpoonup g(t, x(t))$ in X, for all $t \in J$. So, by using $(\mathcal{J}_5)$ and the dominated convergence theorem, we infer that $(Ax_n)(t) \rightharpoonup (Ax)(t)$ in X. Since $(Ax_n)_n$ is bounded $(A(\Omega) \subset \Omega)$, then $Ax_n \rightharpoonup Ax$ in $\mathcal{C}(J, X)$. Consequently, A is weakly sequentially continuous.

Step 3: Let us show that B is a nonlinear contraction.
Let $x,\ y \in \mathcal{C}(J, X)$. Using the assumption $(\mathcal{J}_1)$ (i), we infer that

$$\begin{aligned} \|(Bx)(t) - (By)(t)\| &= \|f(t, x(t)) - f(t, y(t))\| \\ &\leq \varphi\left(\|x(t) - y(t)\|\right) \\ &\leq \varphi\left(\|x - y\|_\infty\right), \end{aligned}$$

for all $t \in J$. So,

$$\|Bx - By\|_\infty \leq \varphi(\|x - y\|_\infty).$$

Hence, B is a nonlinear contraction.

Step 4: $(x = Bx + Ay,\ \ y \in \Omega) \Longrightarrow x \in \Omega$.
First, we claim that $\|(I - B)x(t)\| \geq \|x(t)\| - \varphi(\|x(t)\|)$ for every $x \in \mathcal{C}(J, X)$ and $t \in J$. Indeed, taking into account the assumption $(\mathcal{J}_1)$ (i), we get

$$\begin{aligned} \|(I - B)x(t) - (I - B)y(t)\| &\geq \|x(t) - y(t)\| - \|Bx(t) - By(t)\| \\ &\geq \|x(t) - y(t)\| - \varphi\left(\|x(t) - y(t)\|\right), \end{aligned}$$

for every $x,\ y \in \mathcal{C}(J, X)$ and $t \in J$. In particular, for $y = 0$, the assumption $(\mathcal{J}_1)$ (iii) implies that

$$\|(I - B)x(t)\| \geq \|x(t)\| - \varphi(\|x(t)\|) \text{ for every } x \in \mathcal{C}(J, X) \text{ and } t \in J, \tag{6.4}$$

as claimed. Now, let $x \in \mathcal{C}(J, X)$ and $y \in \Omega$ such that $x = Bx + Ay$ and let us show that $\|x(t)\| \leq \|f(0, 0)\| + b(t)$, for all $t \in J$. Without loss of generality, we may suppose that $x(t) \neq 0$. So, Inequality (6.4) and assumption $(\mathcal{J}_1)$ (ii) imply that $\|(I - B)x(t)\| \geq \lambda \|x(t)\|$, for all $t \in J$. Consequently, we have

$$\begin{aligned} \|x(t)\| &\leq \frac{1}{\lambda} \|((I - B)x)(t)\| \\ &\leq \frac{1}{\lambda} \|(Ay)(t)\|, \end{aligned}$$

for all $t \in J$. Hence, from Eq. (6.3), we deduce that

$$\begin{aligned} \|x(t)\| &\leq b(t) + \frac{1-\lambda}{\lambda}\|f(0,0)\| \\ &< b(t) + \|f(0,0)\|, \end{aligned}$$

for all $t \in J$. Q.E.D.

6.1.2 A general Hammerstein's integral equation

The following integral equation represents a natural generalization of Eq. (6.1).

$$\psi(t) = g(t, L_2((\psi(t))) + \lambda \int_\Omega k(t,s) f(s, L_1(\psi(s)))ds, \tag{6.5}$$

where $\psi \in L^1(\Omega, X)$, represents the space of Lebesgue integrable functions on a measurable subset Ω of $\mathbb{R}^N$ with values in a finite dimensional Banach space X. Here, g is Lipschitzian with respect to the second variable, while $f(.,.)$ (resp. $k(.,.)$) is a nonlinear (resp. measurable) function and $L_i : L^1(\Omega, X) \longrightarrow L^\infty(\Omega, X)$, $i = 1, 2$, are continuous linear maps. Let Ω be a domain of $\mathbb{R}^N$ and let X and Y be two Banach spaces. A function $f : \Omega \times X \longrightarrow Y$ is said to be weak Carathéodory, if:

(i) For any $x \in X$, the map $t \longrightarrow f(t, x)$ is measurable from Ω into Y, and

(ii) for almost all $t \in \Omega$, the map $x \longrightarrow f(t, x)$ is weakly sequentially continuous from X into Y.

Let $m(\Omega, X)$ be the set of all measurable functions $\psi : \Omega \longrightarrow X$. If f is a weak Carathéodory function, then f defines a mapping $N_f : m(\Omega, X) \longrightarrow m(\Omega, Y)$ by $N_f\psi(t) := f(t, \psi(t))$, for all $t \in \Omega$. This mapping is called the Nemytskii's operator associated to f.

Remark 6.1.1 *We should notice that, in general, the Nemytskii's operator N_f is not weakly continuous. In fact, even in the scalar case, only linear functions generate weakly continuous Nemytskii's operators in L^1 spaces (see for example [9, 130]).*

The following result will play a crucial role in our application; for further details and proofs, we may refer to [23].

Lemma 6.1.1 *Let X be a reflexive Banach space, and p, $q \geq 1$ and let $L : L^p(\Omega, X) \longrightarrow L^\infty(\Omega, X)$ be a continuous linear map. Let $f : \Omega \times X \longrightarrow X$ be a weak Carathéodory map satisfying*

$$\|f(t,x)\| \leq A(t)h(\|x\|),$$

where $A \in L^q(\Omega)$ and $h \in L^{loc}_\infty(\mathbb{R}_+)$. Then, if either $q > 1$ or $p = q = 1$, the map $\zeta := N_f \circ L : L^p(\Omega, X) \longrightarrow L^q(\Omega, X)$ is weakly sequentially continuous.

Let X be a reflexive Banach space. First, we observe that Eq. (6.5) may be written in the following form:

$$\psi = \mathcal{A}\psi + \mathcal{B}\psi,$$

where $\mathcal{B} := N_g L_2$ represents the product operator of the linear map L_2 and the Nemytskii's operator associated to the function $g(.,.)$ and $\mathcal{A} := \lambda C N_f L_1$ is the product operator of the linear map L_1 and the Nemytskii's operator associated to $f(.,.)$ and the linear integral operator λC. Note that $\lambda \in \mathbb{C}$ and C is the operator defined from $L^1(\Omega, X)$ into $L^1(\Omega, X)$ by:

$$C\psi(t) = \int_\Omega k(t,s)\psi(s)ds.$$

Now, let us introduce the following assumptions:

($\mathcal{J}_6$) The function $g : \Omega \times X \longrightarrow X$ is a measurable function, $g(.,0) \in L^1(\Omega, X)$ and g is Lipschitzian with respect to the second variable, i.e., there exists an $\alpha \in \mathbb{R}_+$ such that $\|g(t,x) - g(t,y)\| \leq \alpha\|x - y\|$ for all $t \in \Omega$ and x, $y \in X$.

($\mathcal{J}_7$) The functions f, $g : \Omega \times X \longrightarrow X$ satisfy the weak Carathéodory conditions and there exist functions $A_i \in L^1(\Omega, X)$ and $h_i \in L^{loc}_\infty(\mathbb{R}_+)$, $i = 1, 2$, such that

$$\|f(t,x)\| \leq A_1(t)h_1(\|x\|) \quad \text{and} \quad \|g(t,x)\| \leq A_2(t)h_2(\|x\|).$$

($\mathcal{J}_8$) C is a continuous linear operator on $L^1(\Omega, X)$.

($\mathcal{J}_9$) $L_i : L^1(\Omega, X) \longrightarrow L^\infty(\Omega, X)$, $i = 1, 2$, represent continuous linear maps.

($\mathcal{J}_{10}$) $\alpha\mu(\Omega)\|L_2\| \in (0, 1)$, where $\|L_2\|$ denotes the norm of the map L_2 and $\mu(\Omega)$ is the Lebesgue's measure of Ω.

The following definition and lemma give a characterization of $\beta(M)$ for any bounded subset M of $L^1(\Omega, X_0)$ (see [9]).

Definition 6.1.1 *Let Ω be a compact subset of $\mathbb{R}^N$ and let M be a bounded subset of $L^1(\Omega, X_0)$. We call the following real number*

$$\pi_1(M) = \limsup_{\varepsilon \to 0} \left\{ \sup_{\psi \in M} \left\{ \int_D \|\psi(t)\| dt : meas(D) \leq \varepsilon \right\} \right\},$$

the measure of nonequiabsolute continuity of M, where meas(.) denotes the Lebesgue measure.

Lemma 6.1.2 *[9] Let X_0 be a finite-dimensional Banach space and let Ω be a compact subset of $\mathbb{R}^N$. If M is a bounded subset of $L^1(\Omega, X_0)$, then $\beta(M) = \pi_1(M)$.*

The following theorem provides an existence result for Eq. (6.5).

Theorem 6.1.2 *Let X be a finite-dimensional Banach space and let Ω be a compact subset of $\mathbb{R}^N$. Assume that the conditions $(\mathcal{J}_6)$–$(\mathcal{J}_{10})$ are satisfied. Then, Eq. (6.5) has, at least, one solution in $L^1(\Omega, X)$.*

Proof. Notice that Eq. (6.5) can be written in the following form

$$\psi = \mathcal{A}\psi + \mathcal{B}\psi.$$

Claim 1: $\mathcal{A} := \lambda C N_f L_1$ and $\mathcal{B} := N_g L_2$ are weakly sequentially continuous on $L^1(\Omega, X)$. Indeed, from Lemma 6.1.1, we prove that $N_g L_2$ and $N_f L_1$ are weakly sequentially continuous on $L^1(\Omega, X)$. Moreover, using [40], we deduce that C is weakly sequentially continuous on $L^1(\Omega, X)$ which ends the first claim.

Claim 2: Let ψ, $\varphi \in L^1(\Omega, X)$. From the assumption $(\mathcal{J}_6)$, it follows that

$$\begin{aligned}
\|\mathcal{B}\psi - \mathcal{B}\varphi\|_{L^1(\Omega,X)} &= \int_\Omega \|g(t, L_2(\psi(t))) - g(t, L_2(\varphi(t)))\|_X dt \\
&\leq \alpha \int_\Omega \|L_2(\psi(t)) - L_2(\varphi(t))\|_X dt \\
&\leq \alpha\mu(\Omega)\|L_2\|\|\psi - \varphi\|_{L^1(\Omega,X)}.
\end{aligned}$$

Hence, $\mathcal{B}$ is a strict contraction mapping with a constant $\alpha\mu(\Omega)\|L_2\|$ on $L^1(\Omega, X)$.

Claim 3: By using the hypotheses $(\mathcal{J}_6)$–$(\mathcal{J}_{10})$, we have

$$\begin{aligned}\|\psi\| = \|\mathcal{A}\varphi + \mathcal{B}\psi\| &= \|\lambda C N_f L_1\varphi + N_g L_2\psi\| \\ &\leq |\lambda|\|C\|\|A_1\|\|h_1\|_\infty + \|\xi\| + \alpha\|L_2\|\|\psi\|,\end{aligned}$$

where $\xi(t) := \|g(t,0)\|$, $\forall t \in \Omega$. Moreover, we have

$$\begin{aligned}\|\psi\| = \|\mathcal{A}\varphi + \mathcal{B}\psi\| &= \|\lambda C N_f L_1\varphi + N_g L_2\psi\| \\ &\leq |\lambda|\|C\|\|A_1\|\|h_1\|_\infty + \|A_2\|\|h_2\|_\infty.\end{aligned}$$

Let r_0 be the real number defined by:

$$r_0 = \min\left\{\frac{|\lambda|\|C\|\|A_1\|\|h_1\|_\infty + \|\xi\|}{1-\alpha\|L_2\|},\ |\lambda|\|C\|\|A_1\|\|h_1\|_\infty + \|A_2\|\|h_2\|_\infty\right\}. \quad (6.6)$$

Clearly, Eq. (6.6) guarantees that $\psi = \mathcal{A}\varphi + \mathcal{B}\psi$ is in B_r for all $\varphi \in B_r$ and $r \geq r_0$.

Claim 4: $\mathcal{A}B_r$ is relatively weakly compact for all $r \geq r_0$. Indeed, let S be a bounded subset of B_r and let ε be a positive real number. For any nonempty subset D of Ω, and for all $\psi \in S$, we have

$$\begin{aligned}\int_D \|N_f L_1\psi(t)\|dt &\leq \int_D \|A_1(t)\| h_1(\|L_1\psi(t)\|)dt \\ &\leq \|h_1\|_\infty \int_D \|A_1(t)\|dt \\ &= \|h_1\|_\infty \|A_1\|_{L^1(D)}.\end{aligned}$$

Moreover, by using Theorem 1.3.7, we may conclude that

$$\limsup_{\varepsilon\to 0}\left\{\|h_1\|_\infty \int_D \|A_1(t)\|dt : \text{meas}(D) \leq \varepsilon\right\} = 0.$$

Then, we have $\omega(N_f L_1 S) = 0$ and thus, $N_f L_1 S$ is relatively weakly compact. Besides, since λC is bounded, we deduce that $\omega(\mathcal{A}S) = 0$ and then, $\mathcal{A}S$ is relatively weakly compact, too. To end the proof, we may apply Corollary 3.2.2, and we show that the operator $\mathcal{A} + \mathcal{B}$ has, at least, a fixed point in B_r, for all $r \geq r_0$; equivalently, Eq. (6.5) has a solution in B_r. Q.E.D.

Remark 6.1.2 *We should notice that the equality of Lemma 6.1.2, which is fundamental in the proof of Theorem 6.1.2, was established for bounded subsets of the space of Lebesgue integrable functions with values in a finite-dimensional Banach space [9]. Moreover, finite dimensional Banach spaces are reflexive (as required in Lemma 6.1.1). This justifies the assumption that X must be a finite-dimensional Banach space.*

Question 5:

At this point we don't know whether or not Theorem 6.1.2 holds for reflexive infinite-dimensional Banach spaces.

6.2 A Study of Some FIEs in Banach Algebras

In this section, we are concerned with the study of solutions for some functional integral equations (in short FIE). These solutions belong to the Banach algebra $\mathcal{C}(J,X)$. More precisely, let $(X,||.||)$ be a Banach algebra satisfying the condition $(\mathcal{P})$. Let $J=[0,1]$ be the closed and bounded interval in $\mathbb{R}$, the set of all real numbers. Let $\mathcal{C}(J,X)$ be the Banach algebra of all continuous functions from $[0,1]$ to X, endowed with the sup-norm $||\ ||_\infty$, defined by $||f||_\infty=\sup\{||f(t)||\ ;\ t\in[0,1]\}$, for each $f\in\mathcal{C}(J,X)$. Recall that $\mathcal{C}(J,X)$ is also a Banach algebra satisfying the condition $(\mathcal{P})$(see Proposition 1.5.1).

6.2.1 The weak sequential continuity and the weak compactness in FIEs

Let us consider the following functional integral equation:

$$x(t)=a(t)+(Tx)(t)\left[\left(q(t)+\int_0^{\sigma(t)}p(t,s,x(\zeta(s)),x(\eta(s)))ds\right)\cdot u\right] \tag{6.7}$$

for all $t\in J$, where J is the interval $[0,1]$ and X is a Banach algebra satisfying the condition $(\mathcal{P})$. The functions a, q, σ, ζ, η are continuous on J; T, $p(.,.,.,.)$ are nonlinear functions and u is a nonvanishing vector of X.

Remark 6.2.1 *Notice that the FIE (6.7) contains several special types of functional integral equations in $\mathcal{C}(J,\mathbb{R})$:*

(1) *If we take:* $\zeta(s) = s;\ \eta(s) = \lambda s,\quad 0 < \lambda < 1$ *and* $u = 1$, *then the existence results are reduced to those proved in [50] for the nonlinear integral equation:*

$$x(t) = a(t) + (Tx)(t)\left(q(t) + \int_0^{\sigma(t)} p(t,s,x(s),x(\lambda s))ds\right),\quad 0 < \lambda < 1.$$

(2) *If we take:* $q(t) = 0;\ \sigma(t) = 1;\ \zeta(s) = s;\ \eta(s) = \lambda s, 0 < \lambda < 1$ *and* $u = 1$, *then the existence results are reduced to those proved for functional integral equations of Urysohn type:*

$$x(t) = a(t) + (Tx)(t)\int_0^1 p(t,s,x(s),x(\lambda s))ds,\ 0 < \lambda < 1.$$

(3) *If we take* $\sigma(t) = \zeta(t) = t;\ q(t) = 0;\ T = 1;\ p(t,s,x,y) = k(t,s)x$ *and* $u = 1$, *then existence results are reduced to those proved for the classical linear Volterra integral equation on bounded interval in [46]:*

$$x(t) = a(t) + \int_0^t k(t,s)x(s)ds.$$

(4) *If we take* $\sigma(t) = \zeta(t) = t;\ q(t) = 0;\ T = 1;\ p(t,s,x,y) = k(t,s)f(s,x)$ *and* $u = 1$, *then the existence results are reduced to those proved for the nonlinear integral equation of Volterra–Hammerstein type:*

$$x(t) = a(t) + \int_0^t k(t,s)f(s,x(s))ds.$$

(5) *If we take* $a(t) = 0;\ (Tx)(t) = f(t,x(\nu(t)));\ p(t,s,x,y)) = g(s,y)$ *and* $u = 1$, *then the existence results are reduced to those proved in [77] for the nonlinear integral equation:*

$$x(t) = f(t,x(\nu(t)))\left(q(t) + \int_0^{\sigma(t)} g(s,x(\eta(s)))ds\right).$$

(6) *If we take* $\eta(s) = \lambda s,\ 0 < \lambda < 1$, *then existence results are reduced to these proved in [26] for the nonlinear integral equation:*

$$x(t) = a(t) + (Tx)(t)\left(q(t) + \int_0^{\sigma(t)} p(t,s,x(s),x(\lambda s))ds\right)\cdot u,\ 0 < \lambda < 1.$$

Next, we will prove the existence of solutions for the FIE (6.7) under some suitable conditions. For this purpose, let us assume that the functions involved in the FIE (6.7) satisfy the following conditions:

$(\mathcal{J}_{11})$ $a : J \longrightarrow X$ is a continuous function.

$(\mathcal{J}_{12})$ $\sigma, \zeta, \eta : J \longrightarrow J$ are continuous.

$(\mathcal{J}_{13})$ $q : J \longrightarrow \mathbb{R}$ is a continuous function.

$(\mathcal{J}_{14})$ The operator $T : \mathcal{C}(J, X) \longrightarrow \mathcal{C}(J, X)$ is weakly sequentially continuous and weakly compact.

$(\mathcal{J}_{15})$ The function $p : J^2 \times X^2 \longrightarrow \mathbb{R}$ is weakly sequentially continuous such as the partial function $t \longrightarrow p(t, s, x, y)$ is continuous uniformly for $(s, x, y) \in J \times X^2$.

$(\mathcal{J}_{16})$ There exists $r_0 > 0$ such that:

(*a*) $|p(t, s, x, y)| \leq M$ for each t, $s \in J$; x, $y \in X$ such that $\|x\| \leq r_0$ and $\|y\| \leq r_0$,

(*b*) $\|u\|\|Tx\|_\infty \leq 1$ for each $x \in \mathcal{C}(J, X)$ such that $\|x\|_\infty \leq r_0$, and

(*c*) $\|a\|_\infty + \|q\|_\infty + M \leq r_0$.

Theorem 6.2.1 *Under the assumptions* $(\mathcal{J}_{11})$*–*$(\mathcal{J}_{16})$*, the FIE (6.7) has, at least, one solution* $x = x(t)$ *which belongs to the space* $\mathcal{C}(J, X)$*.*

Proof. Recall that $\mathcal{C}(J, X)$ verifies the condition $(\mathcal{P})$. Let us define the subset S of $\mathcal{C}(J, X)$ by:

$$S := \{y \in \mathcal{C}(J, X) : \|y\|_\infty \leq r_0\} = B_{r_0}.$$

Obviously, S is nonempty, closed, convex, and bounded subset of $\mathcal{C}(J, X)$. Let us consider three operators A, L, and U defined on S by:

$$\begin{aligned} (Ax)(t) &= a(t), \\ (Lx)(t) &= (Tx)(t), \ \textit{and} \\ (Ux)(t) &= \left[q(t) + \int_0^{\sigma(t)} p(t, s, x(\zeta(s)), x(\eta(s)))ds \right] \cdot u. \end{aligned}$$

We will prove that the operators A, L, and U satisfy all the conditions of Theorem 3.1.3.

(*i*) Since A is constant, it is weakly sequentially continuous on S and weakly compact.

(*ii*) In view of hypothesis $(\mathcal{J}_{14})$, L is weakly sequentially continuous on S and $L(S)$ is relatively weakly compact.

(*iii*) In order to prove that U satisfies all the conditions of Theorem 3.1.3, we have to demonstrate that U maps S into $\mathcal{C}(J,X)$. For this purpose, let $(t_n)_{n\geq 0}$ be any sequence in J converging to a point t in J. Then, we have

$$\begin{aligned}
&\|(Ux)(t_n)-(Ux)(t)\| \leq |q(t_n)-q(t)|\ \|u\|+\\
&\quad\left\|\left[\int_0^{\sigma(t_n)} p(t_n,s,x(\zeta(s)),x(\eta(s)))ds - \int_0^{\sigma(t)} p(t,s,x(\zeta(s)),x(\eta(s)))ds\right].u\right\|\\
&\leq \left(\Delta_n + \left[\int_0^{\sigma(t_n)} |p(t_n,s,x(\zeta(s)),x(\eta(s))) - p(t,s,x(\zeta(s)),x(\eta(s)))|\,ds\right]\right)\|u\|\\
&\qquad + \left|\int_{\sigma(t)}^{\sigma(t_n)} |p(t,s,x(\zeta(s)),x(\eta(s)))|\,ds\right|\ \|u\|\\
&\leq \left(\Delta_n + \left[\int_0^{1} |p(t_n,s,x(\zeta(s)),x(\eta(s))) - p(t,s,x(\zeta(s)),x(\eta(s)))|\,ds\right]\right)\|u\|\\
&\qquad + M|\sigma(t_n)-\sigma(t)|\ \|u\|,
\end{aligned}$$

where $\Delta_n = |q(t_n)-q(t)|$. Since $t_n \to t$, then $(t_n,s,x(\zeta(s)),x(\eta(s))) \to (t,s,x(\zeta(s)),x(\eta(s)))$, for all $s \in J$. Taking into account the hypothesis $(\mathcal{J}_{15})$, we obtain

$$p(t_n,s,x(\zeta(s)),x(\eta(s))) \to p(t,s,x(\zeta(s)),x(\eta(s))) \text{ in } \mathbb{R}.$$

Moreover, the use of assumption $(\mathcal{J}_{16})$ leads to

$$|p(t_n,s,x(\zeta(s)),x(\eta(s))) - p(t,s,x(\zeta(s)),x(\eta(s)))| \leq 2M$$

for all $t,s \in J$. Now, we can apply the dominated convergence theorem and also the fact that assumption $(\mathcal{J}_{13})$ hold. Hence, we get

$$(Ux)(t_n) \to (Ux)(t) \text{ in } X.$$

It follows that

$$Ux \in \mathcal{C}(J,X).$$

Next, we will prove that U is weakly sequentially continuous on S. For this purpose, let $(x_n)_{n\geq 0}$ be any sequence in S weakly converging to a point x in S. Then, $\{x_n\}_{n=0}^{\infty}$ is bounded. We can apply Dobrakov's theorem (see Theorem 1.4.1) to get

$$\forall t \in J,\ x_n(t) \rightharpoonup x(t).$$

Hence, by using assumptions $(\mathcal{J}_{15})$–$(\mathcal{J}_{16})$ and also the dominated convergence theorem, we obtain

$$\lim_{n\to\infty} \Gamma_n = \Gamma,$$

where

$$\Gamma_n = \int_0^{\sigma(t)} p(t, s, x_n(\zeta(s)), x_n(\eta(s)))ds$$

and

$$\Gamma = \int_0^{\sigma(t)} p(t, s, x(\zeta(s)), x(\eta(s)))ds.$$

Which implies that

$$\lim_{n\to\infty} \Gamma_n \cdot u = \Gamma \cdot u.$$

Hence,

$$(Ux_n)(t) \to (Ux)(t).$$

Thus,

$$(Ux_n)(t) \rightharpoonup (Ux)(t).$$

Since $\{Ux_n\}_{n=0}^{\infty}$ is bounded by $\|u\|(\|q\|_\infty + M)$, then we can again apply Dobrakov's theorem (see Theorem 1.4.1) to obtain

$$Ux_n \rightharpoonup Ux.$$

We conclude that U is weakly sequentially continuous on S. It remains to prove that U is weakly compact. Since S is bounded by r_0, it is sufficient to prove that $U(S)$ is relatively weakly compact.

Step 1: By definition,

$$U(S) := \{Ux : \ \|x\|_\infty \leq r_0\}.$$

For all $t \in J$, we have

$$U(S)(t) := \{(Ux)(t) : \ \|x\|_\infty \leq r_0\}.$$

We claim that $U(S)(t)$ is sequentially relatively weakly compact of X. To show it, let $(x_n)_{n\geq 0}$ be any sequence in S. Then, we have $(Ux_n)(t) = r_n(t) \cdot u$, where

$$r_n(t) = q(t) + \int_0^{\sigma(t)} p(t, s, x_n(\zeta(s)), x_n(\eta(s)))ds.$$

Since $|r_n(t)| \leq (\|q\|_\infty + M)$ and $(r_n(t))_{n\geq 0}$ is an equibounded real sequence, so, there is a renamed subsequence such that

$$r_n(t) \to r(t),$$

which implies that

$$r_n(t) \cdot u \to r(t) \cdot u,$$

and consequently,

$$(Ux_n)(t) \to (q(t) + r(t)) \cdot u.$$

We conclude that $U(S)(t)$ is sequentially relatively compact in X. Then, $U(S)(t)$ is relatively compact in X.

Step 2: We prove that $U(S)$ is weakly equicontinuous on J. If we take $\varepsilon > 0$; $x \in S$; $x^* \in X^*$; t, $t' \in J$ such that $t \leq t'$ and $t' - t \leq \varepsilon$. Then,

$$\begin{aligned}
&|x^*((Ux)(t) - (Ux)(t'))| \leq |q(t) - q(t')||x^*(u)| \\
&\quad + \left| \int_0^{\sigma(t)} p(t, s, x_(\zeta(s)), x(\eta(s)))ds - \int_0^{\sigma(t')} p(t', s, x(\zeta(s)), x(\eta(s)))ds \right| |x^*(u)| \\
&\leq |q(t) - q(t')||x^*(u)| + \left| \left[\int_{\sigma(t)}^{\sigma(t')} |p(t', s, x(\zeta(s)), x(\eta(s)))|ds \right] \right| |x^*(u)| \\
&\quad + \left[\int_0^{\sigma(t)} |p(t, s, x(\zeta(s)), x(\eta(s))) - p(t', s, x(\zeta(s)), x(\eta(s)))|ds \right] |x^*(u)| \\
&\leq [w(q, \varepsilon) + w(p, \varepsilon) + Mw(\sigma, \varepsilon)]|x^*(u)|,
\end{aligned}$$

where

$$w(q, \varepsilon) := \sup\{|q(t) - q(t')| : t, t' \in J; |t - t'| \leq \varepsilon\},$$

$$w(p, \varepsilon) := \sup\{|p(t, s, x, y) - p(t', s, x, y)| : t, t', s \in J; |t - t'| \leq \varepsilon; x, y \in B_{r_0}\},$$

$$w(\sigma, \varepsilon) := \sup\{|\sigma(t) - \sigma(t')| : t, t' \in J; |t - t'| \leq \varepsilon\}.$$

Now, notice that from the above-obtained estimate, taking into account the hypothesis $(\mathcal{J}_{15})$ and in view of the uniform continuity of the functions q, p, and σ on the set J, it follows that $w(q, \varepsilon) \to 0$, $w(p, \varepsilon) \to 0$ and $w(\sigma, \varepsilon) \to 0$ as $\varepsilon \to 0$. Now, we can apply Arzelà–Ascoli's theorem (see Theorem 1.3.9) to get that $U(S)$ is sequentially relatively weakly compact of X. Again an application of Eberlein–Šmulian's theorem (see Theorem 1.3.3) shows that $U(S)$ is relatively weakly compact.

(iv) Finally, it remains to prove that $Ax + LxUx \in S$ for all $x \in S$. To show it, let $x \in S$. Then, by using $(\mathcal{J}_{16})$, for all $t \in J$, one has

$$\begin{aligned} \|(Ax)(t) + (Lx)(t)(Ux)(t)\| &= \|a(t) + (Tx)(t)(Ux)(t)\| \\ &\leq \|a\|_\infty + \|(Ux)(t)\|\|Tx\|_\infty \\ &\leq \|a\|_\infty + (M + \|q\|_\infty)\|u\|\|Tx\|_\infty \\ &\leq r_0. \end{aligned}$$

From the last inequality and taking the supremum over t, we obtain

$$\|Ax + Lx.Ux\|_\infty \leq r_0,$$

and consequently,

$$Ax + Lx.Ux \in S.$$

Thus, all operators A, L, and U fulfill the requirements of Theorem 3.1.3. Hence, the FIE (6.7) has a solution in the space $\mathcal{C}(J, X)$. Q.E.D.

Remark 6.2.2 (i) *When X is infinite-dimensional, the subset $\mathcal{A}_{r_0} = \{x \in X : \|x\| \leq r_0\}$ is not compact. Therefore, the restriction of p on $J^2 \times {\mathcal{A}_{r_0}}^2$ is not uniformly continuous. Thus, we note that the operator U in the FIE (6.7) is not necessarily continuous on S.*
(ii) *When X is finite-dimensional, the subset $U(S) \subset \mathcal{C}(J, X)$ is relatively compact if, and only if, it is weakly equicontinuous on J and $U(S)(J)$ is relatively compact in X (see for instance Corollary A.2.3 in [153, p. 299]) if, and only if, it is weakly equicontinuous on J and $U(S)(J)$ is relatively weakly compact in X if, and only if, $U(S)$ is relatively weakly compact.*

6.2.2 Regular maps in FIEs

In this subsection, we will study two examples using regular mappings.

Example 1 Let us consider the nonlinear functional integral equation:

$$x(t) = a(t) + (T_1x)(t)\left[\left(q(t) + \int_0^{\sigma(t)} p(t, s, x(s), x(\lambda s))ds\right).u\right], \ 0 < \lambda < 1, \tag{6.8}$$

for all $t \in J$, where $u \neq 0$ is a fixed vector of X and the functions a, q, σ, p, T_1 are given, while $x = x(t)$ is an unknown function.

We will prove the existence of solutions for FIE (6.8) under some suitable conditions. Suppose that the functions involved in FIE (6.8) verify the following conditions:

($\mathcal{J}_{17}$) $a : J \longrightarrow X$ is a continuous function.

($\mathcal{J}_{18}$) $\sigma : J \longrightarrow J$ is a continuous and non-decreasing function.

($\mathcal{J}_{19}$) $q : J \longrightarrow \mathbb{R}$ is a continuous function.

($\mathcal{J}_{20}$) The operator $T_1 : \mathcal{C}(J, X) \longrightarrow \mathcal{C}(J, X)$ is such that:

  (*a*) T_1 is Lipschitzian with a Lipschitzian constant α,

  (*b*) T_1 is regular on $\mathcal{C}(J, X)$,

  (*c*) T_1 is weakly sequentially continuous on $\mathcal{C}(J, X)$, and

  (*d*) T_1 is weakly compact.

($\mathcal{J}_{21}$) The function $p : J \times J \times X \times X \longrightarrow \mathbb{R}$ is continuous such that for arbitrary fixed $s \in J$ and $x, y \in X$, the partial function $t \longrightarrow p(t, s, x, y)$ is continuous uniformly for $(s, x, y) \in J \times X \times X$.

($\mathcal{J}_{22}$) There exists $r_0 > 0$ such that

  (*a*) $|p(t, s, x, y)| \leq r_0 - \|q\|_\infty$ for each $t,\ s \in J;\ x,\ y \in X$ such that $||x|| \leq r_0$ and $||y|| \leq r_0$,

  (*b*) $\|T_1 x\|_\infty \leq \left(1 - \frac{\|a\|_\infty}{r_0}\right) \frac{1}{\|u\|}$ for each $x \in \mathcal{C}(J, X)$, and

  (*c*) $\alpha r_0 \|u\| < 1$.

Theorem 6.2.2 *Under the assumptions* ($\mathcal{J}_{17}$)–($\mathcal{J}_{22}$), *FIE (6.8) has, at least, one solution* $x = x(t)$ *which belongs to the space* $\mathcal{C}(J, X)$.

Proof. Let us define the subset S of $\mathcal{C}(J, X)$ by:

$$S := \{y \in \mathcal{C}(J, X),\ \|y\|_\infty \leq r_0\} = B_{r_0}.$$

Obviously, S is nonempty, convex, and closed. Let us consider three operators A, B, and C defined on $\mathcal{C}(J, X)$ by:

$$\begin{aligned} (Ax)(t) &= (T_1 x)(t), \\ (Bx)(t) &= \left[q(t) + \int_0^{\sigma(t)} p(t, s, x(s), x(\lambda s))ds \right].u, \quad 0 < \lambda < 1, and \\ (Cx)(t) &= a(t). \end{aligned}$$

We will prove that the operators A, B, and C satisfy all the conditions of Corollary 3.1.3. (i) From assumption $(\mathcal{J}_{20})(a)$, it follows that A is Lipschitzian with a Lipschitzian constant α. Clearly, C is Lipschitzian with a Lipschitzian constant 0.

(ii) From assumption $(\mathcal{J}_{20})(b)$, it follows that A is regular on $\mathcal{C}(J,X)$.

(iii) Since C is constant, then C is weakly sequentially continuous on S. From assumption $(\mathcal{J}_{20})(c)$, A is weakly sequentially continuous on S. Now, let us show that B is weakly sequentially continuous on S. Firstly, we verify that if $x \in S$, then $Bx \in \mathcal{C}(J,X)$. Let $\{t_n\}$ be any sequence in J converging to a point t in J and denote $\omega_{n,x,t} := \|(Bx)(t_n) - (Bx)(t)\|$. Then,

$$\begin{aligned}
\omega_{n,x,t} &= \left\| \left[\int_0^{\sigma(t_n)} p(t_n,s,x(s),x(\lambda s))ds - \int_0^{\sigma(t)} p(t,s,x(s),x(\lambda s))ds \right] .u \right\| \\
&\leq \left[\int_0^{\sigma(t_n)} |p(t_n,s,x(s),x(\lambda s)) - p(t,s,x(s),x(\lambda s))|ds \right] \|u\| \\
&\quad + \left| \int_{\sigma(t)}^{\sigma(t_n)} |p(t,s,x(s),x(\lambda s))|ds \right| \|u\| \\
&\leq \left[\int_0^{1} |p(t_n,s,x(s),x(\lambda s)) - p(t,s,x(s),x(\lambda s)|ds \right] \|u\| \\
&\quad +(r_0 - \|q\|_\infty)|\sigma(t_n) - \sigma(t)| \; \|u\|.
\end{aligned}$$

Since $t_n \to t$, then $(t_n,s,x(s),x(\lambda s)) \to (t,s,x(s),x(\lambda s))$, for all $s \in J$. Taking into account the assumption $(\mathcal{J}_{21})$, we obtain

$$p(t_n,s,x(s),x(\lambda s)) \to p(t,s,x(s),x(\lambda s)) \text{ in } \mathbb{R}.$$

Moreover, the use of assumption $(\mathcal{J}_{22})$ leads to

$$|p(t_n,s,x(s),x(\lambda s)) - p(t,s,x(s),x(\lambda s))| \leq 2(r_0 - \|q\|_\infty)$$

for all $t, s \in J$, $\lambda \in (0,1)$. Now, let us consider

$$\begin{cases} \varphi : J \longrightarrow \mathbb{R} \\ \qquad s \longrightarrow \varphi(s) = 2(r_0 - \|q\|_\infty). \end{cases}$$

Clearly, $\varphi \in L^1(J)$. Therefore, from the dominated convergence theorem and the assumption $(\mathcal{J}_{18})$, we obtain

$$(Bx)(t_n) \to (Bx)(t).$$

It follows that

$$Bx \in \mathcal{C}(J, X).$$

Next, we have to prove that B is weakly sequentially continuous on S. To show it, let $\{x_n\}$ be any sequence in S weakly converging to a point x in S. So, from the assumptions $(\mathcal{J}_{21})$–$(\mathcal{J}_{22})$ and the dominated convergence theorem, we get

$$\lim_{n\to\infty} \int_0^1 p(t, s, x_n(s), x_n(\lambda s))ds = \int_0^1 p(t, s, x(s), x(\lambda s)),$$

which implies that

$$\lim_{n\to\infty} \left(\int_0^1 p(t, s, x_n(s), x_n(\lambda s))ds \right) .u = \left(\int_0^1 p(t, s, x(s), x(\lambda s))ds \right) .u.$$

Hence,

$$(Bx_n)(t) \to (Bx)(t).$$

Since $(Bx_n)_n$ is bounded by $r_0\|u\|$, then by using Theorem 1.4.1, we get

$$Bx_n \rightharpoonup Bx.$$

We conclude that B is weakly sequentially continuous on S.

(iv) We will prove that $A(S)$, $B(S)$, and $C(S)$ are relatively weakly compact. Since S is bounded by r_0 and taking into account the hypothesis $(\mathcal{J}_{20})(d)$, it follows that $A(S)$ is relatively weakly compact. Now, let us show that $B(S)$ is relatively weakly compact.

Step 1: From the definition,

$$B(S) := \{B(x),\ \|x\|_\infty \leq r_0\}.$$

For all $t \in J$, we have

$$B(S)(t) = \{(Bx)(t),\ \|x\|_\infty \leq r_0\}.$$

We claim that $B(S)(t)$ is sequentially weakly relatively compact in X. To show it, let $\{x_n\}$ be any sequence in S. We have $(Bx_n)(t) = r_n(t).u$, where $r_n(t) = q(t) + \int_0^1 p(t, s, x_n(s), x_n(\lambda s))ds$. Since $|r_n(t)| \leq r_0$ and $(r_n(t))$ is a real sequence, then there is a renamed subsequence such that

$$r_n(t) \to r(t) \text{ in},$$

which implies that

$$r_n(t).u \to r(t).u,$$

and consequently,

$$(Bx_n)(t) \to (q(t) + r(t)).u.$$

We conclude that $B(S)(t)$ is sequentially relatively compact in X. Then, $B(S)(t)$ is sequentially relatively weakly compact in X.

Step 2: We prove that $B(S)$ is weakly equicontinuous on J. If we take $\varepsilon > 0$; $x \in S$; $x^* \in X^*$; $t, t' \in J$ such that $t \leq t'$, $t' - t \leq \varepsilon$ and if we denote $\tau(x,t) := |x^*((Bx)(t) - (Bx)(t'))|$, then

$$\begin{aligned}
\tau(x,t) &= \left| \int_0^{\sigma(t)} p(t,s,x(s),x(\lambda s))ds - \int_0^{\sigma(t')} p(t',s,x(s),x(\lambda s))ds \right| \|x^*(u)\| \\
&\leq \left[\int_0^{\sigma(t)} |p(t,s,x(s),x(\lambda s)) - p(t',s,x(s),x(\lambda s))|ds \right] \|x^*(u)\| \\
&\quad + \left[\int_{\sigma(t)}^{\sigma(t')} |p(t',s,x(s),x(\lambda s))|ds \right] \|x^*(u)\| \\
&\leq [w(p,\varepsilon) + (r_0 - \|q\|_\infty) w(\sigma,\varepsilon)] \|x^*(u)\|,
\end{aligned}$$

where $w(p,\varepsilon) = \sup\{|p(t,s,x,y) - p(t',s,x,y)| : t,t',s \in J; |t-t'| \leq \varepsilon;\ x,\ y \in B_{r_0}\}$, and $w(\sigma,\varepsilon) = \sup\{|\sigma(t) - \sigma(t')| : t,t' \in J; |t-t'| \leq \varepsilon\}$. Taking into account the hypothesis $(\mathcal{J}_{21})$, and in view of the uniform continuity of the function σ on the set J, it follows that $w(p,\varepsilon) \to 0$ and $w(\sigma,\varepsilon) \to 0$ as $\varepsilon \to 0$. An application of Arzelà–Ascoli's theorem (see Theorem 1.3.9), allow us to conclude that $B(S)$ is sequentially weakly relatively compact in X. Again, an application result of Eberlein–Šmulian's theorem (see Theorem 1.3.3) shows that $B(S)$ is relatively weakly compact. Since $C(S) = \{a\}$, hence $C(S)$ is relatively weakly compact.

(v) Finally, it remains to prove the hypothesis (v) of Corollary 3.1.3. For this purpose, let $x \in \mathcal{C}(J,X)$ and $y \in S$ such that

$$x = Ax.By + Cx,$$

or, equivalently for all $t \in J$,

$$x(t) = a(t) + (T_1 x)(t)(By)(t).$$

However, for all $t \in J$, we have

$$\|x(t)\| \leq \|x(t) - a(t)\| + \|a(t)\|.$$

Then,

$$\begin{aligned}\|x(t)\| &\leq \|(T_1x)(t)\|r_0\|u\| + \|a\|_\infty \\ &\leq \left(1 - \frac{\|a\|_\infty}{r_0}\right) r_0 + \|a\|_\infty \\ &= r_0.\end{aligned}$$

From the last inequality and taking the supremum over t, we obtain

$$\|x\|_\infty \leq r_0,$$

and consequently, $x \in S$. We conclude that the operators A, B, and C satisfy all the requirements of Corollary 3.1.3. Thus, FIE (6.8) has a solution in $\mathcal{C}(J, X)$. Q.E.D.

Example 2 Let us consider the following functional integral equation:

$$x(t) = a(t)x(t)+(T_2x)(t)\left[\left(q(t) + \int_0^{\sigma(t)} p(t, s, x(s), x(\lambda s))ds\right).u\right], \; 0 < \lambda < 1, \tag{6.9}$$

for all $t \in J$, where $u \neq 0$ is a fixed vector of X and the functions a, q, σ, p, T_2 are given, while x in $\mathcal{C}(J, X)$ is an unknown function. Suppose that the functions a, q, σ, p, and the operator T_2 verify the following conditions:

($\mathcal{J}_{23}$) $a : J \longrightarrow X$ is a continuous function with $\|a\|_\infty < 1$.

($\mathcal{J}_{24}$) $\sigma : J \longrightarrow J$ is a continuous and non-decreasing function.

($\mathcal{J}_{25}$) $q : J \longrightarrow \mathbb{R}$ is a continuous function.

($\mathcal{J}_{26}$) The operator $T_2 : \mathcal{C}(J, X) \longrightarrow \mathcal{C}(J, X)$ is such that,

(*a*) T_2 is Lipschitzian with a Lipschitzian constant α,

(*b*) T_2 is regular on $\mathcal{C}(J, X)$,

(*c*) $\left(\frac{I}{T_2}\right)^{-1}$ is well defined on $\mathcal{C}(J, X)$, and

(*d*) $\left(\frac{I}{T_2}\right)^{-1}$ is weakly sequentially continuous on $\mathcal{C}(J, X)$.

($\mathcal{J}_{27}$) The function $p : J \times J \times X \times X \longrightarrow \mathbb{R}$ is continuous such that for arbitrary fixed $s \in J$ and $x, y \in X$, the partial function $t \longrightarrow p(t, s, x, y)$ is continuous uniformly for $(s, x, y) \in J \times X \times X$.

$(\mathcal{J}_{28})$ There exists $r_0 > 0$ such that

(a) $|p(t,s,x,y)| \leq r_0 - \|q\|_\infty$ for each $t, s \in J$; x, $y \in X$ such that $\|x\| \leq r_0$ and $\|y\| \leq r_0$,

(b) $\|T_2 x\|_\infty \leq \left(1 - \frac{\|a\|_\infty}{r_0}\right)\frac{1}{\|u\|}$ for each $x \in \mathcal{C}(J,X)$, and

(c) $\alpha r_0 \|u\| < 1$.

Theorem 6.2.3 *Under the assumptions* $(\mathcal{J}_{23})$*–*$(\mathcal{J}_{28})$*, FIE (6.9) has, at least, one solution* $x = x(t)$ *which belongs to the space* $\mathcal{C}(J,X)$*.*

Proof. Let us consider three operators A, B, and C defined on $\mathcal{C}(J,X)$ by:

$$\begin{array}{rcl} (Ax)(t) & = & (T_2 x)(t), \\ (Bx)(t) & = & \left[q(t) + \displaystyle\int_0^{\sigma(t)} p(t,s,x(s),x(\lambda s))ds\right].u, \quad 0 < \lambda < 1, \text{ and} \\ (Cx)(t) & = & a(t)x(t). \end{array}$$

We will check that the operators A, B, and C satisfy all the conditions of Theorem 3.1.5.

(i) From assumption $(\mathcal{J}_{26})(a)$, the map A is Lipschitzian with a constant α. Next, we must show that C is Lipschitzian on $\mathcal{C}(J,X)$. To do it, let us fix arbitrarily x, $y \in \mathcal{C}(J,X)$. If we take an arbitrary $t \in J$, then we get

$$\begin{array}{rcl} \|(Cx)(t) - (Cy)(t)\| & = & \|a(t)x(t) - a(t)y(t)\| \\ & \leq & \|a\|_\infty \, \|x(t) - y(t)\|. \end{array}$$

From the last inequality and taking the supremum over t, we obtain

$$\|Cx - Cy\|_\infty \leq \|a\|_\infty \, \|x - y\|_\infty.$$

This proves that C is Lipschitzian with a Lipschitzian constant $\|a\|_\infty$.

(ii) Arguing as in the proof of Theorem 6.2.2, we show that B is weakly sequentially continuous on S and $B(S)$ is relatively weakly compact.

(iii) From assumption $(\mathcal{J}_{26})(b)$, it follows that A is regular on $\mathcal{C}(J,X)$.

(iv) Let us show that $\left(\frac{I-C}{A}\right)^{-1}$ is weakly sequentially continuous on $B(S)$. For this purpose, let x, $y \in \mathcal{C}(J,X)$ such that

$$\left(\frac{I-C}{A}\right)(x) = y,$$

or equivalently,

$$\frac{(1-a)x}{T_2 x} = y.$$

Since $\|a\|_\infty < 1$, then $(1-a)^{-1}$ exists on $\mathcal{C}(J,X)$. Hence,

$$\left(\frac{I}{T_2}\right)(x) = (1-a)^{-1}y.$$

This implies, from assumption $(\mathcal{J}_{26})(c)$, that

$$x = \left(\frac{I}{T_2}\right)^{-1}((1-a)^{-1}y).$$

Thus,

$$\left(\frac{I-C}{A}\right)^{-1}(x) = \left(\frac{I}{T_2}\right)^{-1}((1-a)^{-1}x)$$

for all $x \in \mathcal{C}(J,X)$. Now, let $\{x_n\}$ be a weakly convergent sequence of $B(S)$ to a point x in $B(S)$. Then,

$$(1-a)^{-1}x_n \rightharpoonup (1-a)^{-1}x,$$

and so, from assumption $(\mathcal{J}_{26})(d)$, it follows that

$$\left(\frac{I}{T_2}\right)^{-1}((1-a)^{-1}x_n) \rightharpoonup \left(\frac{I}{T_2}\right)^{-1}((1-a)^{-1}x).$$

Hence, we conclude that

$$\left(\frac{I-C}{A}\right)^{-1}(x_n) \rightharpoonup \left(\frac{I-C}{A}\right)^{-1}(x).$$

(v) Finally, by using a similar reasoning as in the last point of Theorem 3.1.5, we prove that the condition (v) of Theorem 3.5.1 is fulfilled. As a result, we conclude that the operators A, B, and C satisfy all the requirements of Theorem 3.1.5. Q.E.D.

Remark 6.2.3 *Let us notice that the operator C in FIE (6.9) does not satisfy the condition (iv) of Corollary 3.1.3. In fact, if we take $X = \mathbb{R}$ and $a \equiv \frac{1}{2}$, then $(Cx)(t) = \frac{1}{2}x(t)$. Thus,*

$$C(S) = \left\{\frac{1}{2}x : \|x\|_\infty \leq r_0\right\} = B_{\frac{r_0}{2}}.$$

Since $\mathcal{C}(J,\mathbb{R})$ is infinite-dimensional, then $C(S)$ is not relatively compact. Furthermore, $\mathbb{R}$ is finite-dimensional. Hence, $C(S)$ is not relatively weakly compact [153].

Corollary 6.2.1 *Let $(X, \| \ \|)$ be a Banach algebra satisfying the condition $(\mathcal{P})$, with a positive closed cone X^+. Suppose that the assumptions $(\mathcal{J}_{23})$–$(\mathcal{J}_{28})$ hold. Also, assume that:*

u belongs to X^+, $a(J) \subset X^+$, $q(J) \subset \mathbb{R}_+$, $p(J \times J \times X^+ \times X^+) \subset \mathbb{R}_+$ and $\left(\frac{I}{T_2}\right)^{-1}$ is a positive operator from the positive cone $\mathcal{C}(J, X^+)$ of $\mathcal{C}(J, X)$ into itself.

Then, FIE (6.9) has, at least, one positive solution x in the cone $\mathcal{C}(J, X^+)$.

Proof. Let

$$S^+ := \{x \in S, \ x(t) \in X^+ \text{ for all } t \in J\}.$$

Obviously, S^+ is nonempty, closed, and convex. Similarly to the proof of Theorem 6.2.3, we show that:

(i) A and C are Lipschitzian with a Lipschitzian constant α and $\|a\|_\infty$, respectively.

(ii) A is regular on $\mathcal{C}(J, X)$.

(iii) A, B, and C are weakly sequentially continuous on S^+.

(iv) Because S^+ is a subset of S, then we have $A(S^+), B(S^+)$, and $C(S^+)$ are relatively weakly compact.

(v) Finally, we will show that the hypothesis (v) of Theorem 3.5.3 is satisfied. In fact, let us fix arbitrarily $x \in \mathcal{C}(J, X)$ and $y \in S^+$, such that

$$x = Ax.By + Cx.$$

Arguing as in the proof of Theorem 6.2.3, we get $x \in S$. Moreover, the last equation implies that:

$$\text{for all } t \in J, \ x(t) = a(t)x(t) + (T_2x)(t)(By)(t).$$

Hence,

$$\text{for all } t \in J, \ \frac{x(t)(1 - a(t))}{(T_2x)(t)} = (By)(t).$$

Since for all $t \in J, \|a(t)\| < 1$, it follows that $(1 - a(t))^{-1}$ exists in X, and

$$(1 - a(t))^{-1} = \sum_{n=0}^{+\infty} a^n(t).$$

Since $a(t)$ belongs to the closed positive cone X^+, then $(1 - a(t))^{-1}$ is positive.

Also, we verify that for all $t \in J$, $(By)(t)$ is positive. Therefore, the map ψ defined on J by:

$$\psi(t) = (1 - a(t))^{-1} \left[\left(q(t) + \int_0^{\sigma(t)} p(t,s,x(s),x(\lambda s))ds \right) .u \right]$$

belongs to the positive cone $\mathcal{C}(J, X^+)$ of $\mathcal{C}(J, X)$. Then, B maps $\mathcal{C}(J, X^+)$ into itself. Knowing that

$$\left(\left(\frac{I}{T_2} \right) x \right)(t) = (1-a(t))^{-1} \left[\left(q(t) + \int_0^{\sigma(t)} p(t,s,x(s),x(\lambda s))ds \right) .u \right] = \psi(t),$$

it follows that

$$x = \left(\frac{I}{T_2} \right)^{-1} (\psi).$$

Hence, $x \in \mathcal{C}(J, X^+)$ and consequently, $x \in S^+$. Q.E.D.

Next, let us provide an example of the operator T_2 presented in Theorem 6.2.3.

Let $\mathcal{C}(J, \mathbb{R}) = \mathcal{C}(J)$ denote the Banach algebra of all continuous real-valued functions on J with norm $\|x\|_\infty = \sup_{t \in J} |x(t)|$. Clearly, $\mathcal{C}(J)$ satisfies the condition $(\mathcal{P})$. Let $b : J \longrightarrow \mathbb{R}$ be continuous and nonnegative, and let us define T_2 by:

$$\begin{cases} T_2 : \mathcal{C}(J) \longrightarrow \mathcal{C}(J) \\ \quad x \longrightarrow T_2 x = \dfrac{1}{1 + b|x|}. \end{cases}$$

We obtain the following functional integral equation:

$$x(t) = a(t)x(t) + \frac{1}{1 + b(t)|x(t)|} \left[q(t) + \int_0^{\sigma(t)} p(t,s,x(s),x(\lambda s))ds \right], \; 0 < \lambda < 1. \tag{6.10}$$

Let us check all the conditions of Theorem 6.2.3 for the above FIE (6.10):

(a) Fix x, $y \in \mathcal{C}(J)$. Then, for all $t \in J$, we have

$$\begin{aligned} |(T_2 x)(t) - (T_2 y)(t)| &= \left| \frac{1}{1 + b(t)|x(t)|} - \frac{1}{1 + b(t)|y(t)|} \right| \\ &= \frac{b(t)||y(t)| - |x(t)||}{(1 + b(t)|x(t)|)\,(1 + b(t)|y(t)|)} \\ &\leq \|b\|_\infty \, |x(t) - y(t)|. \end{aligned}$$

Taking the supremum over t, we obtain

$$\|T_2x - T_2y\|_\infty \leq \|b\|_\infty \, \|x - y\|_\infty,$$

which shows that T_2 is Lipschitzian with a Lipschitzian constant $\|b\|_\infty$.

(b) Clearly, T_2 is regular on $\mathcal{C}(J)$.

(c) Let us show that $\left(\frac{I}{T_2}\right)^{-1}$ exists on $\mathcal{C}(J)$. For this purpose, let x, $y \in \mathcal{C}(J)$ such that

$$\left(\frac{I}{T_2}\right)x = y,$$

or equivalently,

$$x(1 + b|x|) = y,$$

which implies

$$|x|(1 + b|x|) = |y|,$$

and then,

$$(\sqrt{b}|x|)^2 + |x| = |y|.$$

For each $t_0 \in J$ such that $b(t_0) = 0$, we have $x = y$. Then, for each $t \in J$ such that $b(t) > 0$, we obtain

$$\left(\sqrt{b(t)}|x(t)| + \frac{1}{2\sqrt{b(t)}}\right)^2 = \frac{1}{4b(t)} + |y(t)|,$$

which further implies that

$$\sqrt{b(t)}|x(t)| = -\frac{1}{2\sqrt{b(t)}} + \sqrt{\frac{1}{4b(t)} + |y(t)|},$$

and then,

$$b(t)|x(t)| = -\frac{1}{2} + \sqrt{\frac{1}{4} + |y(t)|b(t)}.$$

Consequently,

$$x(t) = \frac{y(t)}{1 + b(t)|x(t)|} = \frac{y(t)}{\frac{1}{2} + \sqrt{\frac{1}{4} + b(t)|y(t)|}}.$$

Notice that the equality is also verified for each t such that $b(t) = 0$. Let us consider the function F defined by the expression

$$\begin{cases} F : \mathcal{C}(J) \longrightarrow \mathcal{C}(J) \\ \quad x \longrightarrow F(x) = \dfrac{x}{\frac{1}{2} + \sqrt{\frac{1}{4} + b|x|}}. \end{cases}$$

It is easy to verify that for all $x \in \mathcal{C}(J)$, we have

$$\left(\left(\frac{I}{T_2} \right) \circ F \right)(x) = \left(F \circ \left(\frac{I}{T_2} \right) \right)(x) = x.$$

We conclude that

$$\left(\frac{I}{T_2} \right)^{-1}(x) = \frac{x}{\frac{1}{2} + \sqrt{\frac{1}{4} + b|x|}}.$$

(d) It is an easy exercise to show that T_2 and $\left(\frac{I}{T_2}\right)^{-1}$ are weakly sequentially continuous on $B(S)$.

Remark 6.2.4 *One can easily check that* $\left(\frac{I}{T_2}\right)^{-1}$ *is a positive operator from the positive cone* $\mathcal{C}(J, \mathbb{R}_+)$ *of* $\mathcal{C}(J, \mathbb{R})$ *into itself.*

6.2.3 ω-condensing mappings in FIEs

Let us consider the following example of functional integral equation in $E = \mathcal{C}(J, X)$.

$$x(t) = f(t, x(\nu(t))) + \left[\int_0^{\sigma_1(t)} k(t, s)x(s)ds \right] [(q(t) + \Lambda(t)) \cdot u] \qquad (6.11)$$

for all $t \in J$, where $\Lambda(t) = \displaystyle\int_0^{\sigma_2(t)} p(t, s, x(\zeta(s)), x(\eta(s)))ds$ and u is a nonvanishing vector of X.

Notice that, in 2006, this FIE (6.11) was studied in the Banach algebra $\mathcal{C}(J, \mathbb{R})$ by B. C. Dhage [73].

In what follows, we will assume that the functions involved in FIE (6.11) verify the following conditions:

$(\mathcal{J}_{29})$ $f : J \times X \longrightarrow X$ is continuous such as there exist two nonnegative constants α, β satisfying for all $t \in J$, the operator

$$\begin{cases} f_t : X \longrightarrow X \\ \quad x \longrightarrow f_t(x) = f(t, x) \end{cases}$$

is Lipschitzian with a constant α, and is weakly sequentially continuous, with $|f_t(0)| \leq \beta$.

($\mathcal{J}_{30}$) ν, σ_1, σ_2, ζ, $\eta : J \longrightarrow J$ are continuous on J.

($\mathcal{J}_{31}$) $q : J \longrightarrow \mathbb{R}$ is a continuous function on J.

($\mathcal{J}_{32}$) $k : J^2 \longrightarrow \mathbb{R}$ is a continuous function.

($\mathcal{J}_{33}$) The function $p : J^2 \times X^2 \longrightarrow \mathbb{R}$ is weakly sequentially continuous such as the partial function $t \longrightarrow p(t, s, x, y)$ is continuous uniformly for $(s, x, y) \in J \times X^2$.

($\mathcal{J}_{34}$) There exists $M > 0$ such that:

(*a*) for all $r_0 > 0$, $|p(t, s, x, y)| \leq M$, for each t, $s \in J$; x, $y \in X$, where $||x|| \leq r_0$ and $||y|| \leq r_0$, and

(*b*) $\alpha + (M + \|q\|_\infty)\, \|u\| k_0 < 1$, where $k_0 = \sup\{|k(t, s)| : (t, s) \in J^2\}$.

Theorem 6.2.4 *Under the hypotheses* ($\mathcal{J}_{29}$)*–*($\mathcal{J}_{34}$)*, FIE (6.11) has, at least, one solution* $x = x(t)$ *which belongs to the space* $\mathcal{C}(J, X)$*.*

Proof. Let us define the subset S of $\mathcal{C}(J, X)$ by:

$$S := \{y \in \mathcal{C}(J, X) : \|y\|_\infty \leq r_0\},$$

where

$$r_0 := \frac{\beta}{1 - \alpha - (M + \|q\|_\infty)\, \|u\| k_0}.$$

Obviously, S is a nonempty, closed, convex, and bounded subset of $\mathcal{C}(J, X)$. Let us consider three operators A, L, and U defined on S by:

$$\begin{aligned}
(Ax)(t) &= f(t, x(\nu(t))),\\
(Lx)(t) &= \int_0^{\sigma_1(t)} k(t, s)x(s)ds, \text{ and}\\
(Ux)(t) &= \left[q(t) + \int_0^{\sigma_2(t)} p(t, s, x(\zeta(s)), x(\eta(s)))ds\right] \cdot u.
\end{aligned}$$

We will prove that the operator $F = A + L.U$ satisfies all the conditions of Theorem 2.3.4. (*i*) For all $x \in S$, and by composition, we obtain $Ax \in \mathcal{C}(J, X)$.

Now, let $x, y \in S$. For all $t \in J$, we have

$$\begin{aligned} \|(Ax)(t) - (Ay)(t)\| &= \|f(t, x(\nu(t))) - f(t, y(\nu(t)))\| \\ &\leq \alpha\|x(\nu(t)) - y(\nu(t))\| \\ &\leq \alpha\|x - y\|_\infty. \end{aligned}$$

From the last inequality and taking the supremum over t, we get

$$\|Ax - Ay\|_\infty \leq \alpha\|x - y\|_\infty.$$

This shows that A is Lipschitzian with Lipschitz constant α. Next, we will show that A is weakly sequentially continuous on S. To do so, let $(x_n)_{n\geq 0}$ be any sequence in S weakly converging to a point x in S. Then, $\{x_n\}_{n=0}^{\infty}$ is bounded. We can apply Dobrakov's theorem (see Theorem 1.4.1) in order to get

$$\forall t \in J,\ x_n(t) \rightharpoonup x(t).$$

In view of the hypothesis $(\mathcal{J}_{29})$, we have

$$\forall t \in J,\ Ax_n(t) \rightharpoonup Ax(t).$$

Since $\{Ax_n\}_{n=0}^{\infty}$ is bounded by $\beta + \alpha r_0$, then by again using Dobrakov's theorem (see Theorem 1.4.1), we infer that A is weakly sequentially continuous on S.

(ii) We will verify that L maps S into $\mathcal{C}(J, X)$. For this purpose, let $t, t' \in J$ such that $t \leq t'$ and $t' - t \leq \varepsilon$. Then, we have

$$\begin{aligned} \|(Lx)(t) - (Lx)(t')\| &= \left\| \int_0^{\sigma_1(t)} k(t,s)x(s)ds - \int_0^{\sigma_1(t')} k(t',s)x(s)ds \right\| \\ &\leq \left\| \int_0^{\sigma_1(t)} k(t,s)x(s)ds - \int_0^{\sigma_1(t')} k(t,s)x(s)ds \right\| \\ &+ \left\| \int_0^{\sigma_1(t')} k(t,s)x(s)ds - \int_0^{\sigma_1(t')} k(t',s)x(s)ds \right\| \\ &\leq r_0 \int_0^1 |k(t,s) - k(t',s)|ds + k_0 r_0 |\sigma_1(t) - \sigma_1(t')| \\ &\leq r_0[w(k,\epsilon) + k_0 w(\sigma_1, \varepsilon)], \end{aligned}$$

where

$$\begin{aligned} w(q,\varepsilon) &:= \sup\{|k(t,s) - k(t',s)| : t, t', s \in J; |t - t'| \leq \varepsilon\}, \text{ and} \\ w(\sigma_1,\varepsilon) &:= \sup\{|\sigma_1(t) - \sigma_1(t')| : t, t' \in J; |t - t'| \leq \varepsilon\}. \end{aligned}$$

This shows that L maps S into $\mathcal{C}(J, X)$. Next, since $\|Lx - Ly\|_\infty \leq k_0\|x - y\|_\infty$ and L is linear, then L is weakly sequentially continuous on S.
(iii) Similarly to the proof of Theorem 6.2.4, we can prove that U maps S into $\mathcal{C}(J, X)$ and is weakly sequentially continuous on S. Thus, F maps S into $\mathcal{C}(J, X)$ and F is weakly sequentially continuous on S.
(iv) Knowing that A is Lipschitzian with a constant α, L is Lipschitzian with a constant k_0 and that U is weakly compact [27] and bounded by $(M + \|q\|_\infty)\|u\|$. In view of Lemmas 1.5.2 and 3.1.3, we can prove, for any bounded subset V of S, that

$$\omega(F(V)) \leq [\alpha + (M + \|q\|_\infty)\|u\|k_0]\,\omega(V).$$

By using hypothesis $(\mathcal{J}_{34})$, it follows that F is a condensing map with respect to the measure of weak noncompactness ω.
(v) Finally, it remains to prove that $Ax + Lx.Ux \in S$, for all $x \in S$. To show it, let $x \in S$. Then, by using $(\mathcal{J}_{29})$ and $(\mathcal{J}_{34})$, for all $t \in J$, we have

$$\begin{aligned}
\|(Fx)(t)\| &= \|f(t, x(\nu(t))) + (Lx)(t)(Ux)(t)\| \\
&\leq \|f(t, x(\nu(t))) - f(t, 0)\| + \|f(t, 0)\| + \|(Ux)(t)\|\|Lx\|_\infty \\
&\leq \beta + \alpha r_0 + \|(Ux)(t)\|\|Lx\|_\infty \\
&\leq \beta + [(M + \|q\|_\infty)\|u\|k_0 + \alpha]r_0 \\
&\leq r_0.
\end{aligned}$$

From the last inequality and taking the supremum over t, we obtain

$$\|Ax + Lx.Ux\|_\infty \leq r_0$$

and consequently,

$$Ax + Lx.Ux \in S.$$

We conclude that the operator F fulfills all the requirements of Theorem 2.3.4. Hence, FIE (6.11) has a solution in the space $\mathcal{C}(J, X)$. Q.E.D.

6.2.4 ω-convex-power-condensing mappings in FIEs

Let us consider the following example of nonlinear functional integral equation in $\mathcal{C}(J, X)$.

$$x(t) = a(t) + f(t, x(t))\left[q(t) + \int_0^t b(t-s)p(t, s, x(s), x(\lambda s))ds\right], \quad 0 < \lambda \leq 1. \tag{6.12}$$

In what follows, we will assume that the functions involved in FIE (6.12) verify the following conditions:

($\mathcal{J}_{35}$) The functions $a, b, q : J \longrightarrow X$ are continuous on J.

($\mathcal{J}_{36}$) $f : J \times X \longrightarrow X$ is bounded such as for all $r > 0$, f is uniformly continuous on $J \times B_r$. Moreover, we assume that, for all $t \in J$, the operator

$$\begin{cases} f_t : X \longrightarrow X \\ \qquad x \longrightarrow f_t(x) = f(t, x) \end{cases}$$

is weakly sequentially continuous, and weakly compact on X.

($\mathcal{J}_{37}$) The function $p : J^2 \times X^2 \longrightarrow X$ verifies the following

(*a*) For all $(t, s) \in J^2$, the operator

$$\begin{cases} p_{t,s} : X^2 \longrightarrow X \\ \qquad (x, y) \longrightarrow p_{t,s}(x, y) = p(t, s, x, y) \end{cases}$$

is strongly continuous, and

(*b*) For all $r > 0$, p is bounded and uniformly continuous on $J^2 \times B_r^2$.

($\mathcal{J}_{38}$) $\lim_{r \to +\infty} \sup \dfrac{M_r}{r} < \dfrac{1}{\|f\|_\infty \|b\|_\infty}$, where

$$M_r := \sup\{\|p(t, s, x, y)\| : t, s \in J \text{ and } x, y \in B_r\}.$$

($\mathcal{J}_{39}$) There exist two constants $\alpha_i > 0$ $(i = 1, 2)$ such as, for any bounded and equicontinuous sets $V_i \subset \mathcal{E}$ $(i = 1, 2)$, $(t, s) \in J^2$ and $\lambda \in (0, 1]$, we have:

$$\omega\left(p\left(t, s, V_1(s), V_2(\lambda s)\right)\right) \leq \alpha_1 \omega(V_1(s)) + \alpha_2 \omega(V_2(\lambda s)).$$

Theorem 6.2.5 *Under the assumptions ($\mathcal{J}_{35}$)–($\mathcal{J}_{39}$), FIE (6.12) has, at least, one solution $x = x(t)$ which belongs to the space $\mathcal{C}(J, X)$.*

Proof. The use of the assumption ($\mathcal{J}_{38}$) leads to the existence of $0 < r' < \frac{1}{\|f\|_\infty \|b\|_\infty}$ and $r_0^* > 0$ such that for all $r \geq r_0^*$, we have $M_r < r'r$. Now, let us define the subset B_{r_0} of $\mathcal{C}(J, X)$ by:

$$B_{r_0} := \{y \in \mathcal{C}(J, X) : \|y\|_\infty \leq r_0\},$$

where

$$r_0 := \max\left[r_0^*, (\|a\|_\infty + \|f\|_\infty\|q\|_\infty)(1 - \|f\|_\infty\|b\|_\infty r')^{-1}\right].$$

Let us consider the three operators A, L, and U, defined on B_{r_0} by:

$$\begin{aligned}(Ax)(t) &= a(t),\\ (Lx)(t) &= f(t, x(t)), \text{ and}\\ (Ux)(t) &= q(t) + \int_0^t b(t-s)p(t, s, x(s), x(\lambda s))ds, \ 0 < \lambda \leq 1.\end{aligned}$$

We will prove that the operator $F = A + L.U$ satisfies all the conditions of Corollary 3.5.1.

(i) Clearly, for all $x \in B_{r_0}$, $Ax \in \mathcal{C}(J, X)$. By using $(\mathcal{J}_{36})$, we show that L maps B_{r_0} into $\mathcal{C}(J, X)$. Now, let us prove that U maps B_{r_0} into $\mathcal{C}(J, X)$. For this purpose, let $x \in B_{r_0}$ and let $(t_n)_{n\geq 0}$ be any sequence in J converging to a point t in J. Then,

$$\begin{aligned}&\|(Ux)(t_n) - (Ux)(t)\| \leq \Delta_n(t)\\ &\quad + \left\|\int_0^{t_n} b(t_n - s)p(t_n, s, x(s), x(\lambda s))ds - \int_0^t b(t-s)p(t, s, x(s), x(\lambda s))ds\right\|\\ &\leq \Delta_n(t) + \int_0^{t_n} \|b(t_n - s)p(t_n, s, x(s), x(\lambda s)) - b(t-s)p(t, s, x(s), x(\lambda s))\|ds\\ &\quad + \left|\int_t^{t_n} \|b(t-s)p(t, s, x(s), x(\lambda s))\|ds\right|\\ &\leq \Delta_n(t) + M_{r_0}\|b\|_\infty|t_n - t|\\ &\quad + \int_0^1 \|b(t_n - s)p(t_n, s, x(s), x(\lambda s)) - b(t-s)p(t, s, x(s), x(\lambda s))\|ds,\end{aligned}$$

where $\Delta_n(t) = \|q(t_n) - q(t)\|$. Since $t_n \to t$, it follows, from the Lebesgue dominated convergence theorem, that

$$\int_0^1 \|b(t_n - s)p(t_n, s, x(s), x(\lambda s)) - b(t-s)p(t, s, x(s), x(\lambda s))\|ds \to 0.$$

Hence, F maps B_{r_0} into $\mathcal{C}(J, X)$.

(ii) Let us show that $F(B_{r_0}) \subset B_{r_0}$. To do it, let $x \in B_{r_0}$. Then, for all $t \in J$, we have

$$\begin{aligned}\|F(x(t))\| &\leq \|a\|_\infty + \|f\|_\infty(\|q\|_\infty + \|b\|_\infty M_{r_0})\\ &\leq \|a\|_\infty + \|f\|_\infty\|q\|_\infty + \|f\|_\infty\|b\|_\infty r' r_0 \leq r_0,\end{aligned}$$

which shows that F maps B_{r_0} into itself.

(iii) Now, let us prove that F is weakly sequentially continuous on B_{r_0}. For this purpose, let $(x_n)_{n\geq 0}$ be any sequence in B_{r_0} weakly converging to a point x in B_{r_0}. Then, $\{x_n\}_{n=0}^{\infty}$ is bounded. We can apply Dobrakov's theorem (see Theorem 1.4.1) which allows us to get

$$\forall t \in J,\ x_n(t) \rightharpoonup x(t).$$

Hence, by using $(\mathcal{J}_{36})$, one has

$$\forall t \in J,\ f_t(x_n(t)) \rightharpoonup f_t(x(t)).$$

Since $\{Lx_n\}_{n=0}^{\infty}$ is bounded by $\|f\|_\infty$, then by again using Dobrakov's theorem (see Theorem 1.4.1), we infer that L is weakly sequentially continuous on B_{r_0}. Moreover, for all $(t,s) \in J^2$ and $\lambda \in (0,1)$, we have

$$(t, s, x_n(s), x_n(\lambda s)) \rightharpoonup (t, s, x(s), x(\lambda s)).$$

Taking into account $(\mathcal{J}_{37})(a)$, we get

$$p(t, s, x_n(s), x_n(\lambda s)) \to p(t, s, x(s), x(\lambda s)).$$

Hence,

$$b(t-s)p(t, s, x_n(s), x_n(\lambda s)) \to b(t-s)p(t, s, x(s), x(\lambda s)).$$

Knowing that

$$\|b(t-s)p(t, s, x_n(s), x_n(\lambda s))\| \leq \|b\|_\infty M_{r_0},$$

it follows from the Lebesgue dominated convergence theorem, that

$$\lim_{n\to\infty} \int_0^t b(t-s)p(t, s, x_n(s), x_n(\lambda s))ds = \int_0^t b(t-s)p(t, s, x(s), x(\lambda s))ds.$$

This means that

$$\forall t \in J,\ (Ux_n)(t) \to (Ux)(t),$$

and so,

$$(Ux_n)(t) \rightharpoonup (Ux)(t).$$

Since $\{Ux_n\}_{n=0}^{\infty}$ is bounded by $\|q\|_\infty + \|b\|_\infty M_{r_0}$, then we can again apply Dobrakov's theorem (see Theorem 1.4.1) to show that U, and consequently F, are weakly sequentially continuous on B_{r_0}.

(iv) Let $S = \overline{\text{co}}\, F(B_{r_0})$. Clearly, $F(S) \subset S$. Let $x_0 \in S$. To finish, we will prove that there exist a positive integer $n_0 \geq 1$ and $k \in (0,1)$ such that, for any subset $V \subset S$,

$$\omega\left(F^{(n_0,x_0)}(V)\right) \leq k\omega(V).$$

Q.E.D.

To do this, we need to recall the following well-known lemmas:

Lemma 6.2.1 *[113] Let $V \subset \mathcal{C}(J,X)$ be bounded and equicontinuous. Then,*

(a) The function $t \longrightarrow m(t) = \omega(V(t))$ is continuous on J, and

(b) $\omega(V) = \omega(V(J)) = \max_{t\in J} \omega(V(t))$, where $V(J) = \{x(t) : x \in V,\ t \in J\}$.

Lemma 6.2.2 *[93] Let $V \subset \mathcal{C}(J,X)$ be equicontinuous and $x_0 \in \mathcal{C}(J,X)$. Then, $\overline{co}\{V, x_0\}$ is equicontinuous on $\mathcal{C}(J,X)$.*

Lemma 6.2.3 *Under assumption $(\mathcal{J}_{37})(b)$ of Theorem 6.2.5, for all $t \in J$,*

$$\{s \longrightarrow b(t-s)p(t,s,B_{r_0}(s),B_{r_0}(\lambda s)) : \ t \geq s\}$$

is equicontinuous on $\mathcal{C}(J,X)$.

Proof. Let t,s_1,$s_2 \in J$, $t \geq s_i$ $(i = 1,2)$ and $x \in B_{r_0}$. Then, for all $\varepsilon > 0$, we have

$$\begin{aligned}
&\|b(t-s_1)p(t,s_1,x(s_1),x(\lambda s_1)) - b(t-s_2)p(t,s_2,x(s_2),x(\lambda s_2))\| \\
&\leq \|b(t-s_1)p(t,s_1,x(s_1),x(\lambda s_1)) - b(t-s_2)p(t,s_1,x(s_1),x(\lambda s_1))\| + \\
&\qquad \|b(t-s_2)p(t,s_1,x(s_1),x(\lambda s_1)) - b(t-s_2)p(t,s_2,x(s_2),x(\lambda s_2))\| \\
&\leq M_{r_0}\|b(t-s_1) - b(t-s_2)\| + \\
&\qquad \|b\|_\infty \|p(t,s_1,x(s_1),x(\lambda s_1)) - p(t,s_2,x(s_2),x(\lambda s_2))\| \\
&\leq M_{r_0} w(b,\varepsilon) + \|b\|_\infty w(p,\varepsilon),
\end{aligned}$$

where

$$w(b,\varepsilon) := \sup\{|b(t) - b(t')| : t,\ t' \in J; |t-t'| \leq \varepsilon\},$$

$w(p,\varepsilon) := \sup\{\|p(t,s,x,y) - p(t,s',x',y')\| : t,\ s,\ s' \in J; |s-s'| \leq \varepsilon;\ x,\ y,\ x',\ y' \in B_{r_0}; \|x-x'\| \leq \varepsilon$ and $\|y-y'\| \leq \varepsilon\}$. Now, we are ready to prove that $F : S \longrightarrow S$ is convex-power condensing with respect to the measure of weak noncompactness ω. We first show that $F(B_{r_0}) \subset \mathcal{C}(J,X)$ is equicontinuous. If we take $\varepsilon > 0$; $x \in B_{r_0}$; $t, t' \in J$ such that $t \leq t'$ and $t' - t \leq \varepsilon$, then

$$\begin{aligned}\|(Fx)(t) - (Fx)(t')\| &\leq n(t,t') + \|f(t,x(t))(Ux)(t) - f(t',x(t'))(Ux)(t')\| \\ &\leq n(t,t') + \|f\|_\infty \|(Ux)(t) - (Ux)(t')\| \\ &\qquad + [\|q\|_\infty + M_{r_0}]\, \|f(t,x(t)) - f(t',x(t'))\| \\ &\leq w(a,\varepsilon) + [\|q\|_\infty + M_{r_0}] w(f,\varepsilon) \\ &\qquad + \|f\|_\infty [w(q,\varepsilon) + M_{r_0} w(b,\varepsilon) + \|b\|_\infty w(p,\varepsilon)],\end{aligned}$$

where $n(t,t') = \|a(t) - a(t')\|$. Hence, $F(B_{r_0})$ is equicontinuous on $\mathcal{C}(J,X)$. By using Lemma 6.2.2, we deduce that S is equicontinuous. Notice that, for any $V \subset S$, $F^{(n,x_0)}(V) \subset S$ is bounded and equicontinuous. Then, by using Lemma 6.2.1, we infer that

$$\omega\left(F^{(n,x_0)}(V)\right) = \max_{t \in J} \omega\left(F^{(n,x_0)}(V)(t)\right), \quad n = 1, 2, \ldots$$

For all $t \in J$, and taking into account the assumption $(\mathcal{J}_{36})$, one has

$$\omega(f(t,V)) = \omega(f_t(V)) \leq \omega\left(\overline{f_t(V)}^w\right) = 0.$$

Now, from $(\mathcal{J}_{39})$ and Lemmas 6.2.1, 6.2.2, 6.2.3, and for all $t \in J$, it follows that

$$\begin{aligned}\Theta(x_0,t) &= \omega(F(V)(t)) \\ &\leq \|f\|_\infty \omega\left(q(t) + \int_0^t b(t-s)p(t,s,V(s),V(\lambda s))ds\right) \\ &\leq \|f\|_\infty \left[\omega\{q(t)\} + \omega\left(\int_0^t b(t-s)p(t,s,V(s),V(\lambda s))ds\right)\right] \\ &\leq \|f\|_\infty \omega\left(\int_0^t b(t-s)p(t,s,V(s),V(\lambda s))ds\right) \\ &\leq \|f\|_\infty \|b\|_\infty \omega\left(\int_0^t p(t,s,V(s),V(\lambda s))ds\right),\end{aligned}$$

where $\Theta(x_0,t) = \omega\left(\left(F^{(1,x_0)}(V)\right)(t)\right)$. The use of the mean value theorem [7, Theorem V. 10.4.] leads to

$$\int_0^t p(t,s,V(s),V(\lambda s))ds \subset t\, \overline{\text{co}}\{p(t,s,V(s),V(\lambda s)),\ s \in J\}.$$

Therefore,

$$\begin{aligned}\omega\left(\int_0^t p(t,s,V(s),V(\lambda s))ds\right) &\leq t\omega\left(\overline{co}\{p(t,s,V(s),V(\lambda s)),\ s\in J\}\right)\\ &\leq t\omega\left(\{p(t,s,V(s),V(\lambda s)),\ s\in J\}\right)\\ &\leq t(\alpha_1+\alpha_2)\omega(V).\end{aligned}$$

As a result,

$$\omega\left(\left(F^{(1,x_0)}(V)\right)(t)\right) \leq t\|f\|_\infty\|b\|_\infty(\alpha_1+\alpha_2)\omega(V).$$

Therefore, by using the method of mathematical induction for all positive integers n and $t \in J$, and since $F^{(n,x_0)}(V) \subset S \subset B_{r_0}$ is bounded and equicontinuous, then f_t is weakly compact and the use of assumption $(\mathcal{J}_{39})$ and Lemmas 6.2.1, 6.2.2, 6.2.3 leads to

$$\omega\left(\left(F^{(n,x_0)}(V)\right)(t)\right) \leq \frac{\|f\|_\infty^n\|b\|_\infty^n(\alpha_1+\alpha_2)^n}{n!}\omega(V).$$

From the last inequality and taking the maximum over t, we obtain

$$\omega\left(F^{(n,x_0)}(V)\right) \leq \frac{\|f\|_\infty^n\|b\|_\infty^n(\alpha_1+\alpha_2)^n}{n!}\omega(V).$$

However, it is easy to show that

$$\frac{\|f\|_\infty^n\|b\|_\infty^n(\alpha_1+\alpha_2)^n}{n!} \to 0.$$

Then, there exists a positive integer n_0 such that

$$k = \frac{\|f\|_\infty^{n_0}\|b\|_\infty^{n_0}(\alpha_1+\alpha_2)^{n_0}}{n_0!} < 1.$$

We conclude that the operator F fulfills all the requirements of Corollary 3.5.1. Thus, FIE (6.12) has a solution in the space $\mathcal{C}(J,X)$. Q.E.D.

6.3 Existence Results for FDEs in Banach Algebras

We consider the following nonlinear functional differential equation (in short, FDE) in $\mathcal{C}(J) := \mathcal{C}(J,\mathbb{R})$. For $t \in J$ and $0 < \lambda < 1$

$$\left(\left(\frac{x}{T_2 x}\right) - q_1\right)'(t) = \int_0^t \frac{\partial p}{\partial t}(t,s,x(s),x(\lambda s))ds + p(t,t,x(t),x(\lambda t)) \quad (6.13)$$

satisfying the initial condition

$$x(0) = \zeta \in \mathbb{R}, \tag{6.14}$$

where the functions q_1, p and the operator T_2 are given with $q_1(0) = 0$, while $x = x(t)$ is an unknown function. By a solution of FDE (6.13)–(6.14), we mean an absolutely continuous function $x : J \longrightarrow \mathbb{R}$ that satisfies the two equations (6.13)–(6.14) on J. The existence result for FDE (6.13)–(6.14) is:

Theorem 6.3.1 *We consider the following assumptions:*

($\mathcal{J}_{40}$) *The $q_1 : J \longrightarrow \mathbb{R}$ is a continuous function.*

($\mathcal{J}_{41}$) *The operator $T_2 : \mathcal{C}(J) \longrightarrow \mathcal{C}(J)$ is such that*

(a) *T_2 is Lipschitzian with a Lipschitzian constant α,*

(b) *T_2 is regular on $\mathcal{C}(J)$,*

(c) $\left(\frac{I}{T_2}\right)^{-1}$ *is well defined on $\mathcal{C}(J)$,*

(d) $\left(\frac{I}{T_2}\right)^{-1}$ *is weakly sequentially continuous on $\mathcal{C}(J)$, and*

(e) *for all $x \in \mathcal{C}(J)$, we have $\|(T_2 x)\|_\infty \leq 1$.*

($\mathcal{J}_{42}$) *The function $p : J \times J \times X \times X \longrightarrow \mathbb{R}$ is continuous such that for an arbitrary fixed $s \in J$ and for $x, y \in \mathbb{R}$; the partial function $t \longrightarrow p(t, s, x, y)$ is $\mathcal{C}^1$ on J.*

($\mathcal{J}_{43}$) *There exists $r_0 > 0$ such that*

(a) *For all $t, s \in J$; $y, z \in [-r_0, r_0]$ and $x \in \mathcal{C}(J)$, we have*

$$|p(t, s, y, z)| \leq r_0 - \|q_1\|_\infty - \frac{|\zeta|}{|(T_2 x)(0)|}, \quad \textit{and}$$

(b) $\alpha r_0 < 1$.

Then, FDE (6.13)–(6.14) has, at least, one solution in $\mathcal{C}(J)$.

Proof. Notice that FDE (6.13)–(6.14) is equivalent to the functional integral equation:

$$x(t) = (T_2 x)(t)\left[q_1(t) + \frac{\zeta}{(T_2 x)(0)} + \int_0^t p(t, s, x(s), x(\lambda s))ds\right], t \in J,\ 0 < \lambda < 1. \tag{6.15}$$

Notice also that FIE (6.15) represents a particular case of FIE (6.12) with for all $t \in J$; $\sigma(t) = t, a(t) = 0, u = 1$ and $q(t) = q_1(t) + \frac{\zeta}{(T_2x)(0)}$. Therefore, we have for all $t \in J$; $(Ax)(t) = (T_2x)(t)$, $(Bx)(t) = q(t) + \int_0^t p(t, s, x(s), x(\lambda s))ds$ and $C(x)(t) = 0$. Now, we must prove that the operators A, B, and C satisfy all the conditions of Theorem 3.5.1. Similarly to the proof of the preceding Theorem 6.2.4, we obtain:

(i) A and C are both Lipschitzian with constants α and 0, respectively.

(ii) B is weakly sequentially continuous on S and $B(S)$ is relatively weakly compact, where $S = B_{r_0} := \{x \in \mathcal{C}(J),\ \|x\|_\infty \leq r_0\}$.

(iii) A is regular on $\mathcal{C}(J)$.

(iv) $\left(\frac{I-C}{A}\right)^{-1} = \left(\frac{I}{T_2}\right)^{-1}$ is weakly sequentially continuous on $B(S)$. It remains to prove the assumption (v) of Theorem 3.1.5. First, we have to show that $M = \|B(S)\| \leq r_0$. To do it, let us fix an arbitrary $x \in S$. Then, for $t \in J$, we get

$$\begin{aligned}
|(Bx)(t)| &\leq |q_1(t)| + \frac{|\zeta|}{|(T_2x)(0)|} + \int_0^t |p(t, s, x(s), x(\lambda s))|ds \\
&\leq |q_1(t)| + \frac{|\zeta|}{|(T_2x)(0)|} + \int_0^1 |p(t, s, x(s), x(\lambda s))|ds \\
&\leq \|q_1\|_\infty + \frac{|\zeta|}{|(T_2x)(0)|} + r_0 - \|q_1\|_\infty - \frac{|\zeta|}{|(T_2x)(0)|} \\
&= r_0.
\end{aligned}$$

Taking the supremum over t, we obtain

$$\|Bx\|_\infty \leq r_0.$$

Thus,

$$M \leq r_0.$$

Consequently,

$$\alpha M + \beta = \alpha M \leq \alpha r_0 < 1.$$

Next, let us fix an arbitrary $x \in \mathcal{C}(J)$ and let $y \in S$ such that

$$x = Ax.By + Cx,$$

or, equivalently

$$\text{for all } t \in J,\ x(t) = (T_2x)(t)(By)(t).$$

Then,

$$|x(t)| \leq \|T_2x\|_\infty \|By\|_\infty,$$

and thus, by using the assumption $(\mathcal{J}_{41})(e)$, we deduce that

$$|x(t)| \leq \|By\|_\infty.$$

Since $y \in S$, this implies that

$$|x(t)| \leq r_0,$$

and taking the supremum over t, we obtain

$$\|x\|_\infty \leq r_0.$$

As a result, x is in S. This proves the assumption (v). Now, by applying Theorem 3.1.5, we show that FIE (6.15) has, at least, one solution in $\mathcal{C}(J)$. Q.E.D.

6.4 An Application of Leray–Schauder's Theorem to FIEs

In this section, we are dealing with the following nonlinear functional integral equation:

$$x(t) = a(\nu(t)) + (Tx)(t)\Big[\Big(q(t) + \int_0^{\sigma(t)} p(t, s, g(x(\xi(s))), h(x(\eta(s))))ds\Big).u\Big] \tag{6.16}$$

for all $t \in J$, where $u \neq 0$ is a fixed vector of X and the functions a, ν, q, σ, ξ, η, p, g, h, and T are given while $x = x(t)$ is an unknown function.

We assume the following:

$(\mathcal{J}_{44})$ $a : J \longrightarrow X$ is a continuous function.

$(\mathcal{J}_{45})$ ν, σ, ξ, $\eta : J \longrightarrow J$ are continuous.

$(\mathcal{J}_{46})$ $q : J \longrightarrow \mathbb{R}$ is a continuous function.

$(\mathcal{J}_{47})$ The operator $T : \mathcal{C}(J, X) \longrightarrow \mathcal{C}(J, X)$ is such that:

(*a*) T is Lipschitzian with a constant α,

(*b*) T is regular on $\mathcal{C}(J, X)$,

(*c*) T is weakly sequentially continuous on $\mathcal{C}(J, X)$, *and*

(*d*) T is weakly compact.

($\mathcal{J}_{48}$) The functions g, $h : X \longrightarrow X$ are weakly sequentially continuous on X such that for each $r > 0$, g and h map the bounded subset rB_X into itself,

($\mathcal{J}_{49}$) The function $p : J \times J \times X \times X \longrightarrow \mathbb{R}$ is weakly sequentially continuous such that for an arbitrary fixed $s \in J$ and $x, y \in X$, the partial function $t \longrightarrow p(t, s, x, y)$ is continuous uniformly for $(s, x, y) \in J \times X \times X$, and

($\mathcal{J}_{50}$) there exists $r_0 > 0$ such that:

(*a*) $|p(t, s, x, y)| \leq M$ for each $t,\ s \in J$; $x,\ y \in X$ such that $\|x\| \leq r_0$ and $\|y\| \leq r_0$,

(*b*) $\|u\| \|(Tx)\|_\infty \leq 1$ for each $x \in \mathcal{C}(J, X)$ such that $\|x\|_\infty \leq r_0$,

(*c*) $\|a\|_\infty + \|q\|_\infty + M \leq r_0$,

(*d*) $\alpha r_0 < 1$.

Let us define the subset Ω of $\mathcal{C}(J, X)$ by:

$$\Omega := \{x \in \mathcal{C}(J, X), \|x\|_\infty \leq r_0\} = B_{r_0}.$$

Obviously, Ω is nonempty, convex, and closed. Let U be a weakly open subset of Ω such that $0 \in U$. In order to better understand FIE (6.16), let us consider two operators A and C defined on Ω and another operator B defined on $\overline{U^w}$ as follows:

$$\begin{aligned}
(Ax)(t) &= (Tx)(t),\\
(Bx)(t) &= \left[q(t) + \int_0^{\sigma(t)} p(t, s, g(x(\xi(s))), h(x(\eta(s))))ds\right].u, \text{ and}\\
(Cx)(t) &= a(\nu(t)).
\end{aligned}$$

This means that Eq. (6.16) is equivalent to

$$x = Ax.Bx + Cx.$$

Theorem 6.4.1 *Suppose that the assumptions* $(\mathcal{J}_{44})$–$(\mathcal{J}_{50})$ *hold. Let* $\Omega = B_{r_0}$ *and let* U *be a weakly open subset of* Ω *such that* $0 \in U$. *In addition, suppose that, for any solution* x *to the equation* $x = \lambda A(\frac{x}{\lambda})Bx + \lambda C(\frac{x}{\lambda})$ *for some* $0 < \lambda < 1$, *we have* $x \notin \partial_{\Omega}(U)$. *Then, the FIE (6.16) has a solution in* U.

Proof. We will prove that the operators A, B, and C satisfy all the conditions of Theorem 3.3.3.

(i) From assumption $(\mathcal{J}_{47})(a)$, it follows that A is Lipschitzian with a constant α. It is clear that C is Lipschitzian with a constant 0.

(ii) From assumption $(\mathcal{J}_{47})(b)$, it follows that A is regular on $\mathcal{C}(J, X)$.

(iii) This condition is satisfied by $(\mathcal{J}_{50})(d)$.

(iv) We will show that the hypothesis (iv) of Theorem 3.3.3 is satisfied. In fact, we fix arbitrarily $x \in \mathcal{C}(J, X)$ and $y \in \overline{U^w}$ such that

$$x = Ax.By + Cx,$$

or equivalently, for all $t \in J$,

$$x(t) = (Tx)(t)(By)(t) + a(\nu(t)).$$

In view of hypothesis $(\mathcal{J}_{50})$, and for all $t \in J$, we have

$$\begin{aligned}
\|(Ax)(t)(By)(t) + (Cx)(t)\| &= \|(Tx)(t)(Bx)(t) + a(\nu(t))\| \\
&\leq \|a\|_\infty + \|(By)(t)\| \|(Tx)\|_\infty \\
&\leq \|a\|_\infty + (M + \|q\|_\infty)\|u\| \|(Tx)\|_\infty \\
&\leq r_0.
\end{aligned}$$

From the last inequality, and taking the supremum over t, we obtain

$$\|(Ax).(By) + (Cx)\|_\infty \leq r_0,$$

and consequently,

$$(Ax).(By) + (Cx) \in \Omega.$$

(v) In view of hypothesis $(\mathcal{J}_{47})(c)$, A is weakly sequentially continuous on $\mathcal{C}(J, X)$. Since C is constant, then C is weakly sequentially continuous on Ω. Now, we show that B is weakly sequentially continuous on $\overline{U^w}$. Firstly, we verify that if $x \in \overline{U^w}$, then $Bx \in \mathcal{C}(J, X)$. For this, let $\{t_n\}$ be any sequence in J converging to a point in J. Then,

$$\begin{aligned}
&\|(Bx)(t_n)-(Bx)(t)\| \\
&\leq |q(t_n)-q(t)|\|u\| + \Big[\int_{\sigma(t_n)}^{\sigma(t)} |p(t,s,g(x(\xi(s))),h(x(\eta(s))))|ds\Big]\|u\| \\
&+\Big[\int_0^{\sigma(t_n)} |p(t_n,s,g(x(\xi(s))),h(x(\eta(s))))-p(t,s,g(x(\xi(s))),h(x(\eta(s))))|ds\Big]\|u\| \\
&\leq |q(t_n)-q(t)|\|u\| + M|\sigma(t)-\sigma(t_n)|\|u\| \\
&+\Big[\int_0^{1} |p(t_n,s,g(x(\xi(s))),h(x(\eta(s))))-p(t,s,g(x(\xi(s))),h(x(\eta(s))))|ds\Big]\|u\|.
\end{aligned}$$

Since $t_n \to t$, then for all $s \in J$, we have:

$$(t_n,s,g(x(\xi(s))),h(x(\eta(s)))) \to (t,s,g(x(\xi(s))),h(x(\eta(s)))).$$

Taking into account the hypothesis $(\mathcal{J}_{49})$, we obtain

$$p(t_n,s,g(x(\xi(s))),h(x(\eta(s)))) \to p(t,s,g(x(\xi(s))),h(x(\eta(s)))) \text{ in } \mathbb{R}.$$

Moreover, the use of assumption $(\mathcal{J}_{50})$ leads to

$$|p(t_n,s,g(x(\xi(s))),h(x(\eta(s))))-p(t,s,g(x(\xi(s))),h(x(\eta(s))))| \leq 2M$$

for all $t, s \in J$. Now, we can apply the dominated convergence theorem and since assumption $(\mathcal{J}_{46})$ holds, we get

$$(Bx)(t_n) \to (Bx)(t).$$

It follows that

$$Bx \in \mathcal{C}(J,X).$$

Next, we prove that B is weakly sequentially continuous on $\overline{U^w}$. Let $\{x_n\}$ be any sequence in $\overline{U^w}$ weakly converging to a point $x \in \overline{U^w}$. Then, $\{x_n\}$ is bounded. By applying Theorem 3.5.2, we get

$$x_n(t) \to x(t), \quad \forall t \in J.$$

So, by using assumptions $(\mathcal{J}_{48})$–$(\mathcal{J}_{50})$ and the dominated convergence theorem, we obtain:

$$\int_0^{\sigma(t)} p(t,s,g(x_n(\xi(s))),h(x_n(\eta(s))))ds \to \int_0^{\sigma(t)} p(t,s,g(x(\xi(s))),h(x(\eta(s))))ds$$

as $n \to \infty$, which implies that

$$\lim_{n\to\infty}\Big(q(t)+\int_0^{\sigma(t)} p(t,s,g(x_n(\xi(s))),h(x_n(\eta(s))))ds\Big).u$$
$$=\Big(q(t)+\int_0^{\sigma(t)} p(t,s,g(x(\xi(s))),h(x(\eta(s))))ds\Big).u.$$

Hence,

$$(Bx_n)(t)\to(Bx)(t) \text{ in } X$$

and so,

$$(Bx_n)(t)\rightharpoonup(Bx)(t) \text{ in } X.$$

It is clear that the sequence $\{Bx_n\}$ is bounded by $(\|q\|_\infty+M)\|u\|$. Then, by using Theorem 3.5.2, we get

$$(Bx_n)\rightharpoonup(Bx).$$

Thus, we conclude that B is weakly sequentially continuous on $\overline{U^w}$.

(vi) By using the fact that Ω is bounded by r_0, and in view of assumption $(\mathcal{J}_{47})(d)$, it follows that $A(\Omega)$ is relatively weakly compact. Since $C(\Omega)=\{a\}$, then $C(\Omega)$ is relatively weakly compact. It remains to prove that $B(\overline{U^w})$ is relatively weakly compact. By definition,

$$B(\overline{U^w}):=\{Bx:\|x\|_\infty\le r_0\}.$$

For all $t\in J$, we have

$$B(\overline{U^w})(t):=\{(Bx)(t):\|x\|_\infty\le r_0\}.$$

We claim that $B(\overline{U^w})(t)$ is weakly sequentially relatively compact in X. To show it, let $\{x_n\}$ be any sequence in $\overline{U^w}$. Then, we have $(Bx_n)(t)=r_n(t).u$, where

$$r_n(t)=q(t)+\int_0^{\sigma(t)} p(t,s,g(x_n(\xi(s))),h(x_n(\eta(s))))ds.$$

It is clear that $|r_n(t)|\le(\|q\|_\infty+M)$ and $\{r_n(t)\}$ is a real sequence, so, by using Bolzano–Weirstrass's theorem, there is a renamed subsequence such that

$$r_n(t)\to r(t),$$

which implies

$$r_n(t).u\to r(t).u,$$

and consequently,

$$(Bx_n)(t)\to r(t).u.$$

Hence, we conclude that $B(\overline{U^w})(t)$ is sequentially relatively compact in X. Then, $B(\overline{U^w})(t)$ is sequentially relatively weakly compact in X. Now, we have to prove that $B(\overline{U^w})$ is weakly equicontinuous on J. For this purpose, let $\varepsilon > 0$; $x \in \overline{U^w}$; $x^* \in X^*$; t, $t' \in J$ such that $t \leq t'$ and $t' - t \leq \varepsilon$. Then,

$$\|x^*((Bx)(t) - (Bx)(t'))\| \leq |q(t) - q(t')||x^*(u)|$$

$$+\Big|\int_0^{\sigma(t)} \chi(t,s)ds - \int_0^{\sigma(t')} \chi(t',s)ds\Big||x^*(u)|$$

$$\leq |q(t) - q(t')||x^*(u)|$$

$$+\Big[\int_0^{\sigma(t)} \Big|\chi(t,s) - \chi(t',s)\Big|ds\Big]|x^*(u)| + \Big|\int_{\sigma(t)}^{\sigma(t')} \chi(t',s)ds\Big||x^*(u)|$$

$$\leq [w(q,\varepsilon) + w(p,\varepsilon) + Mw(\sigma,\varepsilon)]|x^*(u)|,$$

where

$$\chi(t,s) := p(t,s,g(x(\xi(s))),h(x(\eta(s)))),$$

$$w(q,\varepsilon) := \sup\{|q(t) - q(t')| : t,t' \in J; |t - t'| \leq \varepsilon\},$$

$$w(p,\varepsilon) := \sup\{|p(t,s,x,y)) - p(t,s,x,y)| : t,t',s \in J; |t - t'| \leq \varepsilon; x,y \in \Omega\},$$

and

$$w(\sigma,\varepsilon) := \sup\{|\sigma(t) - \sigma(t')| : t,t' \in J; |t - t'| \leq \varepsilon\}.$$

Taking into account the hypothesis $(\mathcal{J}_{49})$, and in view of the uniform continuity of the functions q and σ, it follows that $w(q,\varepsilon) \to 0$, $w(p,\varepsilon) \to 0$, and $w(\sigma,\varepsilon) \to 0$ when $\varepsilon \to 0$. Hence, the application of Arzelà–Ascoli's theorem (see Theorem 1.3.9) implies that $B(\overline{U^w})$ is sequentially relatively weakly compact in $\mathcal{E}$. Now, the use of Eberlein–Šmulian's theorem (see Theorem 1.3.3) allows $B(\overline{U^w})$ to be relatively weakly compact. Hence, all the conditions of Theorem 3.3.3 are satisfied and then its application ensures that either conclusion (a) or (b) holds. By using the fact that, for any solution x to the equation $x = \lambda A(\frac{x}{\lambda})Bx + \lambda C(\frac{x}{\lambda})$ for some $0 < \lambda < 1$, $x \notin \partial_\Omega(U)$, we can deduce that conclusion (b) is eliminated and hence Eq. (6.16) has a solution in U. Q.E.D.

Chapter 7

Two-Dimensional Boundary Value Problems

The first aim of this chapter is to give some existence results for a structured problem on L_p-space $(1 \leq p < \infty)$ under abstract boundary conditions of Rotenberg's model type. Next, several coupled systems of nonlinear functional integral equations with bounded or unbounded domains in Banach algebras are considered. Finally, some existence results for coupled systems of perturbed functional differential inclusions of initial and boundary value problems are studied. We should mention that all the involved operator equations are generated by block matrices with the dimensions 2×2. Also considered are both single-valued and multi-valued operators acting in Banach algebras satisfying the so-called condition $(\mathcal{P})$.

7.1 A System of Transport Equations in L_p $(1 < p < \infty)$

Let us consider the following problem in L_p-space $(1 < p < \infty)$ under boundary conditions of Rotenberg's model type [142].

$$\begin{pmatrix} -v\dfrac{\partial}{\partial\mu} - \sigma_1(\mu, v, .) & R_{12} \\ R_{21} & -v\dfrac{\partial}{\partial\mu} - \sigma_2(\mu, v, .) \end{pmatrix} \begin{pmatrix} \psi_1 \\ \psi_2 \end{pmatrix} = \lambda \begin{pmatrix} \psi_1 \\ \psi_2 \end{pmatrix} \tag{7.1}$$

$$\psi_{i}{}_{|\Gamma_0} = K_i\left(\psi_{i}{}_{|\Gamma_1}\right), \quad i = 1, 2, \tag{7.2}$$

where $R_{ij}\psi_j(\mu, v) = \displaystyle\int_a^b r_{ij}(\mu, v, v', \psi_j(\mu, v'))dv'$, $(i, j) \in \{(1, 2), (2, 1)\}$, $\mu \in$

$[0,1]$, $v, v' \in [a,b]$ with $0 \leq a < b < \infty$, $\sigma_i(.,.,.)$, $i = 1,2$, $r_{ij}(.,.,.,.)$ being nonlinear operators, λ is a complex number, $\Gamma_0 = \{0\} \times [a,b]$ and $\Gamma_1 = \{1\} \times [a,b]$. We denote by $\psi_{i_{|\Gamma_0}}$ (resp. $\psi_{i_{|\Gamma_1}}$) the restriction of ψ_i to Γ_0 (resp. Γ_1) while K_i represent nonlinear operators from a suitable function space on Γ_1 to a similar one on Γ_0. The main point in Eq. (7.1) of the proposed model is the nonlinear dependence of the functions $r_{ij}(\mu, v, v', \psi_j(\mu, v'))$ on ψ_j. More specifically, we suppose that

$$r_{ij}(\mu, v, v', \psi(\mu, v')) = k_{ij}(\mu, v, v')f(\mu, v', \psi(\mu, v')); \ (i,j) \in \{(1,2),(2,1)\},$$

where f is a measurable function defined by:

$$\begin{cases} f : [0,1] \times [a,b] \times \mathbb{C} \longrightarrow \mathbb{C} \\ \qquad (\mu, v, u) \longrightarrow f(\mu, v, u) \end{cases}$$

with $k_{ij}(.,.,.)$, $(i,j) \in \{(1,2),(2,1)\}$ representing measurable functions from $[0,1] \times [a,b] \times \mathbb{C}$ into $\mathbb{C}$.

7.1.1 Non-dependence of σ_i on the density of the population

In this subsection, we consider a particular version of (7.1)-(7.2), where each σ_i does not depend on the density of the population i.

$$\begin{pmatrix} -v\dfrac{\partial}{\partial\mu} - \sigma_1(\mu,v)I & R_{12} \\ R_{21} & -v\dfrac{\partial}{\partial\mu} - \sigma_2(\mu,v)I \end{pmatrix}\begin{pmatrix} \psi_1 \\ \psi_2 \end{pmatrix} = \lambda \begin{pmatrix} \psi_1 \\ \psi_2 \end{pmatrix} \tag{7.3}$$

$$\psi_{i_{|\Gamma_0}} = K_i\left(\psi_{i_{|\Gamma_1}}\right), \quad i = 1,2 \tag{7.4}$$

where $\sigma_i(.,.) \in L^\infty([0,1] \times [a,b])$ and λ is a complex number. We will focus on the existence of solutions for the last boundary value problem (7.3)–(7.4). For this purpose, let

$$X_p := L_p([0,1] \times [a,b]; d\mu dv),$$

where $0 \leq a < b < \infty$; $1 < p < \infty$. We denote by X_p^0 and X_p^1, the following boundary spaces

$$X_p^0 := L_p(\{0\} \times [a,b]; vdv)$$

and

$$X_p^1 := L_p(\{1\} \times [a,b]; vdv),$$

endowed with their natural norms. In what follows, $\mathcal{W}_p$ denotes the partial Sobolev space defined by:

$$\mathcal{W}_p = \left\{ \psi \in X_p \text{ such that } v\frac{\partial \psi}{\partial \mu} \in X_p \right\}.$$

Now, let us define the free streaming operator S_{K_i}, $i = 1, 2$ by:

$$\begin{cases} S_{K_i} : \mathcal{D}(S_{K_i}) \subset X_p \longrightarrow X_p \\ \qquad\qquad \psi \longrightarrow S_{K_i}\psi(\mu, v) = -v\frac{\partial \psi}{\partial \mu}(\mu, v) - \sigma_i(\mu, v)\psi(\mu, v) \\ \mathcal{D}(S_{K_i}) = \{\psi \in \mathcal{W}_p \text{ such that } \psi^0 = K_i(\psi^1)\}, \end{cases}$$

where $\psi^0 = \psi_{|\Gamma_0}$, $\psi^1 = \psi_{|\Gamma_1}$ and K_i, $i = 1, 2$ are the following nonlinear boundary operators

$$\begin{cases} K_i : X_p^1 \longrightarrow X_p^0 \\ \qquad u \longrightarrow K_i u \end{cases}$$

satisfying the following conditions:

($\mathcal{R}_1$) There exists $\alpha_i > 0$ such that:

$$\|K_i\varphi_1 - K_i\varphi_2\| \leq \alpha_i \|\varphi_1 - \varphi_2\| \ \ \forall\, \varphi_1, \ \varphi_2 \in X_p^1, \ i = 1, 2.$$

As immediate consequences of ($\mathcal{R}_1$), we have the continuity of the operator K_i from X_p^1 into X_p^0, and the estimate:

$$\|K_i\varphi\| \leq \alpha_i \|\varphi\| + \|K_i(0)\| \ \forall\, \varphi \in X_p^1.$$

Now, let us consider the following equation:

$$(\lambda - S_{K_i})\psi_i = g. \tag{7.5}$$

Our objective is to determine a solution $\psi_i \in \mathcal{D}(S_{K_i})$, where g is given in X_p and $\lambda \in \mathbb{C}$. Let $\underline{\sigma}$ be the real defined by:

$$\underline{\sigma} := \text{ess-}\inf \{\sigma_i(\mu, v), \ \ (\mu, v) \in [0,1] \times [a,b], \ \ i = 1, 2\}.$$

For $Re\lambda > -\underline{\sigma}$, the solution of Eq. (7.5) is formally given by:

$$\psi_i(\mu, v) = \psi_i(0, v)\, e^{-\frac{1}{v}\int_0^{\mu}(\lambda + \sigma_i(\mu', v))d\mu'} + \frac{1}{v}\int_0^{\mu} e^{-\frac{1}{v}\int_{\mu'}^{\mu}(\lambda + \sigma_i(\tau, v))d\tau} g(\mu', v)\, d\mu'.$$

Accordingly, for $\mu = 1$, we get

$$\psi_i(1,v) = \psi_i(0,v)\, e^{-\frac{1}{v}\int_0^1(\lambda+\sigma_i(\mu',v))d\mu'} + \frac{1}{v}\int_0^1 e^{-\frac{1}{v}\int_{\mu'}^1(\lambda+\sigma_i(\tau,v))d\tau} g(\mu',v)\, d\mu'. \tag{7.6}$$

Let us consider the following operators:

$$\begin{cases} P_{i,\lambda} : X_p^0 \longrightarrow X_p^1 \\ \qquad u \longrightarrow (P_{i,\lambda}u)(1,v) := u(0,v)\, e^{-\frac{1}{v}\int_0^1(\lambda+\sigma_i(\mu',v))d\mu'}; \end{cases}$$

$$\begin{cases} Q_{i,\lambda} : X_p^0 \longrightarrow X_p \\ \qquad u \longrightarrow (Q_{i,\lambda}u)(\mu,v) := u(0,v)\, e^{-\frac{1}{v}\int_0^\mu(\lambda+\sigma_i(\mu',v))d\mu'}; \end{cases}$$

$$\begin{cases} \Pi_{i,\lambda} : X_p \longrightarrow X_p^1 \\ \qquad u \longrightarrow (\Pi_{i,\lambda}u)(1,v) := \frac{1}{v}\int_0^1 e^{-\frac{1}{v}\int_{\mu'}^1(\lambda+\sigma_i(\tau,v))d\tau} u(\mu',v)\, d\mu'; \end{cases}$$

and finally,

$$\begin{cases} R_{i,\lambda} : X_p \longrightarrow X_p \\ \qquad u \longrightarrow (R_{i,\lambda}u)(\mu,v) := \frac{1}{v}\int_0^\mu e^{-\frac{1}{v}\int_{\mu'}^\mu(\lambda+\sigma_i(\tau,v))d\tau} u(\mu',v)\, d\mu'. \end{cases}$$

Clearly, for λ satisfying $Re\lambda > -\underline{\sigma}$, the operators $P_{i,\lambda}$, $Q_{i,\lambda}$, $\Pi_{i,\lambda}$, and $R_{i,\lambda}$ are bounded. It is not difficult to check that

$$\|P_{i,\lambda}\| \leq e^{-\frac{1}{b}(Re\lambda+\underline{\sigma})} \tag{7.7}$$

and

$$\|Q_{i,\lambda}\| \leq (p(Re\lambda+\underline{\sigma}))^{-\frac{1}{p}}. \tag{7.8}$$

Moreover, some simple calculations using the Hölder inequality show that

$$\|\Pi_{i,\lambda}\| \leq (Re\lambda+\underline{\sigma})^{-\frac{1}{q}}, \tag{7.9}$$

and

$$\|R_{i,\lambda}\| \leq (Re\lambda+\underline{\sigma})^{-1}. \tag{7.10}$$

Hence, Eq. (7.6) may be abstractly written as

$$\psi_i^1 = P_{i,\lambda}\psi_i^0 + \Pi_{i,\lambda}g.$$

Moreover, ψ_i must satisfy the boundary condition (7.4). Thus, we obtain

$$\psi_i^1 = P_{i,\lambda} K_i \psi_i^1 + \Pi_{i,\lambda} g. \tag{7.11}$$

Notice that the operator $P_{i,\lambda}K_i$ in Eq. (7.11) is defined from X_p^1 into X_p^1. Let $\varphi_1,\ \varphi_2 \in X_p^1$. From both $(\mathcal{R}_1)$ and Eq. (7.7), it follows that

$$\|P_{i,\lambda}K_i\varphi_1 - P_{i,\lambda}K_i\varphi_2\| \leq \alpha_i\ e^{-\frac{Re\lambda+\underline{\sigma}}{b}} \|\varphi_1 - \varphi_2\| \quad \forall \varphi_1,\ \varphi_2 \in X_p^1. \tag{7.12}$$

Now, let us consider the following equation

$$u = P_{i,\lambda}K_i u + \varphi, \quad \varphi \in X_p^1, \tag{7.13}$$

where u is an unknown function and let us define the operator $A_{(i,\lambda,\varphi)}$ on X_p^1 by:

$$\begin{cases} A_{(i,\lambda,\varphi)} : X_p^1 \longrightarrow X_p^1, \\ \qquad\qquad u \longrightarrow (A_{(i,\lambda,\varphi)}u)(1,v) := P_{i,\lambda}K_i u + \varphi. \end{cases}$$

From Eq.(7.12), it follows that

$$\|A_{(i,\lambda,\varphi)}\varphi_1 - A_{(i,\lambda,\varphi)}\varphi_2\| \ = \ \|P_{i,\lambda}K_i\varphi_1 - P_{i,\lambda}K_i\varphi_2\| \leq \alpha_i\ e^{-\frac{Re\lambda+\underline{\sigma}}{b}} \|\varphi_1 - \varphi_2\|.$$

Consequently, for $Re\lambda > -\underline{\sigma} + b\log(\alpha_i)$, the operator $A_{(i,\lambda,\varphi)}$ is a contraction mapping and therefore, by using Theorem 1.2.1, Eq. (7.13) has a unique solution

$$u_{(i,\lambda,\varphi)} = u_i.$$

Let $W_{i,\lambda}$ be the nonlinear operator defined by:

$$W_{i,\lambda}\varphi = u_i, \tag{7.14}$$

where u_i is the solution of Eq. (7.13). Now, we have the following result:

Lemma 7.1.1 *Assume that $(\mathcal{R}_1)$ holds. Then,*
(i) for every λ satisfying $Re\lambda > -\underline{\sigma} + b\log(\alpha_i)$, $i = 1, 2,$ the operator $W_{i,\lambda}$ is continuous and map bounded sets into bounded ones and satisfies the following estimate

$$\|W_{i,\lambda}\varphi_1 - W_{i,\lambda}\varphi_2\| \leq (1 - \alpha_i e^{-(\frac{Re\lambda+\underline{\sigma}}{b})})^{-1} \|\varphi_1 - \varphi_2\|; \quad \varphi_1, \varphi_2 \in X_p^1, \quad i = 1, 2.$$

(ii) If $Re\lambda >$ max$(-\underline{\sigma}, -\underline{\sigma} + b\log(\alpha_i))$, then the operators $(\lambda - S_{K_i})$ are invertible and $(\lambda - S_{K_i})^{-1}$ is given by:

$$(\lambda - S_{K_i})^{-1} = Q_{i,\lambda}K_iW_{i,\lambda}\Pi_{i,\lambda} + R_{i,\lambda}.$$

Moreover, $(\lambda - S_{K_i})^{-1}$ are continuous on X_p and map bounded sets into bounded ones.

Proof. By the definition of $W_{i,\lambda}$, Eq. (7.14) may be written in the form:

$$W_{i,\lambda}\varphi = P_{i,\lambda}K_iW_{i,\lambda}\varphi + \varphi$$

and thus

$$\|W_{i,\lambda}\varphi_1 - W_{i,\lambda}\varphi_2\| \leq \|P_{i,\lambda}K_i(W_{i,\lambda}\varphi_1) - P_{i,\lambda}K_i(W_{i,\lambda}\varphi_2)\| + \|\varphi_1 - \varphi_2\|$$

$$\leq \alpha_i\, e^{-\frac{Re\lambda+\underline{\sigma}}{b}}\|W_{i,\lambda}\varphi_1 - W_{i,\lambda}\varphi_2\| + \|\varphi_1 - \varphi_2\|$$

for any φ_1, $\varphi_2 \in X_p^1$. This leads to the estimate:

$$\|W_{i,\lambda}\varphi_1 - W_{i,\lambda}\varphi_2\| \leq \left(1 - \alpha_i e^{-(\frac{Re\lambda+\underline{\sigma}}{b})}\right)^{-1}\|\varphi_1 - \varphi_2\|, \qquad i = 1,2$$

which proves the continuity of $W_{i,\lambda}$.

The second part of the assertion (i) follows from the estimate:

$$\|W_{i,\lambda}\varphi\| \leq \left(1 - \alpha_i e^{-(\frac{Re\lambda+\underline{\sigma}}{b})}\right)^{-1}\|\varphi\| + \|W_{i,\lambda}(0)\|.$$

(ii) Since $Re\lambda > \max(-\underline{\sigma}, -\underline{\sigma} + b\log(\alpha_i))$, the solution of Eq. (7.11) is given by:

$$\psi_i^1 = W_{i,\lambda}\Pi_{i,\lambda}g. \tag{7.15}$$

Observe that the solution of Eq. (7.5) may be written as:

$$\psi_i = Q_{i,\lambda}K_i\psi_i^1 + R_{i,\lambda}g.$$

Substituting Eq. (7.15) into the equation above, we obtain:

$$\psi_i = Q_{i,\lambda}K_iW_{i,\lambda}\Pi_{i,\lambda}g + R_{i,\lambda}g$$

from which we infer that $(\lambda - S_{K_i})$ is invertible and

$$(\lambda - S_{K_i})^{-1} = Q_{i,\lambda}K_iW_{i,\lambda}\Pi_{i,\lambda} + R_{i,\lambda}.$$

The second part of assertion (ii) follows from the boundedness of the linear operators $Q_{i,\lambda}$, $\Pi_{i,\lambda}$, $R_{i,\lambda}$ and from assertion (i). Q.E.D.

In what follows, and for our subsequent analysis, we need the following assumption:

$(\mathcal{R}_2)$ $r_{ij}(\mu, v, v', \psi(\mu, v')) = k_{ij}(\mu, v, v')f(\mu, v', \psi(\mu, v'))$; $(i,j) \in \{(1,2),(2,1)\}$,

where f is a measurable function defined by:

$$\begin{cases} f : [0,1] \times [a,b] \times \mathbb{C} \longrightarrow \mathbb{C} \\ \qquad (\mu, v, u) \longrightarrow f(\mu, v, u), \end{cases}$$

where $k_{ij}(.,.,.)$, $(i,j) \in \{(1,2),(2,1)\}$ are measurable functions from $[0,1] \times [a,b] \times \mathbb{C}$ into $\mathbb{C}$ which defines a bounded linear operator B_{ij} by:

$$\begin{cases} B_{ij} : X_p \longrightarrow X_p \\ \psi \longrightarrow \int_a^b k_{ij}(\mu, v, v')\psi(\mu, v')dv'. \end{cases} \tag{7.16}$$

Notice that the operators B_{ij}, $(i,j) = (1,2),(2,1)$ act only on the velocity v, so μ may be simply seen as a parameter in $[0,1]$. Then, we will consider B_{ij} as a function

$$B_{ij}(.) : \mu \in [0,1] \longrightarrow B_{ij}(\mu) \in \mathcal{L}(L_p([a,b];dv)).$$

In the following, we will make the assumptions:

$$(\mathcal{R}_3) \begin{cases} \text{- The function } B_{ij}(.) \text{ is measurable, i.e., if } \mathcal{O} \text{ is an open subset of} \\ \mathcal{L}(L_p([a,b];\,dv)), \text{then}\{\mu \in [0,1] \text{ such that} B_{ij}(\mu) \in \mathcal{O}\} \text{ is measurable,} \\ \text{- there exists a compact subset } \mathcal{C} \subset \mathcal{L}(L_p([a,b];\,dv)) \text{ such that} \\ B_{ij}(\mu) \in \mathcal{C} \text{ a.e. on } [0,1], \text{ and} \\ \text{-}B_{ij}(\mu) \in \mathcal{K}(L_p([a,b];\,dv)) \text{ a.e. on } [0,1], \end{cases}$$

where $\mathcal{K}(L_p([a,b],dv))$ stands for the class of compact operators on $L_p([a,b],dv)$.

Lemma 7.1.2 *Assume that B_{ij} satisfies the hypothesis $(\mathcal{R}_3)$. Then, B_{ij} can by approximated, in the uniform topology, by a sequence $(B_{ij,n})_n$ of operators of the form:*

$$\kappa_{ij,n}(\mu, v, v') = \sum_{s=1}^{n} \eta_s(\mu)\theta_s(v)\beta_s(v'),$$

where $\eta_s(.) \in L_\infty([0,1], d\mu)$, $\theta_s(.) \in L_p([a,b], dv)$ and $\beta_s(.) \in L_q([a,b], dv)$ (q denotes the conjugate of p).

Lemma 7.1.3 *Let $p \in (1,\infty)$ and assume that $(\mathcal{R}_1)$ holds. If B_{ij}, $(i,j) \in \{(1,2),(2,1)\}$ satisfies $(\mathcal{R}_3)$, then for any $\lambda \in \mathbb{C}$ such that $Re\lambda > \max(-\underline{\sigma}, -\underline{\sigma} + b\log(\alpha_i))$, the operators $(\lambda - S_{K_i})^{-1}B_{ij}$ are completely continuous on X_p.*

Proof. Let λ be such that $Re\lambda > \max(-\underline{\sigma}, -\underline{\sigma} + b\log(\alpha_i))$. By using Lemma 7.1.1 and the boundedness of B_{ij} (as a linear operator), we infer that $(\lambda - S_{K_i})^{-1}B_{ij}$ is continuous and maps bounded sets into bounded sets. In view of the assertion (ii) of Lemma 7.1.1 we have

$$(\lambda - S_{K_i})^{-1}B_{ij} = Q_{i,\lambda}K_iW_{i,\lambda}\Pi_{i,\lambda}B_{ij} \ + \ R_{i,\lambda}B_{ij}.$$

In order to complete the proof it is sufficient to show that $Q_{i,\lambda}K_iW_{i,\lambda}\Pi_{i,\lambda}B_{ij}$ and $R_{i,\lambda}B_{ij}$ are completely continuous on X_p. We claim that $\Pi_{i,\lambda}B_{ij}$ and $R_{i,\lambda}B_{ij}$ are compact. Indeed, since B_{ij} satisfies $(\mathcal{R}_3)$ it follows from Lemma 7.1.2 that B_{ij} can be approximated, in the uniform topology by a sequence $(B_{ij,n})_n$ of finite rank operators on $L_p([a,b]; dv)$ which converges, in the operator norm, to B_{ij}. Then it suffices to establish the result for a finite rank operator. So, we infer from the linearity and the stability of the compactness by summation that it suffices to prove the result for an operator B_{ij} whose kernel is in the form:

$$\kappa_{ij,n}(\mu, v, v') = \eta_{ij}(\mu)\theta_{ij}(v)\beta_{ij}(v'),$$

where $\eta_{ij}(.) \in L_\infty([0,1], d\mu)$, $\theta_{ij}(.) \in L_p([a,b], dv)$ and $\beta_{ij}(.) \in L_q([a,b], dv)$ (here, q denotes the conjugate of p). In a way similar to Lemma 5.1.2 we achieve the proof. Q.E.D.

Now, let us recall some facts concerning superposition operators required below. Recall $f : [0,1] \times [a,b] \times \mathbb{C} \longrightarrow \mathbb{C}$ is a Carathéodory function, if the following conditions are satisfied

$$\begin{cases} (\mu, v) \longrightarrow f(\mu, v, u) & \text{is measurable on } [0,1] \times [a,b] \text{ for all } u \in \mathbb{C} \\ u \longrightarrow f(\mu, v, u) & \text{is continuous on } \mathbb{C} \text{ a.e. } (\mu, v) \in [0,1] \times [a,b]. \end{cases}$$

Notice that, if f is a Carathéodory function, then we can define the operator $\mathcal{N}_f$ on the set of functions $\psi : [0,1] \times [a,b] \longrightarrow \mathbb{C}$ by:

$$(\mathcal{N}_f\psi)(\mu, v) = f(\mu, v, \psi(\mu, v)) \text{ for every } (\mu, v) \in [0,1] \times [a,b].$$

We assume that:

$(\mathcal{R}_4)$ f is a Carathéodory map satisfying

$$|f(\mu, v, u_1) - f(\mu, v, u_2)| \leq |h(\mu, v)||u_1 - u_2|,$$

where $h \in L^\infty([0,1] \times [a,b], d\mu dv)$.

Theorem 7.1.1 *Assume that* $(\mathcal{R}_1)$ *and* $(\mathcal{R}_2)$ *hold. If* B_{ij}, $(i,j) \in \{(1,2),(2,1)\}$ *satisfy* $(\mathcal{R}_3)$ *on* X_p, *then for each* $r > 0$, *there is* $\lambda_r > 0$ *such that, for each* λ *satisfying* $Re(\lambda) > \lambda_r$, *the problem (7.3)–(7.4) has, at least, one solution in* $B_r \times B_r$.

Proof. Let λ be a complex number such that $Re\lambda > \max(-\underline{\sigma}, -\underline{\sigma} + b\log(\alpha))$ with $\alpha = \max(\alpha_1, \alpha_2)$. Then, according to Lemma 7.1.1 (ii), we deduce that $\lambda - S_{K_i}$ is invertible and therefore, the problem (7.3)–(7.4) may be transformed into

$$\begin{pmatrix} \psi_1 \\ \psi_2 \end{pmatrix} = \mathcal{L}(\lambda)\begin{pmatrix} \psi_1 \\ \psi_2 \end{pmatrix}, \qquad \psi_i^0 = K_i\psi_i^1, \quad i = 1, 2,$$

where

$$\mathcal{L}(\lambda) = \begin{pmatrix} S_{K_1} - (\lambda - 1)I & B_{12}\mathcal{N}_f \\ B_{21}\mathcal{N}_f & S_{K_2} - (\lambda - 1)I \end{pmatrix}.$$

Let $r > 0$. We first check that, for a suitable λ, $\Upsilon(\lambda) := (\lambda - S_{K_1})^{-1}B_{12}\mathcal{N}_f$ leaves B_r invariant. Let $\|\psi_2\| \le r$. From Lemma 7.1.1 and Eqs (7.7)–(7.10), we have

$$\begin{aligned} \|\Upsilon(\lambda)(\psi_2)\| &\le \|Q_{1,\lambda}K_1W_{1,\lambda}\Pi_{1,\lambda}B_{12}\mathcal{N}_f(\psi_2) + R_{1,\lambda}B_{12}\mathcal{N}_f(\psi_2)\| \\ &\le \left[\frac{1}{Re\lambda + \underline{\sigma}} + \frac{1}{p^{\frac{1}{p}}}\frac{\alpha_1}{1 - \alpha_1 e^{-\frac{Re\lambda + \underline{\sigma}}{b}}}\right]\|B_{12}\|M(r) + \Sigma(Re\lambda), \end{aligned}$$

where $M(r)$ is the upper-bound of $\mathcal{N}_f$ on B_r, and

$$\Sigma(Re\lambda) = \left[\frac{\alpha_1\|W_{1,\lambda}(0)\| + \|K_1(0)\|}{(Re\lambda + \underline{\sigma})^{\frac{1}{p}}p^{\frac{1}{p}}}\right].$$

Let $\varepsilon > \max(-\underline{\sigma}, -\underline{\sigma} + b\log(\alpha))$. For $Re\lambda > \varepsilon$, we have

$$(1 - \alpha_1 e^{-\frac{Re\lambda + \underline{\sigma}}{b}})^{-1} \le (1 - \alpha_1 e^{-\frac{\varepsilon + \underline{\sigma}}{b}})^{-1}.$$

Therefore,

$$\|\Upsilon(\lambda)(\psi_2)\| \le \left[\frac{1}{Re\lambda + \underline{\sigma}} + \frac{1}{p^{\frac{1}{p}}}\frac{\alpha_1}{1 - \alpha_1 e^{-\frac{\varepsilon + \underline{\sigma}}{b}}}\right]\|B_{12}\|M(r) + \Sigma(Re\lambda).$$

By using Eq. (7.14), we have

$$P_{1,\lambda}K_1W_{1,\lambda}(0) = W_{1,\lambda}(0).$$

Let $0 < \delta < \frac{1}{\alpha_1}$. From Eq. (7.8), there exists λ_r such that for any λ satisfying $Re\lambda > \max(-\underline{\sigma}, -\underline{\sigma} + b\log(\alpha), \lambda_r)$, we have $\|P_{1,\lambda}\| \leq \delta$. Then, by using $(\mathcal{R}_1)$, we deduce that

$$\begin{aligned} \|W_{1,\lambda}(0)\| &\leq \|P_{1,\lambda}\|\|K_1 W_{1,\lambda}(0)\| \\ &\leq \delta(\alpha_1\|W_{1,\lambda}(0)\| + \|K_1(0)\|). \end{aligned}$$

It follows that

$$\|W_{1,\lambda}(0)\| \leq \frac{\delta\|K_1(0)\|}{1-\delta\alpha_1}.$$

Therefore,

$$\begin{aligned} \|\Upsilon(\lambda)(\psi_2)\| &\leq \left[\frac{1}{Re\lambda + \underline{\sigma}} + \frac{1}{p^{\frac{1}{p}}} \frac{\alpha_1}{1-\alpha_1 e^{-\frac{\varepsilon+\underline{\sigma}}{b}}}\right] \|B_{12}\|M(r) + \widetilde{\Sigma}(Re\lambda), \\ &\leq Q(Re\lambda), \end{aligned}$$

where

$$Q(t) = \left[\frac{1}{t+\underline{\sigma}} + \frac{1}{p^{\frac{1}{p}}} \frac{\alpha_1}{1-\alpha_1 e^{-\frac{\varepsilon+\underline{\sigma}}{b}}}\right] \|B_{12}\|M(r) + \left[\frac{(\frac{\alpha_1\delta}{1-\delta\alpha_1}+1)\|K_1(0)\|}{(t+\underline{\sigma})^{\frac{1}{p}} p^{\frac{1}{p}}}\right],$$

and

$$\widetilde{\Sigma}(Re\lambda) = \left[\frac{(\frac{\alpha_1\delta}{1-\delta\alpha_1}+1)\|K_1(0)\|}{(Re\lambda+\underline{\sigma})^{\frac{1}{p}} p^{\frac{1}{p}}}\right].$$

Clearly, $Q(.)$ is continuous strictly decreasing in $t > 0$ and satisfies $\lim_{t\to+\infty} Q(t) = 0$. Hence, there exists λ'_r, such that $Q(\lambda'_r) \leq r$. Obviously, if $Re\lambda \geq \lambda'_r$, then $(\lambda - S_{K_1})^{-1}B_{12}\mathcal{N}_f$ maps B_r into itself and $(\mathcal{R}_1)$ is satisfied. Clearly, from Lemma 7.1.3, the operator $S(\lambda) = B_{21}\mathcal{N}_f(\lambda - S_{K_1})^{-1}B_{12}\mathcal{N}_f$ is continuous, and so has a closed graph. Now, we claim that $S(\lambda)(B_r)$ is relatively compact. Indeed, $\mathcal{N}_f(B_r)$ is a bounded subset of X_p. From Lemma 7.1.3, it follows that $O_r := (\lambda - S_{K_1})^{-1}B_{12}\mathcal{N}_f(B_r)$ is relatively compact. Since $B_{21}\mathcal{N}_f$ is continuous, $B_{12}\mathcal{N}_f(\overline{O_r})$ is compact and so, $S(\lambda)(B_r)$ is relatively compact. For $Re\lambda \geq \max(\lambda_r, \lambda'_r)$ and $\|\psi_2\| \leq r$, we have

$$\|S(\lambda)(\psi_2)\| \leq r.$$

By using Lemma 7.1.1, we deduce that there exists $\lambda''_r \geq \max(\lambda_r, \lambda'_r)$ such that, for any $Re\lambda \geq \lambda''_r$, we have $\Delta(\lambda)\psi_2 := (\lambda - S_{K_2})^{-1}S(\lambda)\psi_2 \in B_r$. The result follows from Theorem 1.6.5. Q.E.D.

Theorem 7.1.2 *Assume that* $(\mathcal{R}_1)$, $(\mathcal{R}_2)$, *and* $(\mathcal{R}_4)$ *hold. If* B_{ij}, $(i,j) \in \{(1,2),(2,1)\}$ *satisfy* $(\mathcal{R}_3)$ *on* X_p, *then for each* $r > 0$, *there is* $\lambda_r > 0$ *such that, for each* λ *satisfying* $Re(\lambda) > \lambda_r$, *the problem (7.3)–(7.4) has a unique solution in* $B_r \times B_r$.

Proof. Let λ be a complex number such that $Re\lambda > \max(-\underline{\sigma}, -\underline{\sigma} + b\log(\alpha))$ with $\alpha = \max(\alpha_1, \alpha_2)$ and let $\psi_1, \psi_2 \in X_p$. Using the same notations as in the above proof, we have

$$\|S(\lambda)\psi_1 - S(\lambda)\psi_2\| \leq \|B_{21}\|\|B_{12}\|\|h\|_\infty^2 \mathcal{F}_1(Re\lambda)\|\psi_1 - \psi_2\|,$$

where

$$\mathcal{F}_1(Re\lambda) = \left[\frac{1}{Re\lambda + \underline{\sigma}} + \frac{1}{p^{\frac{1}{p}}} \frac{1}{1 - \alpha_1 e^{-\frac{Re\lambda + \underline{\sigma}}{b}}}\right]$$

is a continuous, positive, and strictly decreasing function on $]0, +\infty[$ satisfying $\lim_{t \to +\infty} \mathcal{F}_1(t) = 0$. By the same way, we have

$$\begin{aligned} \|\Delta(\lambda)\psi_1 - \Delta(\lambda)\psi_2\| &\leq \left[\frac{1}{Re\lambda + \underline{\sigma}} + \frac{1}{p^{\frac{1}{p}}} \frac{1}{1 - \alpha_2 e^{-\frac{Re\lambda + \underline{\sigma}}{b}}}\right] \|S(\lambda)\psi_1 - S(\lambda)\psi_2\| \\ &\leq \mathcal{F}_2(Re\lambda)\|S(\lambda)\psi_1 - S(\lambda)\psi_2\|, \end{aligned}$$

where $\mathcal{F}_1(.)$ is a continuous, positive, and strictly decreasing function on $]0, +\infty[$ satisfying $\lim_{t \to +\infty} \mathcal{F}_1(t) = 0$. Then, we have

$$\|\Delta(\lambda)\psi_1 - \Delta(\lambda)\psi_2\| \leq \|B_{21}\|\|B_{12}\|\|h\|_\infty^2 \mathcal{F}_1(Re\lambda)\mathcal{F}_2(Re\lambda)\|\psi_1 - \psi_2\|. \quad (7.17)$$

The function $\mathcal{F}_1(.)\mathcal{F}_2(.)$ satisfies the same properties as $\mathcal{F}_i$, $i = 1, 2$. We conclude that there exists a complex number λ_1 such that $\lambda_1 > \max(-\underline{\sigma}, -\underline{\sigma} + b\log(\alpha))$ and

$$\|B_{21}\|\|B_{12}\|\|h\|_\infty^2 \mathcal{F}_1(Re\lambda)\mathcal{F}_2(Re\lambda) < 1 \text{ for any } Re\lambda \geq \lambda_1. \quad (7.18)$$

Obviously, from Inequalities (7.17)–(7.18), and for $Re\lambda \geq \lambda_1$, the operator $\Delta(\lambda)$ is a contraction mapping on B_r and maps B_r into itself. Hence, the use of Banach's fixed point theorem (see Theorem 1.2.1) allows us to conclude that there exists a unique ψ_2 in B_r, such that $\Delta(\lambda)\psi_2 = \psi_2$. Let us take $\psi_1 := \Upsilon(\lambda)\psi_2$. By using the same argument as in the above proof, ψ_1 lies in B_r and so, $\begin{pmatrix} \psi_1 \\ \psi_2 \end{pmatrix}$ is a unique fixed point for the problem (7.3)–(7.4) in $B_r \times B_r$. Q.E.D.

7.1.2 Dependence of σ_i on the density of the population

Now, let us discuss the existence of solutions for the more general nonlinear boundary problem (7.1)–(7.2). When dealing with this problem, some technical difficulties arise. So, we need the following assumption:

($\mathcal{R}_5$) $K_i \in \mathcal{L}(X_p^1, X_p^0)$ and, for some $r > 0$, we have

$$|\sigma_i(\mu, v, \psi_1) - \sigma_i(\mu, v, \psi_2)| \leq |\omega_i(\mu, v)| \, \|\psi_1 - \psi_2\| \quad i = 1, 2 \ \ (\psi_1, \psi_2 \in X_p),$$

where $\mathcal{L}(X_p^1, X_p^0)$ denotes the set of all bounded linear operators from X_p^1 into X_p^0, $\omega_i(.,.) \in L^\infty([0,1] \times [a,b], d\mu dv)$ and $\mathcal{N}_{\sigma_i}$ acts from B_r into B_r.

Let us define the free streaming operator $\widehat{S}_{K_i}$ by:

$$\begin{cases} \widehat{S}_{K_i} : \mathcal{D}(\widehat{S}_{K_i}) \subset X_p \to X_p \\ \qquad\qquad \psi \to \widehat{S}_{K_i}\psi(\mu, v) = -v\dfrac{\partial \psi}{\partial \mu}(\mu, v) \\ \mathcal{D}(\widehat{S}_{K_i}) = \left\{\psi \in \mathcal{W}_p \text{ such that } \psi^0 = K_i(\psi^1)\right\}. \end{cases}$$

Theorem 7.1.3 *Assume that ($\mathcal{R}_2$), ($\mathcal{R}_4$), and ($\mathcal{R}_5$) hold. If B_{ij}, $(i,j) \in \{(1,2),(2,1)\}$ satisfy ($\mathcal{R}_3$) on X_p, then for each $r > 0$, there is $\lambda_r > 0$ such that, for each λ satisfying $Re(\lambda) > \lambda_r$, the problem (7.1)–(7.2) has, at least, one solution in $B_r \times B_r$.*

Proof. Since K_i, $i = 1, 2$ are linear in view of ($\mathcal{R}_5$), the operators $\widehat{S}_{K_i}$ are linear, too. By using Lemma 7.1.1, we deduce that

$$\{\lambda \in \mathbb{C} \text{ such that } Re\lambda > \max(0, b\log\|K_1\|, b\log\|K_2\|)\} \subset \varrho(\widehat{S}_{K_i}),$$

where $\varrho(\widehat{S}_{K_i})$ denotes the resolvent set of $\widehat{S}_{K_i}$. Let λ be a complex number such that $Re\lambda > \max(0, b\log\|K_1\|, b\log\|K_2\|)$. Then, by using the linearity of the operator $(\lambda - \widehat{S}_{K_i})^{-1}$, the problem (7.1)–(7.2) may be written in the following form:

$$\begin{pmatrix} \psi_1 \\ \psi_2 \end{pmatrix} = \widehat{\mathcal{L}}(\lambda)\begin{pmatrix} \psi_1 \\ \psi_2 \end{pmatrix}, \qquad \psi_i \in \mathcal{D}(\widehat{S}_{K_i}), \ \ i = 1, 2,$$

where

$$\widehat{\mathcal{L}}(\lambda) = \begin{pmatrix} \widehat{S}_{K_1} - (\lambda - 1)I + \mathcal{N}_{\sigma_1} & B_{12}\mathcal{N}_f \\ B_{21}\mathcal{N}_f & \widehat{S}_{K_2} - (\lambda - 1)I + \mathcal{N}_{\sigma_2} \end{pmatrix}.$$

The same problem may also be transformed into the form:

$$\begin{pmatrix} \psi_1 \\ \psi_2 \end{pmatrix} = \mathcal{G}_1(\lambda)\begin{pmatrix} \psi_1 \\ \psi_2 \end{pmatrix} + \mathcal{G}_2(\lambda)\begin{pmatrix} \psi_1 \\ \psi_2 \end{pmatrix}, \quad \psi_i \in \mathcal{D}(\widehat{S}_{K_i}), \quad i = 1,2 \tag{7.19}$$

where

$$\mathcal{G}_1(\lambda) = \begin{pmatrix} (\lambda - \widehat{S}_{K_1})^{-1}\mathcal{N}_{\sigma_1} & 0 \\ 0 & (\lambda - \widehat{S}_{K_2})^{-1}\mathcal{N}_{\sigma_2} \end{pmatrix},$$

and

$$\mathcal{G}_2(\lambda) = \begin{pmatrix} 0 & (\lambda - \widehat{S}_{K_1})^{-1}B_{12}\mathcal{N}_f \\ (\lambda - \widehat{S}_{K_2})^{-1}B_{21}\mathcal{N}_f & 0 \end{pmatrix}.$$

Let us check that, for a suitable λ, the operator $\mathcal{G}_1(\lambda)$ is a contraction mapping. Indeed, let $\begin{pmatrix} \psi_1 \\ \psi_2 \end{pmatrix}, \begin{pmatrix} \varphi_1 \\ \varphi_2 \end{pmatrix} \in X_p$. Then, for each $i = 1, 2$, we have

$$\|(\lambda - \widehat{S}_{K_i})^{-1}(\mathcal{N}_{\sigma_i}\varphi_i - \mathcal{N}_{\sigma_i}\psi_i)\| \leq \|(\lambda - \widehat{S}_{K_i})^{-1}\|\|\mathcal{N}_{\sigma_i}\varphi_i - \mathcal{N}_{\sigma_i}\psi_i\|.$$

A simple calculation, using the estimates (7.7)–(7.10), leads to

$$\|(\lambda - \widehat{S}_{K_i})^{-1}\| \leq \frac{1}{Re\lambda}\left[1 + \frac{\gamma}{1 - \gamma e^{-\frac{Re\lambda}{b}}}\right], \quad i = 1, 2, \tag{7.20}$$

where $\gamma = \max(\|K_1\|, \|K_2\|)$. Moreover, by taking into account the assumption on $\sigma_i(.,.,.)$, we get

$$\|\mathcal{N}_{\sigma_i}\varphi_i - \mathcal{N}_{\sigma_i}\psi_i\| \leq \|\omega\|_\infty\|\varphi_i - \psi_i\|,$$

where $\|\omega\|_\infty = \max(\|\omega_1\|_\infty, \|\omega_2\|_\infty)$. By using the relationships (7.19) and (7.20), we infer that, for $V = \begin{pmatrix} \varphi_1 \\ \varphi_2 \end{pmatrix}$ and $W = \begin{pmatrix} \psi_1 \\ \psi_2 \end{pmatrix}$, we have

$$\begin{aligned} \Delta(\lambda, V, W) &= \left\|\begin{pmatrix} (\lambda - \widehat{S}_{K_1})^{-1}\mathcal{N}_{\sigma_1}\varphi_1 \\ (\lambda - \widehat{S}_{K_2})^{-1}\mathcal{N}_{\sigma_1}\varphi_2 \end{pmatrix} - \begin{pmatrix} (\lambda - \widehat{S}_{K_1})^{-1}\mathcal{N}_{\sigma_1}\psi_1 \\ (\lambda - \widehat{S}_{K_2})^{-1}\mathcal{N}_{\sigma_1}\psi_2 \end{pmatrix}\right\| \\ &\leq \frac{1}{Re\lambda}\left[1 + \frac{\gamma}{1 - \gamma e^{-\frac{Re\lambda}{b}}}\right]\|\omega\|_\infty\left\|\begin{pmatrix} \varphi_1 \\ \varphi_2 \end{pmatrix} - \begin{pmatrix} \psi_1 \\ \psi_2 \end{pmatrix}\right\| \\ &\leq \Xi(Re\lambda)\left\|\begin{pmatrix} \varphi_1 \\ \varphi_2 \end{pmatrix} - \begin{pmatrix} \psi_1 \\ \psi_2 \end{pmatrix}\right\|, \end{aligned}$$

where

$$\Delta(\lambda, V, W) := \|\mathcal{G}_1(\lambda)V - \mathcal{G}_1(\lambda)W\|$$

and

$$\Xi(t) := \frac{1}{t}\left[1 + \frac{\gamma}{1-\gamma e^{-\frac{t}{b}}}\right]\|\omega\|_\infty.$$

Let us notice that Ξ is a continuous and strictly decreasing function defined on $]0,\infty[$ and

$$\lim_{t\to\infty}\Xi(t) = 0.$$

Hence, there exists $\lambda_1 \in]\max(0, b\log\|K_1\|, b\log\|K_2\|), \infty[$ such that $\Xi(\lambda_1) < 1$ and so, for $Re\lambda \geq \lambda_1$, $\mathcal{G}_1(\lambda)$ is a contraction mapping. By using Lemma 7.1.1 and arguing as in the proof of Theorem 7.1.1, we can show that the operator $\mathcal{G}_2(\lambda)$ is completely continuous on X_p. Theorem 4.1.4 achieves the proof. Q.E.D.

Question 6:

What happens if the reproduction rules are not generated by a bounded linear operator K_i from X_p^1 to X_p^0? To our knowledge, this question is not yet developed.

Now, let us discuss the existence of positive solutions for our boundary value problem. Let B_{ij} be defined by Eq. (7.16) and let $k_{ij}^+(.,.,.)$ (resp. $k_{ij}^-(.,.,.)$) denote the positive part (resp. the negative part) of $k_{ij}(.,.,.)$:

$$k_{ij}(\mu,v,v') = k_{ij}^+(\mu,v,v') - k_{ij}^-(\mu,v,v') \qquad (\mu,v,v') \in [0,1]\times[a,b]\times[a,b].$$

Let us define the following nonnegative operators :

$$B_{ij}^\pm : \psi \longrightarrow B_{ij}^\pm\psi(\mu,v) := \int_a^b k_{ij}^\pm(\mu,v,v')\psi(\mu,v')\,dv'.$$

Clearly,

$$B_{ij} = B_{ij}^+ - B_{ij}^-.$$

Now, let $|B_{ij}|$ denote the following nonnegative operator:

$$|B_{ij}| := B_{ij}^+ + B_{ij}^-$$

i.e.,

$$|B_{ij}|\psi(\mu,v) = \int_a^b |k_{ij}|(\mu,v,v')\psi(\mu,v')\,dv', \quad \psi \in X_p.$$

Assume that

$(\mathcal{R}_6)$ $$K_i[(X_p^1)^+)] \subset (X_p^0)^+,$$

where $(X_p^1)^+$ (resp. $(X_p^0)^+$) denotes the positive cone of the space X_p^1 (resp. X_p^0). Let $r > 0$. We define the set B_r^+ by $B_r^+ := B_r \cap X_p^+$.

Theorem 7.1.4 *Assume that* $(\mathcal{R}_1)$, $(\mathcal{R}_2)$, $(\mathcal{R}_3)$, $(\mathcal{R}_4)$, *and* $(\mathcal{R}_6)$ *hold. If* B_{ij} *is a positive operator and if* $\mathcal{N}_f(X_p^+) \subset X_p^+$, *then for each* $r > 0$, *there is* $\lambda_r > 0$ *such that for all* $\lambda > \lambda_r$, *the problem (7.1)–(7.2) has, at least, one solution in* B_r^+.

Proof. Obviously, the operators $P_{i,\lambda}$, $\Pi_{i,\lambda}$, $Q_{i,\lambda}$, and $R_{i,\lambda}$ are bounded and positive. Accordingly, by using arguments similar to those used in the proof of Theorem 5.2.4, we can reach the desired result. Q.E.D.

7.2 A Study of a Biological Coupled System in L_1

The aim of this section is to apply Theorems 4.2.4 and 4.2.6 in order to discuss the existence results for the two-dimensional boundary value problem (7.3)–(7.4) in the Banach space $L_1 \times L_1$. For this purpose, let us first specify the functional setting of the problem. Let us consider

$$X := L_1([0,1] \times [a,b]; d\mu dv),$$

where $0 \leq a < b < \infty$. Let us denote by X^0 and X^1 the following boundary spaces

$$X^0 := L_1(\{0\} \times [a,b]; vdv),$$
$$X^1 := L_1(\{1\} \times [a,b]; vdv),$$

endowed with their natural norms. Let $\mathcal{W}$ be the space defined by:

$$\mathcal{W} = \left\{\psi \in X \text{ such that } v\frac{\partial \psi}{\partial \mu} \in X\right\}.$$

It is well known (see, for example, [52, 53, 63]) that any ψ in $\mathcal{W}$ has traces on the spatial boundary $\{0\}$ and $\{1\}$ which belong, respectively, to the spaces X^0 and X^1.

As in Section 7.1.1, we define the free streaming operator S_{K_i}, $i = 1,\ 2$, by:

$$\begin{cases} S_{K_i} : \mathcal{D}(S_{K_i}) \subset X \longrightarrow X, \\ \qquad\qquad \psi_i \longrightarrow S_{K_i}\psi_i(\mu, v) = -v\dfrac{\partial \psi_i}{\partial \mu}(\mu, v) - \sigma_i(\mu, v)\psi_i(\mu, v), \\ \mathcal{D}(S_{K_i}) = \{\psi_i \in \mathcal{W} \text{ such that } \psi_i^0 = K_i(\psi_i^1)\}, \end{cases}$$

where $\sigma_i(.,.) \in L^\infty([0,1] \times [a,b])$, $\psi_i^0 = \psi_{i|\Gamma_0}$, $\psi_i^1 = \psi_{i|\Gamma_1}$ and K_i, $i = 1, 2$, represent the following nonlinear boundary operators

$$\begin{cases} K_i : X^1 \longrightarrow X^0, \\ \qquad u \longrightarrow K_i u, \end{cases}$$

satisfying the following conditions:

$(\mathcal{R}_7)$ There exists $\alpha_i > 0$ such that

$$\|K_i\varphi_1 - K_i\varphi_2\| \leq \alpha_i \|\varphi_1 - \varphi_2\| \quad \text{for all } \varphi_1,\ \varphi_2 \in X^1, \quad i = 1,\ 2.$$

$(\mathcal{R}_8)$ K_2 is a weakly compact operator on X^1.

As immediate consequences of $(\mathcal{R}_7)$, we have the continuity of the operator K_i from X^1 into X^0, and the following estimate:

$$\|K_i\varphi\| \leq \alpha_i\|\varphi\| + \|K_i(0)\| \quad \text{for all } \varphi \in X^1.$$

Let us consider the following equation

$$(\lambda - S_{K_i})\psi_i = g. \tag{7.21}$$

Following the same reasoning as in the previous subsection, our first task is to determine a solution $\psi_i \in \mathcal{D}(S_{K_i})$, where g is given in X and $\lambda \in \mathbb{C}$.
Let $\underline{\sigma}$ be the real defined by:

$$\underline{\sigma} := \text{ess-}\inf\{\sigma_i(\mu, v),\ (\mu, v) \in [0,1] \times [a,b],\ i = 1,\ 2\}.$$

For $Re\lambda > -\underline{\sigma}$, the solution of Eq. (7.21) is formally given by:

$$\psi_i(\mu, v) = \psi_i(0, v)\, e^{-\frac{1}{v}\int_0^\mu (\lambda+\sigma_i(\mu',v))d\mu'} + \frac{1}{v}\int_0^\mu e^{-\frac{1}{v}\int_{\mu'}^\mu (\lambda+\sigma_i(\tau,v))d\tau} g(\mu', v)\, d\mu'.$$

Accordingly, for $\mu = 1$, we get

$$\psi_i(1, v) = \psi_i(0, v)\, e^{-\frac{1}{v}\int_0^1 (\lambda+\sigma_i(\mu',v))d\mu'} + \frac{1}{v}\int_0^1 e^{-\frac{1}{v}\int_{\mu'}^1 (\lambda+\sigma_i(\tau,v))d\tau} g(\mu', v)\, d\mu'. \tag{7.22}$$

Let us introduce the following operators:

$$\begin{cases} P_{i,\lambda} : X^0 \longrightarrow X^1 \\ \qquad u \longrightarrow (P_{i,\lambda}u)(1, v) := u(0, v)\, e^{-\frac{1}{v}\int_0^1 (\lambda+\sigma_i(\mu',v))d\mu'}, \end{cases}$$

$$\begin{cases} Q_{i,\lambda} : X^0 \longrightarrow X \\ \qquad u \longrightarrow (Q_{i,\lambda}u)(\mu, v) := u(0, v)\, e^{-\frac{1}{v}\int_0^\mu (\lambda+\sigma_i(\mu',v))d\mu'}, \end{cases}$$

$$\begin{cases} \Pi_{i,\lambda} : X \longrightarrow X^1 \\ \qquad u \longrightarrow (\Pi_{i,\lambda} u)(1, v) := \dfrac{1}{v} \displaystyle\int_0^1 e^{-\frac{1}{v}\int_{\mu'}^1 (\lambda+\sigma_i(\tau,v))d\tau} u(\mu', v)\, d\mu', \end{cases}$$

and finally,

$$\begin{cases} R_{i,\lambda} : X \longrightarrow X \\ \qquad u \longrightarrow (R_{i,\lambda} u)(\mu, v) := \dfrac{1}{v} \displaystyle\int_0^\mu e^{-\frac{1}{v}\int_{\mu'}^\mu (\lambda+\sigma_i(\tau,v))d\tau} u(\mu', v)\, d\mu'. \end{cases}$$

Clearly, for λ satisfying $Re\lambda > -\underline{\sigma}$, the operators $P_{i,\lambda}$, $Q_{i,\lambda}$, $\Pi_{i,\lambda}$, and $R_{i,\lambda}$, $i = 1,\ 2$, are bounded. It is not difficult to check that

$$\|P_{i,\lambda}\| \leq e^{-\frac{1}{b}(Re\lambda+\underline{\sigma})}, \tag{7.23}$$

and

$$\|Q_{i,\lambda}\| \leq (Re\lambda + \underline{\sigma})^{-1}. \tag{7.24}$$

Moreover, some simple calculations show that

$$\|\Pi_{i,\lambda}\| \leq 1, \tag{7.25}$$

and

$$\|R_{i,\lambda}\| \leq (Re\lambda + \underline{\sigma})^{-1}. \tag{7.26}$$

Hence, Eq. (7.22) may be abstractly written as

$$\psi_i^1 = P_{i,\lambda}\psi_i^0 + \Pi_{i,\lambda} g.$$

Moreover, ψ_i must satisfy the boundary condition (7.2). Thus, we obtain

$$\psi_i^1 = P_{i,\lambda} K_i \psi_i^1 + \Pi_{i,\lambda} g. \tag{7.27}$$

Notice that the operator $P_{i,\lambda} K_i$ appearing in Eq. (7.27), is defined from X^1 into X^1. Let $\varphi_1,\ \varphi_2 \in X^1$. From $(\mathcal{R}_7)$ and the estimate (7.23), we have

$$\|P_{i,\lambda} K_i \varphi_1 - P_{i,\lambda} K_i \varphi_2\| \leq \alpha_i\, e^{-\frac{Re\lambda+\underline{\sigma}}{b}} \|\varphi_1 - \varphi_2\|. \tag{7.28}$$

Now, let us consider the following equation

$$u = P_{i,\lambda} K_i u + \varphi, \quad \varphi \in X^1, \tag{7.29}$$

where u is an unknown function and let us define the operator $A_{(i,\lambda,\varphi)}$ on X^1 by:

$$\begin{cases} A_{(i,\lambda,\varphi)} : X^1 & \longrightarrow X^1, \\ \qquad\qquad u & \longrightarrow (A_{(i,\lambda,\varphi)}u)(1,v) := P_{i,\lambda}K_i u + \varphi. \end{cases}$$

From the estimate (7.28), it follows that

$$\|A_{(i,\lambda,\varphi)}\varphi_1 - A_{(i,\lambda,\varphi)}\varphi_2\| = \|P_{i,\lambda}K_i\varphi_1 - P_{i,\lambda}K_i\varphi_2\| \leq \alpha_i\, e^{-\frac{Re\lambda+\underline{\sigma}}{b}}\|\varphi_1 - \varphi_2\|.$$

Consequently, for $Re\lambda > -\underline{\sigma} + b\log(\alpha_i)$, the operator $A_{(i,\lambda,\varphi)}$ is a contraction mapping and therefore, Eq. (7.29) has a unique solution

$$u_{(i,\lambda,\varphi)} = u_i.$$

Let $W_{i,\lambda}$ be the nonlinear operator defined by:

$$W_{i,\lambda}\varphi = u_i, \tag{7.30}$$

where u_i is the solution of Eq. (7.29). Arguing as in the proof of Lemma 7.1.1, we have the following result:

Lemma 7.2.1 *Assume that* $(\mathcal{R}_7)$ *holds. Then,*
(i) For every λ *satisfying* $Re\lambda > -\underline{\sigma} + b\log(\alpha_i)$, $i = 1,\ 2$, *the operator* $W_{i,\lambda}$ *is continuous and maps bounded sets into bounded ones and satisfying the following estimate*

$$\|W_{i,\lambda}\varphi_1 - W_{i,\lambda}\varphi_2\| \leq \left(1 - \alpha_i e^{-(\frac{Re\lambda+\underline{\sigma}}{b})}\right)^{-1}\|\varphi_1 - \varphi_2\| \quad \textit{for all } \varphi_1,\ \varphi_2 \in X^1.$$

(ii) If $Re\lambda > \max(-\underline{\sigma}, -\underline{\sigma}+b\log(\alpha_i))$, *then the operator* $(\lambda - S_{K_i})$ *is invertible and* $(\lambda - S_{K_i})^{-1}$ *is given by:*

$$(\lambda - S_{K_i})^{-1} = Q_{i,\lambda}K_iW_{i,\lambda}\Pi_{i,\lambda} + R_{i,\lambda}.$$

Moreover, $(\lambda - S_{K_i})^{-1}$ *is continuous on* X *and maps bounded sets into bounded ones.*

In what follows and for our subsequent analysis, we need the following hypothesis:

$(\mathcal{R}_9)$ $r_{ij}(\mu, v, v', \psi_j(\mu, v')) = k_{ij}(\mu, v, v')f(\mu, v', \psi_j(\mu, v'))$,

where f and $k_{ij}(.,.,.)$, $(i,j) \in \{(1,2),(2,1)\}$, are defined as in the previous subsection. We should also recall that the linear operator B_{ij} is defined by:

$$\begin{cases} B_{ij} : X \longrightarrow X \\ \qquad \psi_j \longrightarrow \displaystyle\int_a^b k_{ij}(\mu, v, v')\psi_j(\mu, v')dv'. \end{cases} \tag{7.31}$$

The following useful definition of regularity can be found in [126].

Definition 7.2.1 *Let B_{ij}, $(i,j) \in \{(1,2),(2,1)\}$, be the operator defined by Eq. (7.31). Then, B_{ij} is said to be a regular operator if $\{k_{ij}(\mu,.,v')$ such that $(\mu,v') \in [0,1]\times[a,b]\}$ is a relatively weakly compact subset of $L_1([a,b];d\mu)$.*

Let us recall the following result, which states a basic fact for the theory of these operators on L_1 spaces (see [57]).

Lemma 7.2.2 *Assume that g satisfies the Carathéodory conditions. If the operator $\mathcal{N}_g$ acts from L_1 into L_1, then $\mathcal{N}_g$ is continuous and takes bounded sets into bounded ones. Moreover, there is a constant $k > 0$ and a positive function $h(.) \in L_1$ such that*

$$|g(x,y)| \leq h(x) + k|y| \text{ a.e. in } x, \text{ for all } y \in \mathbb{R}.$$

We will also assume that:

$(\mathcal{R}_{10})$ f satisfies the Carathéodory conditions and $\mathcal{N}_f$ acts from X into X.

We recall the following lemma established in [119] which will play a crucial role below.

Lemma 7.2.3 *If condition $(\mathcal{R}_{10})$ holds true, then for every weakly convergent sequence $(\psi_n)_n$, there exists a weakly convergent subsequence of $\mathcal{N}_f\psi_n$.*

Proof. Let $(\psi_n)_n \in \mathbb{N}$ be a weakly convergent sequence in X. Then, the set $G := \{(\psi_n)_n,\ n \in \mathbb{N}\}$ is sequentially weakly compact and therefore $\omega(G) = 0$. On the other hand Lemma 7.2.2 implies that there exists $k > 0$ and $h(.) \in X$ such that

$$|f(x,\xi,\psi_n(x,\xi))| \leq h(x,\xi) + k|\psi_n(x,\xi)|.$$

So,

$$\int_\Delta \mathcal{N}_f\psi_n(x,\xi)dxd\xi \leq \int_\Delta h(x,\xi)dxd\xi \leq k\int_\Delta |\psi_n(x,\xi)|dxd\xi$$

for all measurable subsets of Δ of $[-a,a]\times[-1,1]$. This together with Lemma 6.1.2 implies that $\omega(\mathcal{N}_f(G)) \leq k\omega(G)$ and therefore, by Lemma 1.4.1, $\mathcal{N}_f(G)$ is relatively weakly compact. This completes the proof. Q.E.D.

Now, we are ready to state the first existence result of this section.

Theorem 7.2.1 *Assume that* $(\mathcal{R}_7)$*–*$(\mathcal{R}_{10})$ *hold. If* B_{12} *is a regular collision operator on* X*, then for each* $r > 0$*, there is* $\lambda_r > 0$ *such that, for each* λ *satisfying* $Re(\lambda) > \lambda_r$*, the problem*

$$\begin{pmatrix} -v\dfrac{\partial}{\partial\mu} - \sigma_1(\mu, v)I & R_{12} \\ R_{21} & -v\dfrac{\partial}{\partial\mu} - \sigma_2(\mu, v)I \end{pmatrix} \begin{pmatrix} \psi_1 \\ \psi_2 \end{pmatrix} = \lambda \begin{pmatrix} \psi_1 \\ \psi_2 \end{pmatrix}, \quad (7.32)$$

$$\psi_{i_{|\Gamma_0}} = K_i\left(\psi_{i_{|\Gamma_1}}\right), \quad i = 1,\ 2, \quad (7.33)$$

has, at least, one solution in $B_r \times B_r$*.*

Proof. Let λ be a complex number such that $Re\lambda > \max(-\underline{\sigma}, -\underline{\sigma} + b\log(\alpha))$ with $\alpha = \max(\alpha_1, \alpha_2)$. Then, according to Lemma 7.2.1, we infer that $\lambda - S_{K_i}$ is invertible and therefore, the problem (7.32)–(7.33) may be transformed into

$$\mathcal{L}_\lambda \begin{pmatrix} \psi_1 \\ \psi_2 \end{pmatrix} = \begin{pmatrix} \psi_1 \\ \psi_2 \end{pmatrix}, \quad \psi_i^0 = K_i\psi_i^1, \quad i = 1,\ 2,$$

where

$$\mathcal{L}_\lambda = \begin{pmatrix} S_{K_1} - (\lambda - 1)I & B_{12}\mathcal{N}_f \\ B_{21}\mathcal{N}_f & S_{K_2} - (\lambda - 1)I \end{pmatrix}.$$

Claim 1: Let $r > 0$. First, we check that, for a suitable λ, the operator $S_\lambda := (\lambda - S_{K_1})^{-1}B_{12}\mathcal{N}_f$ leaves B_r invariant. Let $\psi \in B_r$. From both Lemma 7.2.1 and the estimates (7.23)–(7.26), we have

$$\begin{aligned} \|S_\lambda\psi\| &\leq \|Q_{1,\lambda}K_1W_{1,\lambda}\Pi_{1,\lambda}B_{12}\mathcal{N}_f\psi + R_{1,\lambda}B_{12}\mathcal{N}_f\psi\| \\ &\leq \left[1 + \frac{\alpha_1}{1 - \alpha_1 e^{-\frac{Re\lambda+\underline{\sigma}}{b}}}\right]\frac{\|B_{12}\|M(r)}{Re\lambda + \underline{\sigma}} + \frac{\alpha_1\|W_{1,\lambda}(0)\| + \|K_1(0)\|}{Re\lambda + \underline{\sigma}}, \end{aligned}$$

where $M(r)$ is the upper-bound of $\mathcal{N}_f$ on B_r. Let $\varepsilon > \max(-\underline{\sigma}, -\underline{\sigma} + b\log(\alpha))$. For $Re\lambda > \varepsilon$, we have

$$(1 - \alpha_1 e^{-\frac{Re\lambda+\underline{\sigma}}{b}})^{-1} \leq (1 - \alpha_1 e^{-\frac{\varepsilon+\underline{\sigma}}{b}})^{-1}.$$

Therefore,

$$\|S_\lambda\psi\| \leq \left[1 + \frac{\alpha_1}{1 - \alpha_1 e^{-\frac{\varepsilon+\underline{\sigma}}{b}}}\right]\frac{\|B_{12}\|M(r)}{Re\lambda + \underline{\sigma}} + \frac{\alpha_1\|W_{1,\lambda}(0)\| + \|K_1(0)\|}{Re\lambda + \underline{\sigma}}.$$

By using Eq. (7.30), we deduce that

$$P_{1,\lambda}K_1W_{1,\lambda}(0) = W_{1,\lambda}(0).$$

Let $0 < \delta < \frac{1}{\alpha_1}$. From the estimate (7.23), there exists λ_1 such that for any λ satisfying $Re\lambda > \max(-\underline{\sigma}, -\underline{\sigma} + b\log(\alpha), \lambda_1)$, we have $\|P_{1,\lambda}\| \leq \delta$. Then, by using the assumption $(\mathcal{R}_7)$, we deduce that

$$\begin{aligned}\|W_{1,\lambda}(0)\| &\leq \|P_{1,\lambda}\|\|K_1W_{1,\lambda}(0)\| \\ &\leq \delta(\alpha_1\|W_{1,\lambda}(0)\| + \|K_1(0)\|).\end{aligned}$$

It follows that

$$\|W_{1,\lambda}(0)\| \leq \frac{\delta\|K_1(0)\|}{1-\delta\alpha_1}.$$

Therefore,

$$\begin{aligned}\|S_\lambda\psi\| &\leq \left[1 + \frac{\alpha_1}{1-\alpha_1 e^{-\frac{\varepsilon+\underline{\sigma}}{b}}}\right]\frac{\|B_{12}\|M(r)}{Re\lambda+\underline{\sigma}} + \frac{\left(\frac{\alpha_1\delta}{1-\delta\alpha_1}+1\right)\|K_1(0)\|}{Re\lambda+\underline{\sigma}} \\ &\leq Q(Re\lambda),\end{aligned}$$

where

$$Q(t) = \left[1 + \frac{\alpha_1}{1-\alpha_1 e^{-\frac{\varepsilon+\underline{\sigma}}{b}}}\right]\frac{\|B_{12}\|M(r)}{t+\underline{\sigma}} + \frac{\left(\frac{\alpha_1\delta}{1-\delta\alpha_1}+1\right)\|K_1(0)\|}{t+\underline{\sigma}}.$$

Clearly, $Q(.)$ is continuous strictly decreasing in $t > 0$ and satisfies $\lim_{t\to+\infty} Q(t) = 0$. Hence, there exists λ_2, such that $Q(\lambda_2) \leq r$. Obviously, if $Re\lambda \geq \max(\lambda_1, \lambda_2)$, then $(\lambda - S_{K_1})^{-1}B_{12}\mathcal{N}_f$ maps B_r into itself.

Now, let us recall the following conditions (introduced in Chapter 4), which will be needed in the sequel.

$$(\mathcal{A}_1)\quad \begin{cases}\text{If } (\psi_n)_{n\in\mathbb{N}} \text{ is a weakly convergent sequence in } L_1, \text{ then} \\ (A\psi_n)_{n\in\mathbb{N}} \text{ has a strongly convergent subsequence in } L_1\end{cases}$$

and

$$(\mathcal{A}_2)\quad \begin{cases}\text{if } (\psi_n)_{n\in\mathbb{N}} \text{ is a weakly convergent sequence in } L_1, \text{ then} \\ (A\psi_n)_{n\in\mathbb{N}} \text{ has a weakly convergent subsequence in } L_1.\end{cases}$$

Claim 2: It is immediate that the operator S_λ is continuous and weakly compact on X. Now, let us check that S_λ satisfies the condition $(\mathcal{A}_1)$. For this, let $(\psi_n)_{n\in\mathbb{N}}$ be a weakly convergent sequence of X. Using the fact that $\mathcal{N}_f$ satisfies $(\mathcal{A}_2)$, we deduce that $(\mathcal{N}_f\psi_n)_{n\in\mathbb{N}}$ has a weakly convergent subsequence, say $(\mathcal{N}_f\psi_{n_k})_{k\in\mathbb{N}}$. Moreover, by using the fact that $\Pi_{1,\lambda}B_{12}$ $R_{1,\lambda}B_{12}$ are weakly compact, the assumption $(\mathcal{A}_1)$, and Lemma 7.2.1, we deduce that $(\lambda - S_{K_1})^{-1}B_{12}$ is weakly-strongly sequentially continuous. Hence, $(S_\lambda\psi_{n_k})_{k\in\mathbb{N}}$ converges strongly in X and so, S_λ satisfies the assumption $(\mathcal{A}_1)$.

Claim 3: Clearly, $B_{21}\mathcal{N}_f$ is continuous on X. Now, let us check that $B_{21}\mathcal{N}_f$ satisfies the condition $\mathcal{A}_2$. To do so, let $(\psi_n)_{n\in\mathbb{N}}$ be a weakly convergent sequence of X. By using the fact that $\mathcal{N}_f$ satisfies the condition $(\mathcal{A}_2)$, we infer that $(\mathcal{N}_f\psi_n)_{n\in\mathbb{N}}$ has a weakly convergent subsequence, say $(\mathcal{N}_f\psi_{n_k})_{k\in\mathbb{N}}$. Moreover, the continuity of the linear operator B_{21} shows that it is weakly continuous on X (see [40]). So, $(B_{21}\mathcal{N}_f\psi_{n_k})_{k\in\mathbb{N}}$ converges weakly in X. Then, $B_{21}\mathcal{N}_f$ satisfies the assumption $(\mathcal{A}_2)$.

Claim 4: Clearly, from Lemma 7.2.1, we deduce that $(\lambda - S_{K_2})^{-1}$ exists and is continuous on X. Now, let us check that $(\lambda - S_{K_2})^{-1}$ satisfies the condition $(\mathcal{A}_2)$. For this, let $(\psi_n)_{n\in\mathbb{N}}$ be a weakly convergent sequence of X. By using the fact that $(W_{2,\lambda}\Pi_{2,\lambda}\psi_n)_{n\in\mathbb{N}}$ is a bounded sequence and that K_2 is a weakly compact operator on X^1, we infer that $(K_2W_{2,\lambda}\Pi_{2,\lambda}\psi_n)_{n\in\mathbb{N}}$ has a weakly convergent subsequence, say $(K_2W_{2,\lambda}\Pi_{2,\lambda}\psi_{n_k})_{k\in\mathbb{N}}$. Besides, by using the continuity of the linear operators $Q_{2,\lambda}$ and $R_{2,\lambda}$, we show that $\left((\lambda - S_{K_2})^{-1}\psi_{n_k}\right)_{k\in\mathbb{N}}$ converge weakly in X. Then the operator $(\lambda - S_{K_2})^{-1}$ satisfies $(\mathcal{A}_2)$.

Arguing as in claim 1 for $S_\lambda\psi$, there exists λ_r such that, for $Re\lambda \geq \lambda_r$, we have $\Gamma\psi := (\lambda - S_{K_2})^{-1}B_{21}\mathcal{N}_fS_\lambda\psi \in B_r$. Finally, Γ has, at least, a fixed point in B_r, or equivalently, the problem (7.32)–(7.33) has a solution in $B_r \times B_r$. Q.E.D.

Now, we may discuss the existence of solutions for the more general nonlinear boundary problem (7.1)–(7.2). For this purpose, we need the following assumption:

$(\mathcal{R}_{11})$ $K_i \in \mathcal{L}(X^1, X^0)$, and for each $r > 0$, the function $\sigma_i(.,.,.)$, $i = 1,\ 2$, satisfies

$$|\sigma_i(\mu, v, \psi_1) - \sigma_i(\mu, v, \psi_2)| \ \leq \ |\omega_i(\mu, v)|\ \|\psi_1 - \psi_2\| \quad \text{for all } \psi_1, \psi_2 \in B_r,$$

where $\mathcal{L}(X^1, X^0)$ denotes the set of all bounded linear operators from X^1 into X^0 and $\omega_i(.,.) \in L^\infty([0,1] \times [a,b])$, and $\mathcal{N}_{\sigma_i}$ acts from X into X.

Let us define the free streaming operator $\widehat{S}_{K_i}$, $i = 1,\ 2$, by:

$$\begin{cases} \widehat{S}_{K_i} : \mathcal{D}(\widehat{S}_{K_i}) \subset X \longrightarrow X, \\ \qquad\qquad \psi_i \ \longrightarrow \widehat{S}_{K_i}\psi_i(\mu, v) = -v\dfrac{\partial \psi_i}{\partial \mu}(\mu, v), \\ \mathcal{D}(\widehat{S}_{K_i}) = \left\{\psi_i \in \mathcal{W} \text{ such that } \psi_i^0 = K_i(\psi_i^1)\right\}. \end{cases}$$

Theorem 7.2.2 *Assume that* $(\mathcal{R}_9)$*–*$(\mathcal{R}_{11})$ *hold. If* B_{ij}, $(i,j) \in \{(1,2),(2,1)\}$, *are regular collision operators on* X*, then for each* $r > 0$*, there is* $\lambda_r > 0$ *such that, for each* λ *satisfying* $Re(\lambda) > \lambda_r$*, the problem* (7.1)*–*(7.2) *has, at least, one solution in* $B_r \times B_r$*.*

Proof. Since K_i, $i = 1,\ 2$, is linear, the operator $\widehat{S}_{K_i}$ is linear, too. By using Lemma 7.2.1, we deduce that

$$\{\lambda \in \mathbb{C} \ \text{ such that } \ Re\lambda > \max(0, b\log\|K_1\|, b\log\|K_2\|)\} \subset \varrho(\widehat{S}_{K_i}),$$

where $\varrho(\widehat{S}_{K_i})$ denotes the resolvent set of $\widehat{S}_{K_i}$. Let $\lambda \in \mathbb{C}$ such that $Re\lambda > \max(0, b\log\|K_1\|, b\log\|K_2\|)$. Then, by using the linearity of the operator $(\lambda - \widehat{S}_{K_i})^{-1}$, the problem (7.1)–(7.2) may be written in the form:

$$\widehat{\mathcal{L}}_\lambda \begin{pmatrix} \psi_1 \\ \psi_2 \end{pmatrix} = \begin{pmatrix} \psi_1 \\ \psi_2 \end{pmatrix}, \quad \psi_i^0 = K_i\psi_i^1, \quad i = 1,\ 2,$$

where

$$\widehat{\mathcal{L}}_\lambda = \begin{pmatrix} \widehat{S}_{K_1} - (\lambda - 1)I + \mathcal{N}_{-\sigma_1} & B_{12}\mathcal{N}_f \\ B_{21}\mathcal{N}_f & \widehat{S}_{K_2} - (\lambda - 1)I + \mathcal{N}_{-\sigma_2} \end{pmatrix}.$$

This problem may be transformed into the form:

$$\mathcal{G}_{1,\lambda} \begin{pmatrix} \psi_1 \\ \psi_2 \end{pmatrix} + \mathcal{G}_{2,\lambda} \begin{pmatrix} \psi_1 \\ \psi_2 \end{pmatrix} = \begin{pmatrix} \psi_1 \\ \psi_2 \end{pmatrix}, \quad \psi_i^0 = K_i\psi_i^1, \quad i = 1,\ 2,$$

where

$$\mathcal{G}_{1,\lambda} = \begin{pmatrix} (\lambda - \widehat{S}_{K_1})^{-1}\mathcal{N}_{-\sigma_1} & 0 \\ 0 & (\lambda - \widehat{S}_{K_2})^{-1}\mathcal{N}_{-\sigma_2} \end{pmatrix},$$

and

$$\mathcal{G}_{2,\lambda} = \begin{pmatrix} 0 & (\lambda - \widehat{S}_{K_1})^{-1}B_{12}\mathcal{N}_f \\ (\lambda - \widehat{S}_{K_2})^{-1}B_{21}\mathcal{N}_f & 0 \end{pmatrix}.$$

Claim 1: Let us check that, for a suitable λ, the operator $\mathcal{G}_{1,\lambda}$ is a contraction mapping. Indeed, let $\begin{pmatrix} \psi_1 \\ \psi_2 \end{pmatrix}, \begin{pmatrix} \varphi_1 \\ \varphi_2 \end{pmatrix} \in X \times X$. For $i = 1,\ 2$, we have

$$\|(\lambda - \widehat{S}_{K_i})^{-1}(\mathcal{N}_{-\sigma_i}\varphi_i - \mathcal{N}_{-\sigma_i}\psi_i)\| \leq \|(\lambda - \widehat{S}_{K_i})^{-1}\|\|\mathcal{N}_{-\sigma_i}\varphi_i - \mathcal{N}_{-\sigma_i}\psi_i\|.$$

A simple calculation leads to

$$\|(\lambda - \widehat{S}_{K_i})^{-1}\| \leq \frac{1}{Re\lambda}\left[1 + \frac{\gamma}{1 - \gamma e^{-\frac{Re\lambda}{b}}}\right], \quad i = 1,\ 2, \tag{7.34}$$

where $\gamma = \max(\|K_1\|, \|K_2\|)$. Moreover, by taking into account the assumption on $\sigma_i(.,.,.)$, we get

$$\|\mathcal{N}_{-\sigma_i}\varphi_i - \mathcal{N}_{-\sigma_i}\psi_i\| \leq \|\omega\|_\infty\|\varphi_i - \psi_i\|, \quad i = 1,\ 2,$$

where $\|\omega\|_\infty = \max(\|\omega_1\|_\infty, \|\omega_2\|_\infty)$. Let us denote by $\Phi := \begin{pmatrix} \varphi_1 \\ \varphi_2 \end{pmatrix}$ and $\Psi := \begin{pmatrix} \psi_1 \\ \psi_2 \end{pmatrix}$. By using the estimate (7.34), we have

$$\begin{aligned}
\Delta(\lambda, \Phi, \Psi) &= \left\|\begin{pmatrix} (\lambda - \widehat{S}_{K_1})^{-1}\mathcal{N}_{-\sigma_1}\varphi_1 \\ (\lambda - \widehat{S}_{K_2})^{-1}\mathcal{N}_{-\sigma_2}\varphi_2 \end{pmatrix} - \begin{pmatrix} (\lambda - \widehat{S}_{K_1})^{-1}\mathcal{N}_{-\sigma_1}\psi_1 \\ (\lambda - \widehat{S}_{K_2})^{-1}\mathcal{N}_{-\sigma_2}\psi_2 \end{pmatrix}\right\| \\
&\leq \frac{\|\omega\|_\infty}{Re\lambda}\left[1 + \frac{\gamma}{1 - \gamma e^{-\frac{Re\lambda}{b}}}\right]\left\|\begin{pmatrix} \varphi_1 \\ \varphi_2 \end{pmatrix} - \begin{pmatrix} \psi_1 \\ \psi_2 \end{pmatrix}\right\| \\
&\leq \Xi(Re\lambda)\left\|\begin{pmatrix} \varphi_1 \\ \varphi_2 \end{pmatrix} - \begin{pmatrix} \psi_1 \\ \psi_2 \end{pmatrix}\right\|,
\end{aligned}$$

where

$$\Delta(\lambda, \Phi, \Psi) := \left\|\mathcal{G}_{1,\lambda}\Phi - \mathcal{G}_{1,\lambda}\Psi\right\|.$$

By the same arguments as those used in Section 7.1.2, we deduce that, there exists $\lambda_1 \in]\max(0, b\log\|K_1\|, b\log\|K_2\|), \infty[$ such that $\Xi(\lambda_1) < 1$. Hence, for $Re\lambda \geq \lambda_1$, $\mathcal{G}_{1,\lambda}$ is a contraction mapping.

Claim 2: By using Theorem 5.2.2, and arguing as in the proof of Theorem 7.2.1, we show that $\mathcal{G}_{1,\lambda}$ satisfies $(\mathcal{A}_2)$, and that $\mathcal{G}_{2,\lambda}$ is continuous, weakly compact on $X \times X$ and satisfies $(\mathcal{A}_1)$.

Claim 3: Let $r > 0$ and $\varphi_1, \psi_2 \in B_r$. According to the estimation (7.34), we obtain

$$\begin{aligned} \|(\lambda - \widehat{S}_{K_1})^{-1}\mathcal{N}_{-\sigma_1}\varphi_1 + (\lambda - \widehat{S}_{K_1})^{-1}B_{12}\mathcal{N}_f\psi_2\| &\leq \tfrac{\|B_{12}\|M(r)+M'(r)}{Re\lambda}\Lambda(\lambda) \\ &\leq T(Re\lambda), \end{aligned}$$

where $T(.)$ has the same properties as $\Xi(.)$,

$$\Lambda(\lambda) := \left[1 + \frac{\|K_1\|}{1 - \|K_1\|e^{-\frac{Re\lambda}{b}}}\right]$$

and $M(r)$, $M'(r)$ are the upper-bounds of $\mathcal{N}_f$ and $\mathcal{N}_\sigma$ on B_r. Arguing as above, we show that there exists λ_2 such that, for $Re\lambda \geq \lambda_2$, we have

$$(\lambda - \widehat{S}_{K_1})^{-1}\mathcal{N}_{-\sigma_1}B_r + (\lambda - \widehat{S}_{K_1})^{-1}B_{12}\mathcal{N}_f B_r \subset B_r.$$

By using a similar reasoning, we may prove that there exists λ_3 such that, for $Re\lambda \geq \lambda_3$, we have

$$(\lambda - \widehat{S}_{K_2})^{-1}\mathcal{N}_{-\sigma_2}B_r + (\lambda - \widehat{S}_{K_2})^{-1}B_{21}\mathcal{N}_f B_r \subset B_r.$$

Finally, if $\lambda_r = \max(\lambda_1, \lambda_2, \lambda_3)$, then for all λ satisfying $Re\lambda \geq \lambda_r$, the operators $\mathcal{G}_{1,\lambda}$ and $\mathcal{G}_{2,\lambda}$ satisfy the conditions of Theorem 4.2.6. Consequently, the problem (7.1)–(7.2) has a solution in $B_r \times B_r$ for all λ such that $Re\lambda \geq \lambda_r$. Q.E.D.

7.3 A Coupled Functional Integral System in Banach Algebras

Let us consider the following nonlinear functional integral system:

$$\begin{cases} x(t) = a(t)x(t) + y(t)\Big[q(t) + \displaystyle\int_0^{\sigma(t)} h\big(t, y(\eta(s))\big)ds\Big] \\ y(t) = \dfrac{1}{1 + |x(\theta(t))|} - f\big(t, x(\theta(t))\big) + g(t, y(t)) \end{cases} \tag{7.35}$$

for $t \in J$, where J is the interval $[0, 1]$ and x, y are unknown functions in the Banach algebra $\mathcal{C}(J, \mathbb{R}) := \mathcal{C}(J)$ of all real-valued continuous functions on J. Here, g is a contraction condition with respect to the second variable while $f(.,.)$ (resp. $h(.,.)$, $\sigma(.)$, $\eta(.)$, $\theta(.)$, $a(.)$, $q(.)$) is a nonlinear (resp. are continuous) function(s). The system (7.35) can be written as a fixed point problem:

$$\begin{pmatrix} x(t) \\ y(t) \end{pmatrix} = \begin{pmatrix} a(t) & I.K(t,.) \\ f_1(t,.) & g(t,.) \end{pmatrix} \begin{pmatrix} x(t) \\ y(t) \end{pmatrix},$$

where

$$K(t, y(t)) = q(t) + \int_0^{\sigma(t)} h\big(t, y(\eta(s))\big)ds,$$

$$f_1(t, x(t)) = \frac{1}{1 + |x(\theta(t))|} - f\big(t, x(\theta(t))\big)$$

and I represents the identity operator on $\mathcal{C}(J)$. We equip the space $\mathcal{C}(J)$ with the norm $\|x\|_\infty = \sup_{t\in J}|x(t)|$. Clearly, $\mathcal{C}(J)$ is a complete normed algebra with respect to this supremum norm. Assume that the functions involved in Eq. (7.35) verify the following conditions:

$(\mathcal{R}_{12})$ The function $f : J \times \mathbb{R} \longrightarrow \mathbb{R}$ satisfies the Lipschitzian condition with a constant $\alpha \in]1, +\infty[$.

$(\mathcal{R}_{13})$ The function $g : J \times \mathbb{R} \longrightarrow \mathbb{R}$ is such that:

(a) g is continuous and satisfies the Lipschitzian condition with a constant k in $]\frac{2}{1+\alpha}, 1[$,

(b) $|g(t, x(t))| \leq (1 - k)|x(t)|$ for all $x \in \mathcal{C}(J)$, and

(c) $\|f(., x(.))\|_\infty \leq 1$ for $x \in \mathcal{C}(J)$ such that $\|x\| \leq 1 + k$.

$(\mathcal{R}_{14})$ $a : J \longrightarrow \mathbb{R}$ is a continuous function and $\|a\|_\infty < \frac{1}{2}$.

$(\mathcal{R}_{15})$ $\theta, \sigma, \eta : J \longrightarrow J$ are continuous and nondecreasing functions such that $\sigma(t) \leq t$, for all $t \in J$.

$(\mathcal{R}_{16})$ $q : J \longrightarrow \mathbb{R}$ is a continuous function with $\|q\|_\infty < \frac{1-\|a\|_\infty}{2(1+\alpha)}(1 - k)$.

$(\mathcal{R}_{17})$ The operator $h : J \times \mathbb{R} \longrightarrow \mathbb{R}$ is such that:

(a) h is continuous, and

(b) $\|h(t, x(.))\|_\infty \leq \frac{1-\|a\|_\infty}{2(1+\alpha)}(1 - k) - \|q\|$, $t \in J$, $x \in \mathcal{C}(J)$, and $\|x\| \leq 1 + \alpha$.

Theorem 7.3.1 *Under the assumptions* $(\mathcal{R}_{12})$–$(\mathcal{R}_{17})$*, the functional integral system* (7.35) *has, at least, one solution in* $\mathcal{C}(J)$.

Proof. Consider the mapping A, B, C, D, and B' on $\mathcal{C}(J)$ defined by:

$$\begin{cases} (Ax)(t) = a(t)x(t) \\ (Bx)(t) = x(t) \\ (Cx)(t) = \dfrac{1}{1+|x(\theta(t))|} - f\left(t, x(\theta(t))\right) \\ (Dx)(t) = g\left(t, x(t)\right) \\ (B'x)(t) = q(t) + \displaystyle\int_0^{\sigma(t)} h(t, x(\eta(s)))ds. \end{cases}$$

Then, the problem (7.35) is equivalent to the system:

$$\begin{cases} x(t) = Ax(t) + By(t).B'y(t) \\ y(t) = Cx(t) + Dy(t). \end{cases}$$

We will show that A, B, C, D, and B' satisfy all the conditions of Theorem 4.3.3. For this purpose, let us define the subsets S and S' of $\mathcal{C}(J)$ by:

$$S := \{y \in \mathcal{C}(J), \|y\| \leq 1+k\} \text{ and } S' := \{y \in \mathcal{C}(J), \|y\| \leq 1+\alpha\}.$$

Obviously, S and S' are nonempty, bounded, convex, and closed subsets of the Banach algebra $\mathcal{C}(J)$.
First, let us begin by showing that C is Lipschitzian with the constant $1+\alpha$ on S. To see this, let x, $y \in S$. So,

$$\begin{aligned} \Delta(t) &= \left|\frac{1}{1+|x(\theta(t))|} - f\left(t, x(\theta(t))\right) - \frac{1}{1+|y(\theta(t))|} - f\left(t, y(\theta(t))\right)\right| \\ &\leq \left|\frac{|x(\theta(t))| - |y(\theta(t))|}{(1+|x(\theta(t))|)(1+|y(\theta(t))|)}\right| + \left|f\left(t, y(\theta(t))\right) - f\left(t, x(\theta(t))\right)\right| \\ &\leq (1+\alpha)\|x-y\|, \end{aligned}$$

where

$$\Delta(t) := \|Cx(t) - Cy(t)\|.$$

Accordingly, we have

$$\|Cx - Cy\| \leq (1+\alpha)\|x-y\|.$$

Clearly, A, B are Lipschitzian with the constants $\|a\|_\infty$ and 1, respectively, and the operator D satisfies the contraction condition with a constant k. We claim that $C(S)$ is a relatively compact subset in $\mathcal{C}(J)$. Let $\{y_n\}$ be any sequence of $C(S)$. Then, there exists $\{x_n\}$ of S such that $y_n = C(x_n)$. By taking into account the hypothesis $(\mathcal{R}_{13})$, we obtain

$$|y_n(t)| = \left|\frac{1}{1+\big|x_n(\theta(t))\big|} - f\big(t, x_n(\theta(t))\big)\right| \leq 2$$

which shows that $\{y_n(t)\}$ has a subsequence $\{y_{n_k}(t)\}$ that converges to $y(t)$. Consequently, $C(S)$ is a sequentially relatively compact subset in $\mathcal{C}(J)$. By applying Eberlein–Šmulian's theorem (see Theorem 1.3.3), we infer that $C(S)$ is relatively compact. Next, let us show that $C(S) \subset (I - D)(S')$. To do it, let $x \in S$ be a fixed point. Let us define the mapping

$$\begin{cases} \varphi_x : \mathcal{C}(J) \longrightarrow \mathcal{C}(J) \\ \qquad\quad y \longrightarrow Cx + Dy. \end{cases}$$

From the hypothesis $(\mathcal{R}_{13})(a)$, we deduce that the operator φ_x is a contraction with a constant k. Hence, by applying Banach's theorem, we show that there exists a unique point $y \in \mathcal{C}(J)$ such that $Cx + Dy = y$. Therefore,

$$C(S) \subset (I - D)(\mathcal{C}(J)).$$

Since $y \in \mathcal{C}(J)$, then there is $t^* \in J$ such that

$$\begin{aligned} \|y\|_\infty = |y(t^*)| = |Cx(t^*) + Dy(t^*)| &\leq \frac{1}{1+|x(\theta(t^*))|} + 1 + (1-k)|y(t^*)| \\ &\leq 2 + (1-k)|y(t^*)|, \end{aligned}$$

or, equivalently $|y(t^*)| < 1 + \alpha$. We conclude that $C(S) \subset (I - D)(S')$.

Now, let us claim that B' is continuous on S'. First, we begin by showing that if $x \in S'$, then $B'x \in S'$. To see this, let $\{t_n\}$ be any sequence in J converging to a point $t \in J$ and denote $\Delta_n(t) = |B'x(t_n) - B'x(t)|$. Then,

$$\begin{aligned}
\Delta_n(t) &= \left| q(t_n) + \int_0^{\sigma(t_n)} h(t_n, x(\eta(s)))ds - q(t) - \int_0^{\sigma(t)} h(t, x(\eta(s)))ds \right| \\
&\leq |q(t_n) - q(t)| + \left| \int_0^{\sigma(t_n)} h(t_n, x(\eta(s)))ds - \int_0^{\sigma(t)} h(t, x(\eta(s)))ds \right| \\
&\leq |q(t_n) - q(t)| + \left| \int_0^{\sigma(t_n)} \big| h(t_n, x(\eta(s))) - h(t, x(\eta(s)))\big| ds \right| \\
&\quad + \left| \int_{\sigma(t)}^{\sigma(t_n)} h\big(t, x(\eta(s))\big) ds \right| \\
&\leq |q(t_n) - q(t)| + \int_0^1 |h(t_n, x(\eta(s))) - h(t, x(\eta(s)))|\, ds \\
&\quad + (1 - \|q\|)\, |\sigma(t_n) - \sigma(t)|\, .
\end{aligned}$$

Since $t_n \to t$, so $(t_n, x(\eta(s))) \to (t, x(\eta(s)))$ for all $s \in J$. By taking into account the assumption $(\mathcal{R}_{17})$, we get

$$h(t_n, x(\eta(s))) \to h(t, x(\eta(s))).$$

Moreover, the use of hypothesis $(\mathcal{R}_{17})$ leads to

$$|h(t_n,\ x(\eta(s))) - h(t,\ x(\eta(s)))| \leq 2(1 - \|q\|_\infty) \quad \text{for all} \quad t, s \in J.$$

Now, let us consider

$$\begin{cases} \varphi : J \longrightarrow \mathbb{R} \\ \qquad s \longrightarrow \varphi(s) = 2(1 - \|q\|_\infty). \end{cases}$$

Clearly, $\varphi \in L^1(J)$. Therefore, by using the dominated convergence theorem as well as the assumption $(\mathcal{R}_{15})$, we obtain

$$B'x_n(t) \to B'x(t).$$

So, $B'x \in \mathcal{C}(J)$ and consequently, there is a scalar $t^* \in J$ such that

$$\|B'x\| = |B'x(t^*)| \leq 1.$$

Let $\{x_n\}$ be any sequence in S' such that $x_n \to x$. Then,

$$(t, x_n(\eta(s))) \to (t, x(\eta(s))).$$

The use of both hypothesis $(\mathcal{R}_{15})$ and the dominated convergence theorem shows that the operator B' is continuous. Now, from the hypothesis $(\mathcal{R}_{17})(b)$, it follows that

$$\begin{aligned}
M &= \|T'(S)\| = \sup_{x\in S}\|T'x\| \\
&\leq \sup\Big\{\sup_{t\in J}\{|q(t)| + \int_0^{\sigma(t)} \big|h(t,\ x(\eta(s)))\big|ds\}\ ;\ \ x\in S\Big\} \\
&\leq \sup_{t\in J}\Big\{\sup_{t\in J}|q(t)| + \sup_{t\in J}\int_0^{\sigma(t)} \big|h(t,\ x(\eta(s)))\big|ds;\ x\in S\Big\} \\
&\leq \sup_{t\in J}\Big\{\sup_{t\in J}|q(t)| + \sup_{t\in J}\int_0^{\sigma(t)} \sup_{t\in J}\big|h(t,\ x(\eta(s)))\big|ds;\ x\in S\Big\} \\
&\leq \sup_{t\in J}\Big\{\sup_{t\in J}|q(t)| + \sup_{t\in J}\int_0^{t} \sup_{s\in J}\big|h(t,\ x(\eta(s)))\big|ds;\ x\in S\Big\} \\
&< \frac{(1-\|a\|_\infty)(1-k)}{2(1+\alpha)}.
\end{aligned}$$

Therefore,

$$\frac{1}{1-k}M + \|a\|_\infty < 1.$$

Then, it remains to verify the hypothesis (vi) of Theorem 4.3.3. Let $x,\ z \in S$. Then, for all $t \in J$, we have

$$\begin{aligned}
|Ax(t) + \Pi x(t)\Pi' z(t)| &\leq \|a\|_\infty\|x\| + |x_1(t)||B'(z_1(t))| \\
&\leq \tfrac{1+k}{2} + (1+\alpha)\big(\tfrac{(1-\|a\|_\infty)(1-k)}{2(1+\alpha)}\big) \\
&\leq \tfrac{1+k}{2} + \tfrac{1+k}{2} \\
&\leq 1+k,
\end{aligned}$$

where $\Pi := B(I-D)^{-1}C$, $\Pi' := B'(I-D)^{-1}C$, $x_1 = (I-D)^{-1}Cx$, and

$$z_1 = (I-D)^{-1}Cz.$$

This implies that

$$Ax + B(I-D)^{-1}Cx.B'(I-D)^{-1}Cz \in S \ \text{ for any } x,\ z\in S.$$

Finally, we conclude that the operators A, B, C, D, and B' satisfy all the requirements of Theorem 4.3.3, and the proof is achieved. Q.E.D.

7.4 A Coupled System in Banach Algebras under the Condition $(\mathcal{P})$

In this section, we illustrate the applicability of Theorem 4.3.7 and Theorem 4.4.1 by considering the following examples of nonlinear functional integral equations.

Example 1. Let X be a Banach algebra satisfying the condition $(\mathcal{P})$. Let us consider the following system of nonlinear integral equations occurring in some biological problems, and also in ones dealing with physics:

$$\begin{cases} x(t) = f(t,x(t)) + [a(t)y(t)] \cdot \left[\left(\int_0^{\sigma_1(t)} k(t,s)f_1(s,y(\eta(s)))ds\right) u\right] \\ y(t) = \left[\left(q(t) + \int_0^{\sigma_2(t)} p\left(t,s,x(s),x(\lambda s)\right) ds\right) v\right] + g(t,y(t)), \end{cases} \tag{7.36}$$

where $u \in X\backslash\{0\}$ and $v \in X\backslash\{0\}$. We will seek the solutions of the system (7.36) in the space $\mathcal{C}(J,X)$ of all continuous functions on $J = [0,\mathcal{T}], 0 < \mathcal{T} < \infty$ endowed with the norm $\|.\|_\infty$. Let us assume that the functions involved in Eq. (7.36) satisfy the following assumptions:

$(\mathcal{R}_{18})$ The functions a and k are such that:

(a) $a : J \longrightarrow X$ is continuous, and

(b) $k : J \times J \to \mathbb{R}$ is nonnegative and continuous.

$(\mathcal{R}_{19})$ σ_1, σ_2, $\eta : J \longrightarrow J$ are continuous.

$(\mathcal{R}_{20})$ $q : J \longrightarrow \mathbb{R}$ is continuous.

$(\mathcal{R}_{21})$ The function $p : J \times J \times X \times X \longrightarrow \mathbb{R}$ is weakly sequentially continuous such that, for an arbitrary fixed $s \in J$ and x, $y \in X$, the partial function $t \longrightarrow p(t,s,x,y)$ is continuous.

$(\mathcal{R}_{22})$ The function $f : J \times X \longrightarrow X$ is such that:

(a) f is weakly sequentially continuous with respect to the second variable, and

(b) f is a contraction map with a constant k' with respect to the second variable.

$(\mathcal{R}_{23})$ The function $f_1 : J \times X \longrightarrow \mathbb{R}$ is such that:

(a) f_1 is weakly sequentially continuous with respect to the second variable, and

(b) $\|f_1(.,x(.))\| \leq \lambda \|x\|_\infty$.

$(\mathcal{R}_{24})$ The function $g : J \times X \longrightarrow X$ is such that:

(a) g is continuous,

(b) g is weakly sequentially continuous with respect to the second variable,

(c) g is a Φ-nonlinear contraction with respect to the second variable, and

(d) $\Phi(r) < (1-\lambda)r$, for all $r > 0$.

Theorem 7.4.1 *Suppose that the assumptions $(\mathcal{R}_{18})$–$(\mathcal{R}_{24})$ hold. Moreover, if there exists a real number $r_0 > 0$ such that:*

$$\begin{cases} |p(t,s,x(s),x(\lambda s))| \leq r_0, \ \text{for } x \in \mathcal{C}(J,X) \text{ with } \|x\|_\infty \leq r_0 \text{ and } t,s \in J, \\ \\ \|f(t,x(t))\| \leq k'\|x(t)\|, \ \text{ for } t \in J \text{ and } x \in \mathcal{C}(J,X) \text{ with } \|x\|_\infty \leq r_0, \\ \\ \|g(.,x(.))\| \leq \lambda\|x\|_\infty, \ \text{for } x \in \mathcal{C}(J,X) \text{ such that } \|x\|_\infty \leq r_0, \\ \\ \|a\|_\infty \leq \dfrac{(1-k')r_0}{\delta^2 K \mathcal{T} \lambda \|u\|_\infty}, \ \text{ with } u \in X\backslash\{0\}, \\ \\ \text{where } K = \sup\limits_{t,s\in J} k(t,s), \ \lambda\delta = (\|q\|_\infty + \mathcal{T}r_0)\,\|v\|_\infty + r_0, \ \text{ and } v \in X\backslash\{0\}. \end{cases} \tag{7.37}$$

Then the nonlinear system (7.36) has, at least, one solution in $\mathcal{C}(J,X) \times \mathcal{C}(J,X)$.

Proof. Let B_{r_0} be the closed ball in $\mathcal{C}(J,X)$ centered at the origin and of radius r_0, and consider the nonlinear mapping A, B, C, D, and B' on $\mathcal{C}(J,X)$ defined by:

$$\begin{cases} (Ax)(t) = f(t, x(t)), \ t \in J, \\ (Bx)(t) = a(t)x(t), \ t \in J, \\ (B'x)(t) = \left(\displaystyle\int_0^{\sigma_1(t)} k(t,s) f_1(s, x(\eta(s))) ds \right) \cdot u; \ t \in J \text{ and } u \in X \backslash \{0\}, \\ (Cx)(t) = \left(q(t) + \displaystyle\int_0^{\sigma_2(t)} p\left(t, s, x(s), x(\lambda s)\right) ds \right) \cdot v, \\ \text{where } t \in J, \ 0 < \lambda < 1, \ v \in X \backslash \{0\}, \text{ and} \\ (Dx)(t) = g\left(t, x(t)\right), \ t \in J. \end{cases} \tag{7.38}$$

In order to apply Theorem 4.3.7, we have to verify the following steps.

Step 1: We first show that the entries of the block operator matrix (4.6) are all well-defined operators. Obviously, the maps $Ax(.)$, $By(.)$ and $Dy(.)$ are continuous on J in view of the assumptions $(\mathcal{R}_{18})$, $(\mathcal{R}_{22})(c)$, and $(\mathcal{R}_{24})(c)$, for all $(x, y) \in B_{r_0} \times \mathcal{C}(J, X)$. Moreover, we claim that the two maps $Cx(.)$ and $B'y(.)$ are continuous on J for all $(x, y) \in B_{r_0} \times (I - D)^{-1}C(B_{r_0})$. We begin by showing that the set $(I - D)^{-1}C(B_{r_0})$ is bounded. Indeed, from hypothesis $(\mathcal{R}_{24})(c)$ and Theorem 1.6.10, it follows that $(I - D)^{-1}$ exists and is continuous on $(I - D)(\mathcal{C}(J, X))$. Let $y \in \mathcal{C}(J, X)$ with $y = (I - D)^{-1}Cx$, for some $x \in B_{r_0}$. Then, for all $t \in J$, we have

$$y(t) = \left(q(t) + \int_0^{\sigma_2(t)} p\left(t, s, x(s), x(\lambda s)\right) ds \right) \cdot v + g\left(t, y(t)\right).$$

Since $y \in \mathcal{C}(J, X)$, then there is $t^* \in J$ such that

$$\begin{aligned} \|y\|_\infty &= \|y(t^*)\| \\ &\leq \left| q(t^*) + \int_0^{\sigma_2(t^*)} p\left(t^*, s, x(s), x(\lambda s)\right) ds \right| \|v\| \\ &\quad + \|g(t^*, y(t^*)) - g(t^*, x(t^*))\| + \|g(t^*, x(t^*))\| \\ &\leq \left(\|q\|_\infty + \mathcal{T} r_0\right) \|v\| + \Phi\left(\|x(t^*) - y(t^*)\|\right) + \|g(t^*, x(t^*))\| \\ &< \left(\|q\|_\infty + \mathcal{T} r_0\right) \|v\| + (1 - \lambda)\|y(t^*)\| + \|x\|_\infty \\ &\leq \left(\|q\|_\infty + \mathcal{T} r_0\right) \|v\| + r_0 + (1 - \lambda)\|y\|_\infty. \end{aligned}$$

Consequently,

$$\|y\|_\infty < \delta,$$

where

$$\delta = \frac{1}{\lambda}\left[(\|q\|_\infty + \mathcal{T}r_0)\,\|v\| + r_0\right].$$

Hence, $(I-D)^{-1}C(B_{r_0})$ is bounded with a bound δ. Now, let $\{t_n\}$ be any sequence in J converging to a point t in J. We denote

$$\Delta_n := \|(B'y)(t_n) - (B'y)(t)\|.$$

Then,

$$\begin{aligned}
\Delta_n &\leq \left|\int_0^{\sigma_1(t_n)} k(t_n,s)f_1(s,y(\eta(s)))ds - \int_0^{\sigma_1(t)} k(t,s)f_1(s,y(\eta(s)))ds\right| \|u\| \\
&\leq \left[\int_0^{\sigma_1(t_n)} |k(t_n,s)-k(t,s)|\,|f_1(s,y(\eta(s)))|\,ds\right] \|u\| \\
&\quad + \left|\int_{\sigma_1(t_n)}^{\sigma_1(t)} k(t,s)f_1(s,y(\eta(s)))ds\right| \|u\|.
\end{aligned}$$

Moreover, by taking into account that $(I-D)^{-1}C(B_{r_0})$ is bounded with a bound δ and by using the assumption $(\mathcal{R}_{23})(b)$, we get

$$\begin{aligned}
\Delta_n &\leq \left[\int_0^{\mathcal{T}} |k(t_n,s)-k(t,s)|\,\lambda\delta ds\right] \|u\| + \left|\int_{\sigma_1(t_n)}^{\sigma_1(t)} K\lambda\delta ds\right| \|u\| \\
&\leq \left[\int_0^{\mathcal{T}} |k(t_n,s)-k(t,s)|\,\lambda\delta ds\right] \|u\| + K\lambda\delta\,|\sigma_1(t_n)-\sigma_1(t)|\,\|u\|.
\end{aligned}$$

The continuity of k and σ_1 on the compact interval $[0,\mathcal{T}]$ implies that $B'y(.)$ is continuous. Similarly, the use of the first inequality in (7.37) as well as the dominated convergence theorem implies that the operator C is well defined.

Step 2: Let us prove that the entries of the block operator matrix (4.6) are weakly sequentially continuous. In order to apply Theorem 4.3.7, we need to show that B and B' are weakly sequentially continuous on $(I-D)^{-1}C(B_{r_0})$ and that A and C are weakly sequentially continuous on B_{r_0}. Let $\{x_n\}_{n=0}^{\infty}$ be a weakly convergent sequence of $(I-D)^{-1}C(B_{r_0})$ to a point x. Since

$(I - D)^{-1}C(B_{r_0})$ is bounded, then we can apply Dobrakov's theorem (see Theorem 1.4.1) in order to get

$$x_n(t) \rightharpoonup x(t) \text{ in } X.$$

By using the condition $(\mathcal{P})$, we get

$$a(t)x_n(t) \rightharpoonup a(t)x(t) \text{ in } X$$

i.e.,

$$(Bx_n)(t) \rightharpoonup (Bx)(t) \text{ in } X.$$

Since $\{Bx_n\}_{n=0}^{\infty}$ is bounded with a bound $\|a\|_\infty \delta$, then we can again apply Dobrakov's theorem to obtain $Bx_n \rightharpoonup Bx$. Consequently, B is weakly sequentially continuous. Now, the use of assumption $(\mathcal{R}_{23})(a)$ and Dobrakov's theorem, (see Theorem 1.4.1) allows us to get

$$f_1(t, x_n(t)) \rightharpoonup f_1(t, x(t)).$$

Moreover, the use of the dominated convergence theorem leads to

$$\lim_{n\to\infty} \int_0^{\sigma_1(t)} k(t,s) f_1(s, x_n(\eta(s)))ds = \int_0^{\sigma_1(t)} k(t,s) f_1(s, x(\eta(s)))ds.$$

So,

$$\left(\int_0^{\sigma_1(t)} k(t,s) f_1(s, x_n(\eta(s)))ds\right) \cdot u \to \left(\int_0^{\sigma_1(t)} k(t,s) f_1(s, x(\eta(s)))ds\right) \cdot u.$$

This means that $B'x_n \rightharpoonup B'x$ and consequently, B' is weakly sequentially continuous on $(I-D)^{-1}C(B_{r_0})$. Moreover, since g is weakly sequentially continuous with respect to the second variable and since $g(.,x_n(.))$ is bounded with a bound $\lambda\|x_n\|_\infty$, it follows that the operator D already defined in Eq. (7.38) is also weakly sequentially continuous in view of hypothesis $(\mathcal{R}_{22})(a)$. Besides, by taking into account that B_{r_0} is bounded and using Dobrakov's theorem, (see Theorem 1.4.1) we deduce that A is a weakly sequentially continuous operator on B_{r_0}. Next, let us show that C is weakly sequentially continuous on B_{r_0}. To do it, let $\{x_n\}_{n=0}^{\infty}$ be any sequence in B_{r_0} weakly converging to a point $x \in B_{r_0}$. Then, by using Dobrakov's theorem (see Theorem 1.4.1), we get for all $t \in J$, $x_n(t) \rightharpoonup x(t)$. Then,

$$p(t, s, x_n(s), x_n(\lambda s)) \rightharpoonup p(t, s, x_n(s), x_n(\lambda s)).$$

Knowing that

$$p(t, s, x_n(s), x_n(\lambda s)) \leq r_0,$$

it follows, from the dominated convergence theorem, that $(Cx_n)(t) \rightharpoonup (Cx)(t)$. Since the sequence $\{Cx_n\}$ is bounded with a bound $\|q\|_\infty + \mathcal{T} r_0$, we can again apply Dobrakov's theorem to deduce that C is weakly sequentially continuous on B_{r_0}.

Step 3: Next, let us show that C is weakly compact and that A is condensing on B_{r_0}. We should prove that $C(B_{r_0})$ is relatively weakly compact. By definition, we have

$$\text{for all } t \in J, \quad C(B_{r_0})(t) = \{(Cx)(t) \; ; \; \|x\|_\infty \leq r_0\}.$$

Then, $C(B_{r_0})(t)$ is sequentially relatively weakly compact in X. To see this, let $\{x_n\}$ be any sequence in B_{r_0}. Then, we have $(Cx_n)(t) = r_n(t) \cdot v$, where

$$r_n(t) = q(t) + \int_0^{\sigma_2(t)} p\left(t, s, x_n(s), x_n(\lambda s)\right) ds.$$

So, by using the first inequality in (7.37), $|r_n(t)| \leq \|q\|_\infty + \mathcal{T} r_0$, which shows that $\{r_n\}$ is a uniformly bounded sequence in $\mathcal{C}(J, \mathbb{R})$. Next, we show that $\{r_n\}$ is an equicontinuous set. Let t_1, $t_2 \in J$. Then, we have

$$\begin{aligned}
|r_n(t_1) - r_n(t_2)| \;&\leq\; |q(t_1) - q(t_2)| + \left| \int_{\sigma_2(t_1)}^{\sigma_2(t_2)} \left|p\left(t_2, s, x(s), x(\lambda s)\right)\right| ds \right| \\
&\quad + \left| \int_0^{\sigma_2(t_1)} \left|p\left(t_1, s, x(s), x(\lambda s)\right) - p\left(t_2, s, x(s), x(\lambda s)\right)\right| ds \right| \\
&\leq \left| \int_0^{\mathcal{T}} \left|p\left(t_1, s, x(s), x(\lambda s)\right) - p\left(t_2, s, x(s), x(\lambda s)\right)\right| ds \right| \\
&\quad + |q(t_1) - q(t_2)| + r_0 \left|\sigma_2(t_1) - \sigma_2(t_2)\right|.
\end{aligned}$$

Since p, q, and σ_2 are uniformly continuous functions, we conclude that $\{r_n\}$ is an equicontinuous set. As a result, $C(B_{r_0})(t)$ is sequentially relatively weakly compact. Next, we will show that $C(B_{r_0})$ is a weakly equicontinuous set. If we take $\varepsilon > 0$, $x \in B_{r_0}$, $x^* \in X^*$ and $t, t' \in J$ such that $t \leq t'$, $t' - t \leq \varepsilon$, and using the first inequality in (7.37), we obtain

$$\begin{aligned} G(t,t') &\leq \left[\int_0^{\sigma_2(t)} |p(t,s,x(s),x(\lambda s)) - p(t',s,x(s),x(\lambda s))|\, ds\right] \|x^*(v)\| \\ &+ |q(t)-q(t')|\|x^*(v)\| + \left[\int_{\sigma_2(t)}^{\sigma_2(t')} |p(t',s,x(s),x(\lambda s))|\, ds\right] \|x^*(v)\| \\ &\leq [w(q,\varepsilon) + \mathcal{T}w(p,\varepsilon) + r_0 w(\sigma_2,\varepsilon)]\, \|x^*(v)\|, \end{aligned}$$

where

$$G(t,t') := |x^*((Cx)(t) - (Cx)(t'))|,$$

$$\begin{cases} w(q,\varepsilon) = \sup\{|q(t)-q(t')| \ : \ t,t' \in J \ ; \ |t-t'| \leq \varepsilon\}, \\ w(p,\varepsilon) = \sup\limits_{t,t',\ s\in J,\ x,y\in S} \{|p(t,s,x,y) - p(t',s,x,y)| \ : |t-t'| \leq \varepsilon\}, \text{ and} \\ w(\sigma_2,\varepsilon) = \sup\{|\sigma_2(t)-\sigma_2(t')| \ : \ t,t' \in J \ ; \ |t-t'| \leq \varepsilon\}. \end{cases}$$

By taking into account the assumption $(\mathcal{R}_{21})$, and in view of the uniform continuity of the functions q and σ on the set J, it follows that $w(q,\varepsilon) \to 0$, $w(p,\varepsilon) \to 0$ and $w(\sigma_2,\varepsilon) \to 0$ as $\varepsilon \to 0$. By applying Arzelà–Ascoli's theorem (see Theorem 1.3.9), we conclude that $C(B_{r_0})$ is sequentially weakly relatively compact in X. Again, an application of Eberlein–Šmulian's theorem (see Theorem 1.3.3) implies that $C(B_{r_0})$ is relatively weakly compact. As a result, C is weakly compact. Now, the use of the assumption $(\mathcal{R}_{22})$ and Lemma 3.1.3 allows us to deduce that the operator A is condensing.

Step 4: To finish, it is sufficient to show that

$$Ax + B(I-D)^{-1}Cx \cdot B'(I-D)^{-1}Cx \in S \text{ for all } x \in B_{r_0}.$$

Let $y \in \mathcal{C}(J,X)$ be arbitrary, with

$$y = Ax + B(I-D)^{-1}Cx \cdot B'(I-D)^{-1}Cx,$$

for some $x \in B_{r_0}$. Then, for all $t \in J$, we have

$$\|y(t)\| \leq \|(Ax)(t)\| + \left\|B(I-D)^{-1}Cx(t) \cdot B'(I-D)^{-1}Cx(t)\right\|.$$

We should notice that, for all $x \in (I-D)^{-1}C(B_{r_0})$, there exists a unique $z \in \mathcal{C}(J, X)$ such that $z = x$, with $\|z\| \leq \delta$. Therefore,

$$\begin{aligned} \|y(t)\| &\leq \|f(t,x(t))\| + \|a(t)z(t)\| \left\| \left(\int_0^{\sigma_1(t)} k(t,s) f_1(s, z(\eta(s)))ds \right) \cdot u \right\| \\ &\leq k'\|x(t)\| + \|a(t)\|\|z(t)\| \left(\int_0^{\mathcal{T}} |k(t,s)\lambda|\|z(\eta(s))\|ds \right) \|u\|_\infty \\ &\leq k'r_0 + \lambda K \mathcal{T} \delta^2 \|a\|_\infty \|u\|_\infty, \end{aligned}$$

where

$$K = \sup_{t,s\in J} |k(t,s)|.$$

Since $y \in \mathcal{C}(J, X)$, there is $t^* \in J$ such that $\|y\|_\infty = \|y(t^*)\|$ and so, $\|y\|_\infty \leq r_0$ in view of the last inequality in (7.37). Hence, the hypothesis (iii) of Theorem 4.3.7 is satisfied, which achieves the proof. Q.E.D.

Example 2. Let $J = [0,1]$ be the closed and bounded interval in $\mathbb{R}$. Let $\mathcal{C}(J,\mathbb{R})$ be the Banach algebra of all continuous functions from J to $\mathbb{R}$ endowed with the sup-norm $\|\cdot\|_\infty$ defined by $\|f\|_\infty = \sup_{t\in J}|f(t)|$, for each $f \in \mathcal{C}(J,\mathbb{R})$. In the sequel, we need the following definition that can be found in [72].

Definition 7.4.1 *A mapping $f : J \times \mathbb{R} \longrightarrow \mathbb{R}$ is said to satisfy L^1-Carathéodory condition or simply is called L^1-Carathéodory, if*

(a) $t \longrightarrow f(t,x)$ is measurable for each $x \in \mathbb{R}$.

(b) $x \longrightarrow f(t,x)$ is almost everywhere continuous for $t \in J$, and

(c) for each real number $r > 0$, there exists a function $h_r \in L^1(J,\mathbb{R})$, such that

$$|f(t,x)| \leq h_r(t) \ ; \ t \in J$$

for all $x \in \mathbb{R}$ with $|x| \leq r$.

Let us consider the following nonlinear system of functional integral equations:

$$\begin{cases} x(t) = \displaystyle\int_0^{\sigma_1(t)} k_1(t,s) f_1(s, x(\eta_1(s)))ds + y(t) \cdot \left[\int_0^{\sigma_2(t)} k_2(t,s) f_2(s, y(\eta_2(s)))ds \right] \\ y(t) = \dfrac{1}{1+b(t)|x(t)|} - g(t, \dfrac{1}{1+b(t)|x(t)|}) + g(t, y(t)) \end{cases} \tag{7.39}$$

for all $t \in J$, where the functions $\sigma_1, \sigma_2, \eta_1, \eta_2, k_1, k_2, b, f_1, f_2, g$ are given while $x = x(t)$ and $y = y(t)$ are unknown functions.

In the remaining part of this example, we will assume that:

($\mathcal{R}_{25}$) The functions $\sigma_i, \eta_i : J \longrightarrow J$ are continuous for $i = 1, 2$.

($\mathcal{R}_{26}$) The function $b : J \longrightarrow \mathbb{R}$ is continuous and nonnegative.

($\mathcal{R}_{27}$) The functions $k_i : J \times J \longrightarrow \mathbb{R}$ are continuous and nonnegative for $i = 1, 2$.

($\mathcal{R}_{28}$) The function $f_1 : J \times \mathbb{R} \longrightarrow \mathbb{R}$ is generalized Lipschitz with a Lipschitz function l_1.

($\mathcal{R}_{29}$) The function $f_2 : J \times \mathbb{R} \longrightarrow \mathbb{R}$ is L^1-Carathéodory and it is generalized Lipschitz with a Lipschitz function l_2.

($\mathcal{R}_{30}$) The function $g : J \times \mathbb{R} \longrightarrow \mathbb{R}$ is such that:

(a) The function $x \longrightarrow g(t, x)$ is a contraction with a constant k, for $t \in J$.

(b) The function $t \longrightarrow g(t, x)$ is a continuous mapping on J, for all $x \in \mathcal{C}(J, \mathbb{R})$.

($\mathcal{R}_{31}$) There exists $r > 2$ such that:

(a) $|f_1(t, x(t))| \leq |x(t)|$, for $x \in \mathcal{C}(J, \mathbb{R})$ such that $\|x\|_\infty \leq r$.

(b) $K_2\|h_r\|_{L^1} \leq (1 - K_1)r$, where $K_1 = \sup\limits_{t,s\in J} k_1(t, s) < 1$ and $K_2 = \sup\limits_{t,s\in J} k_2(t, s)$.

(c) $\dfrac{\|b\|_\infty(1 + k)}{1 - k} K_2\|h_r\|_{L^1} + K_1\|l_1\|_{L^1} < k$.

(d) $0 \leq \dfrac{rK_2\|l_2\|_{L^1}}{(1 - k)^2} < 1$.

Theorem 7.4.2 *Under the assumptions* ($\mathcal{R}_{25}$)–($\mathcal{R}_{31}$), *the* FIE (7.39) *has, at least, one solution in* $\mathcal{C}(J, \mathbb{R}) \times \mathcal{C}(J, \mathbb{R})$.

Proof. Let us define the subset S on $\mathcal{C}(J, \mathbb{R})$ by:

$$S := \{y \in \mathcal{C}(J, \mathbb{R}), \|y\| \leq r\}.$$

Obviously, S is a nonempty, bounded, convex, and closed subset of $\mathcal{C}(J, \mathbb{R})$.

Consider the mapping A, B, C, D, and B' on $\mathcal{C}(J,\mathbb{R})$ defined by:

$$\begin{cases} (Ax)(t) = \displaystyle\int_0^{\sigma_1(t)} k_1(t,s)f_1(s,x(\eta_1(s)))ds \ ; \ t \in J \\ (Bx)(t) = x(t) \ ; \ t \in J \\ (Cx)(t) = \dfrac{1}{1+b(t)|x(t)|} - g(t, \dfrac{1}{1+b(t)|x(t)|}) \ ; \ t \in J \\ (Dx)(t) = g\left(t, x(t)\right) \ ; \ t \in J \\ (B'x)(t) = \displaystyle\int_0^{\sigma_2(t)} k_2(t,s)f_2(s,x(\eta_2(s)))ds \ ; \ t \in J. \end{cases} \tag{7.40}$$

We will show that A, B, C, D, and B' satisfy all conditions of Theorem 4.4.1. Since the two maps f_2 and k_2 are continuous and since f_2 is L^1-Carathéodory, we deduce from the dominated convergence theorem, that B' is continuous on S. Moreover, it is easy to verify that, by composition, the operators A, B, C, D, and B' defined in Eqs. (7.40) are well defined. Now, let us show that A is Lipschitzian on S. For this purpose, let x, $y \in S$. So,

$$\begin{aligned} \|Ax - Ay\| &\leq \sup_{t\in J} \int_0^{\sigma_1(t)} k_1(t,s) \left| f_1(s, x(\eta_1(s))) - f_1(s, y(\eta_1(s))) \right| ds \\ &\leq \sup_{t\in J} \int_0^{\sigma_1(t)} K_1 l_1(s) \left| x(\eta_1(s)) - y(\eta_1(s)) \right| ds \\ &\leq K_1 \|x-y\| \int_0^1 l_1(s)ds. \end{aligned}$$

This shows that A is Lipschitzian with the constant $K_1\|l_1\|_{L^1}$. By using the same argument, we conclude that B' is Lipschitzian with the constant $K_2\|l_2\|_{L^1}$. Again from assumption $(\mathcal{R}_{30})(a)$, we deduce that the operator C is Lipschitzian with the constant $(1+k)\|b\|_\infty$. Next, we will show that $C(S) \subset (I-D)(S)$. Let $x \in S$ and $t \in J$. Then, it is easy to justify that

$$(Cx)(t) = (I-D)\left(\frac{1}{1+b|x|}\right)(t).$$

So, there exists a point $y \in S$ such that $(Cx)(t) = (I-D)(y)(t)$. Consequently, $C(S) \subset (I-D)(S)$ and, from our assumptions, we have $(I-D)^{-1}$ exists on $C(S)$ and $(I-D)^{-1}C$ is Lipschitzian on S, with the constant $\frac{1+k}{1-k}\|b\|_\infty$. Next, let us prove that C is a strongly continuous mapping on S. To do it, let

$\{x_n\}_{n=0}^{\infty}$ be any sequence in S weakly converging to a point x. Then, $x \in S$ since S is weakly closed in $\mathcal{C}(J, \mathbb{R})$, and by using Dobrakov's theorem (see Theorem 1.4.1), we have for all $t \in J$

$$x_n(t) \rightharpoonup x(t) \quad \text{in } \mathbb{R}.$$

Since C is Lipschitzian, then $Cx_n(t) \to Cx(t)$ and consequently, $Cx_n \to Cx$. This shows that C is a strongly continuous operator on S. Clearly, $B(I - D)^{-1}C$ is regular on $S \supseteq B'(I-D)^{-1}C(S)$, since $(I-D)^{-1}Cx = \dfrac{1}{1+b|x|}$, for all $x \in S$. Therefore, $\left(\frac{I}{T}\right)^{-1}$ exists on $B'(S)$. To see it, let $y \in S$ be arbitrary, with

$$\left(\frac{I}{B(I-D)^{-1}C}\right)(x) = y, \quad \text{for some } x \in S,$$

or equivalently, for all $t \in J$,

$$x(t)(1 + b(t)|x(t)|) = y(t),$$

which implies that

$$|x(t)||(1 + b(t)|x(t)|) = |y(t)|.$$

For each $t \in J$ such that $b(t) = 0$, we have $x = y$. Then, for each $t \in J$ such that $b(t) > 0$, we obtain

$$\left(\sqrt{b(t)}|x(t)| + \frac{1}{2\sqrt{b(t)}}\right)^2 = \frac{1}{4b(t)} + |y(t)|$$

which further implies that,

$$\sqrt{b(t)}|x(t)| = \frac{-1}{2\sqrt{b(t)}} + \sqrt{\frac{1}{4b(t)} + |y(t)|}.$$

Hence,

$$b(t)|x(t)| = \frac{-1}{2} + \sqrt{\frac{1}{4} + b(t)|y(t)|}$$

and consequently,

$$x(t) = \frac{y(t)}{1 + b(t)|x(t)|} = \frac{y(t)}{\dfrac{1}{2} + \sqrt{\dfrac{1}{4} + b(t)|y(t)|}}.$$

Let us consider the function G defined by the expression

$$\begin{cases} F : \mathcal{C}(J,\mathbb{R}) \longrightarrow \mathcal{C}(J,\mathbb{R}) \\ \quad x \longrightarrow G(x) = \dfrac{x}{\frac{1}{2}+\sqrt{\frac{1}{4}+b|x|}}. \end{cases}$$

It is easy to verify that, we have for all $x \in \mathcal{C}(J,\mathbb{R})$

$$\left(\left(\frac{I}{T}\right)\circ G\right)(x) = \left(G\circ\left(\frac{I}{T}\right)\right)(x) = x.$$

We conclude that

$$\left(\frac{I}{T}\right)^{-1} x = \frac{x}{\frac{1}{2}+\sqrt{\frac{1}{4}+b|x|}}.$$

Moreover, by taking into account that $K_1\|l_1\|_{L^1} < 1$ and $A(S) \subset S$, and by using the fixed point theorem of Boyd and Wong (see Theorem 1.6.10), we deduce that $(I-A)^{-1}$ exists on $(I-A)(S)$. Consequently, by referring to B. C. Dhage in [73], we have

$$\left(\frac{I-A}{T}\right)^{-1} = \left(\frac{I}{T}\right)^{-1}(I-A)^{-1}.$$

So, the operator $\left(\frac{I-A}{T}\right)^{-1}$ exists on $B'(S)$. Again, by using assumption $(\mathcal{R}_{29})$,

$$\begin{aligned} M_1 &= \sup_{x\in S}\|B'(I-D)^{-1}Cx\| \\ &\leq \sup_{x\in S}\|B'x\| \\ &\leq \sup_{x\in S}\left\{\sup_{t\in J}\left|\int_0^{\sigma_2(t)} k_2(t,s)f_2(s,x(\eta_2(s)))ds\right|\right\} \\ &\leq K_2\|h_r\|_{L^1}. \end{aligned}$$

Consequently, in view of assumption $(\mathcal{R}_{31})$, we have

$$\frac{(1+k)\|b\|_\infty}{1-k}M_1 + K_1\|l_1\|_{L^1} < k.$$

Now, since

$$M_2 = \sup_{x\in S}\|B(I-D)^{-1}Cx\| \leq \sup_{x\in S}\|Bx\| \leq r,$$

then we have

$$0 \leq \frac{M_2K_2\|l_2\|_{L^1}}{(1-k)^2} < 1.$$

Next, let $x \in \mathcal{C}(J, \mathbb{R})$ be arbitrary, with

$$x = Ax + Tx \cdot T'y, \quad \text{for some } y \in S.$$

Then, for all $t \in J$, we have

$$\begin{aligned}
|x(t)| &\leq |Ax(t)| + |Tx(t)||T'y(t)| \\
&\leq \int_0^{\sigma_1(t)} |k_1(t,s) f_1(s, x(\eta_1(s)))|\, ds \\
&\quad + \frac{1}{1 + b(t)|x(t)|} \left| \int_0^{\sigma_2(t)} k_2(t,s) f_2\left(s, \frac{1}{1 + b(\eta_2(s))|x(\eta_2(s))|}\right) ds \right| \\
&\leq \int_0^{\sigma_1(t)} k_1(t,s)\, |f_1(s, x(\eta_1(s)))|\, ds \\
&\quad + \frac{1}{1 + b(t)|x(t)|} \int_0^1 k_2(t,s) \left| f_2\left(s, \frac{1}{1 + b(\eta_2(s))|x(\eta_2(s))|}\right)\right| ds \\
&\leq \int_0^1 K_1\, |x(\eta_1(s))|\, ds + \int_0^1 K_2 h_r(s) ds \\
&\leq K_1 \|x\|_\infty + K_2 \|h_r\|_{L^1} \\
&\leq r.
\end{aligned}$$

The desired conclusion follows from a direct application of Theorem 4.4.1. The proof is complete. Q.E.D.

Remark 7.4.1 *By applying a fixed point theorem of the Krasnosel'skii type and by giving the suitable assumptions, Ntouyas et al. in [76] obtained some results on the existence of solutions to the following nonlinear functional integral equation:*

$$x(t) = K(t, x(t)) + \int_0^{\sigma(t)} v(t,s) g(s, x(\theta(s))) ds.$$

7.5 Nonlinear Equations with Unbounded Domain

In this section, we will study the existence of solutions for the following coupled system:

$$\begin{cases} x(t) = h(x(t)) + f_1(t, y(t)) \int_I g(t, s, y(s))ds \\ y(t) = f_2(t, x(t)) + g_2(t, y(t)), \end{cases} \tag{7.41}$$

where $t \in I$ and the functions h, f_1, f_2, g, and g_2 are given, and x, $y \in \mathcal{C}_b(I)$, the space of all bounded, continuous, and real valued functions φ defined on I such that I is a real, closed, and unbounded interval. Let us mention that the space $X := \mathcal{C}_b(I)$, equipped with the standard supremum norm, $\|\varphi\| = \sup_{t\in I} |\varphi(t)|$, is a Banach algebra.

Before studying Eq. (7.41), we show that a non-compact map of the form

$$Fx = Hx + Lx.Kx$$

can be, in some cases, a strict set-contraction with respect to Kuratowskii's measure of noncompactness μ.

Theorem 7.5.1 *Let S be a subset of a Banach algebra X, and suppose that $F : S \longrightarrow X$ is of the form $Fx = Hx + Lx.Kx$, where*

(i) $H : S \longrightarrow X$ is $\mathcal{D}$-Lipschitzian with a $\mathcal{D}$-function ϕ_H,

(ii) $L : S \longrightarrow X$ is $\mathcal{D}$-Lipschitzian with a $\mathcal{D}$-function ϕ_L, and

(iii) $K : S \longrightarrow X$ is a compact mapping.

Suppose that $\beta = \sup_{x\in S} \|Kx\| < \infty$. If $\phi_H(r) + \beta\phi_L(r) < r$ for all $r > 0$, then F is a strict-set-contraction.

Proof. Let C be a bounded subset of S, let $\varepsilon > 0$ be fixed and $\delta = \sup_{x\in C} \|Lx\|$. We may assume that $\alpha > 0$, $\beta > 0$ and $\delta > 0$. Since $K(C)$ is relatively compact, there exist finitely many sets D_1,D_2,...,D_m in B such that diam $D_i < (\delta)^{-1}\varepsilon$, $i = 1, 2, ..., m$, and $K(C) = \cup_{i=1}^{m} D_i$. Let us choose the sets $C_1, C_2, ..., C_n$ such that diam $C_i \leq \mu(C)$ and $C = \cup_{i=1}^{n} C_i$, and then let us define the sets $S_{i,j}$, $i = 1, ..., n$, $j = 1, ..., m$, by:

$$S_{i,j} = \{t + x.y \text{ such that } t \in H(C_i),\ x \in L(C_i) \text{ and } y \in D_j\}.$$

Obviously,

$$F(C) \subset \bigcup_{i,j} S_{i,j}.$$

Notice that, if w, $z \in S_{i,j}$, then there exist u, v, a, $b \in C_i$ and x, $y \in D_j$ such that $w = Ha + Lu.x$, and $z = Hb + Lv.y$. Therefore,

$$\begin{aligned}
\|w - z\| &= \|Ha - Hb + Lu.x - Lv.y\| \\
&\leq \|Ha - Hb\| + \|Lu.x - Lv.y\| \\
&\leq \phi_H(\|a - b\|) + \|x\|\|Lu. - Lv\| + \|Lv\|\|x - y\| \\
&\leq \phi_H(\|a - b\|) + \|x\|\phi_L(\|u. - v\|) + \|Lv\|\|x - y\| \\
&\leq \phi_H(\mu(C)) + \beta\phi_L(\mu(C)) + \delta(\delta)^{-1}\varepsilon \\
&\leq \phi_H(\mu(C)) + \beta\phi_L(\mu(C)) + \varepsilon \\
&\leq \mu(C) + \varepsilon.
\end{aligned}$$

We have proved that diam $S_{i,j} \leq \mu(C) + \varepsilon$, which implies that

$$\mu(F(C)) \leq \mu(C),$$

and that F is a strict-set-contraction. Q.E.D.

The following proposition can be proved by the same reasoning.

Proposition 7.5.1 *Let S be a subset of a Banach algebra $(X, \|.\|_X)$, and let H, K, and L be three maps from S into X. Assume that:*

(i) H maps bounded sets into bounded ones and there exists a constant λ such that $\mu(H(C)) \leq \lambda\mu(C)$ for every bounded set $C \in S$,

(ii) L maps bounded sets into bounded ones and there exists a constant α such that $\mu(L(C)) \leq \alpha\mu(C)$ for every bounded set $C \in S$,

(iii) K is a compact operator, and

(iv) $\lambda + \alpha. \sup_{z \in S} \|Kz\| < 1$.

Then, the mapping $F : S \longrightarrow X$ defined by $Fz = Hz + Lz.Kz$, $z \in S$, is a strict-set-contraction.

Now, the following theorem shows that, under some conditions, an operator of the form $F := A + T.T'$ may have a fixed point.

Theorem 7.5.2 *Let S be a closed, bounded, and convex subset of a Banach algebra X, let S' be a closed, bounded, and convex subset of a Banach algebra Y, and let $A : S \longrightarrow X$, $B, B' : S' \longrightarrow X$, $C : S \longrightarrow Y$, and $D : S' \longrightarrow S'$ be five operators such that:*

(i) A, B, and C are Lipschitzian with the constants α, β, and λ, respectively,

(ii) C is compact and B' is continuous,

(iii) D is a k-contraction,

(iv) $C(S) \subset (I - D)(S')$, and

(v) $Ax + Tx.T'x \in S$ for all $x \in S$, where $T = B(I - D)^{-1}C$, and $T' = B'(I - D)^{-1}C$.

Then, the mapping $F := A + T.T'$ has, at least, a fixed point in S whenever $\alpha + \beta\frac{\lambda}{1-k}\delta < 1$, where $\delta = \sup_{x\in S} \|T'(S)\|$.

Proof. Since D is a k-contraction on S', B and C are β, λ-Lipschitzian, respectively, and $C(S) \subset (I - D)(S')$, then the operator $T = B(I - D)^{-1}C$ exists and is Lipschitzian with the constant $\beta\frac{\lambda}{1-k}$. Also, the operator $T' = B'(I - D)^{-1}C$ exists and is compact. Indeed, let Ω be a bounded subset of X and $\alpha_n \in T'(\Omega) := B'(I - D)^{-1}C(\Omega)$. Then, there exists a sequence $(\psi_n)_n \in C(\Omega)$ such that $\alpha_n = B'(I-D)^{-1}\psi_n$ for all $n \in \mathbb{N}$. Due to the fact that C is compact, we deduce that $C(\Omega)$ is relatively compact and hence, $(\psi_n)_n$ has a convergent subsequence $(\psi_{\varphi_n})_n \longrightarrow \psi$. Therefore, $\alpha_{\varphi_n} := B'(I - D)^{-1}\psi_{\varphi_n}$ is convergent, since $B'(I - D)^{-1}$ is continuous. So, Theorem 7.5.1 establishes that F is a strict-set-contraction. Hence, by applying Darbo's theorem, we deduce that F has, at least, a fixed point. Q.E.D.

Next, we recall the following useful compactness criterion:

Proposition 7.5.2 *[127] Let C be a bounded subset of $\mathcal{C}_b(I)$. Assume that C is pointwise equicontinuous in I, and*

$$\lim_{a\to+\infty} \sup_{\varphi\in C} \Big\{ \sup\{|\varphi(t)|;\quad t \in I,\ |t| \geq a\}\Big\} = 0.$$

Then, C is relatively compact in $\mathcal{C}_b(I)$.

For the following, let us define the subsets S and S' on $\mathcal{C}_b(I)$ by:

$$S := \{x \in \mathcal{C}_b(I), \|x\| \leq r\}$$

and

$$S' := \Big\{y \in \mathcal{C}_b(I), \|y\| \le p := \frac{\theta r + \|f_2(.,0)\| + \|g_2(.,0)\|}{1-k}\Big\}.$$

Now, let us specify the assumptions under which the equations of (7.41) will be investigated. These assumptions are as follows:

($\mathcal{R}_{32}$) $h : \mathbb{R} \longrightarrow \mathbb{R}$ is α-Lipschitzian, i.e., $\|hx - hy\| \le \alpha\|x-y\|$ for all x, $y \in S$, and $t \longrightarrow h(0)(t)$ is bounded on I.

($\mathcal{R}_{33}$) $f_1 : I \times \mathbb{R} \longrightarrow \mathbb{R}$ satisfies :

(a) f_1 is η-Lipschitzian with respect to the second variable, i.e.,

$$\|f_1(.,x) - f_1(.,y)\| \le \eta\|x-y\| \text{ for all } x,\ y \in S',$$

(b) the function $t \longrightarrow f_1(t,0)$ is bounded on I, and

(c) $\|f_1(.,0)\| > \dfrac{[(1-\alpha)r - \|h(0)\|]\eta\theta}{(1-\alpha)(1-k)} - \eta p.$

($\mathcal{R}_{34}$) $f_2 : I \times \mathbb{R} \longrightarrow \mathbb{R}$ satisfies:

(a) f_2 is θ-Lipschitzian with respect to the second variable, i.e.,

$$\|f_2(.,x) - f_2(.,y)\| \le \theta\|x-y\| \text{ for all } x,\ y \in S'.$$

(b) f_2 is 1-Lipschitzian with respect to the first variable, i.e.,

$$\|f_2(t,.) - f_2(s,.)\| \le \|t-s\| \text{ for all } t,\ s \in I,$$

(c) the function $t \longrightarrow f_2(t,x)$ is bounded on I for all $x \in S$, and

(d) $\lim_{a\longrightarrow\infty} \sup_{f_2(.,x)\in f_2(.,S)}\{\sup\{|f_2(t,x)|; t \in I, |t| \ge a\}\} = 0.$

($\mathcal{R}_{35}$) $g_2 : I \times \mathbb{R} \longrightarrow \mathbb{R}$ satisfies :

(a) g_2 is a k-contraction with respect to the second variable, i.e.,

$$\|g_2(.,x) - g_2(.,y)\| \le k\|x-y\| \text{ for all } x,\ y \in S',$$

(b) the function $t \longrightarrow g_2(t,0)$ is bounded on I, and

(c) $\|g_2(.,0)\| \le p(1-k)$.

($\mathcal{R}_{36}$) $g : I \times I \times \mathbb{R} \longrightarrow \mathbb{R}$ satisfies the following conditions:

(i) for every $t \in I$, the function $s \longrightarrow g(t,s,x)$ is measurable for all $x \in \mathbb{R}$ and the function $x \longrightarrow g(t,s,x)$ is continuous for almost all $s \in I$,

(ii) for every $r > 0$ and for all $t \in I$, the function $s \longrightarrow \sup_{|x|\le r}|g(t,s,x)|$

belongs to $L^1(I)$,

(iii) for every $r > 0$, $\lim_{t\in I,|t|\to+\infty} \int_I \sup_{|x|\leq r} |g(t,s,x)|ds = 0$, and

(iv) for every $r > 0$ and for every $\tau \in I$, $\lim_{t\to\tau} \int_I \sup_{|x|\leq r} |g(t,s,x) - g(\tau,s,x)|ds = 0$.

$(\mathcal{R}_{37})$ $\| \int_I g(t,s,x(s)ds\| \leq \dfrac{(1-\alpha)r - \|h(0)\|}{\eta p + \|f_1(.,0)\|}$.

Now, we can prove the following theorem:

Theorem 7.5.3 *Under the assumptions* $(\mathcal{R}_{32})$–$(\mathcal{R}_{37})$*, the problem (7.41) has, at least, one solution in* $\mathcal{C}_b(I)$ *whenever* $\alpha + \frac{\eta\theta}{1-k}\frac{(1-\alpha)r-\|h(0)\|}{\eta p+\|f_1(.,0)\|} < 1$.

Proof. Consider the mapping A, B, B', C, and D on $\mathcal{C}_b(I)$ defined by:

$$\begin{cases} Ax(t) = h(x(t)) \\ By(t) = f_1(t,y(t)) \\ B'y(t) = \int_I g(t,s,y(s))ds \\ Cx(t) = f_2(t,x(t)) \\ Dy(t) = g_2(t,y(t)). \end{cases}$$

Observe that the problem (7.41) can be written in the following form:

$$\begin{cases} x(t) = Ax(t) + By(t).B'y(t) \\ y(t) = Cx(t) + Dy(t). \end{cases}$$

We will prove that operators A, B, B', C, and D satisfy all the conditions of Theorem 7.5.2. Obviously, S and S' are nonempty, closed, bounded, and convex subsets of $\mathcal{C}_b(I)$.

(i) It is clear that the operator A maps $\mathcal{C}_b(I)$ into itself and is α-Lipschitzian in view of assumption $(\mathcal{R}_{32})$. Let us consider the superposition operators B, C, and D. The hypotheses on the functions f_1, f_2, and g_2 ensure that operators B, C, and D map $\mathcal{C}_b(I)$ into itself and that B and C are Lipschitzian with constants η, and θ, respectively.

(ii) Now, let us show that the set $C(S)$ is relatively compact. Knowing that $C(S) := \{Cx : x \in S\}$, this subset is nothing else than

$$f_2(.,S) := \{f_2(.,x) : x \in \mathcal{C}_b(I); \|x\| \leq r\}.$$

First, let us show that $C(S)$ is bounded. For this purpose, let $x \in S$. Then,

$$\begin{aligned} \|Cx\| &= \|f_2(.,x)\| \\ &= \sup_{t\in I} \|f_2(t,x(t))\| \\ &\leq \sup_{t\in I} \|f_2(t,x(t)) - f_2(t,0)\| + \sup_{t\in I} \|f_2(t,0)\| \\ &\leq \theta \sup_{t\in I} \|x(t)\| + \|f_2(.,0)\| \\ &\leq \theta\|x\| + \|f_2(.,0)\| \\ &\leq \theta.r + \|f_2(.,0)\| \end{aligned}$$

Since f_2 is a Lipschitzian map, it is easy to verify that $C(S)$ is a pointwise equicontinuous set. By taking into account the hypothesis $(\mathcal{R}_{33})$, the result is obtained by a simple application of Proposition 7.5.2. Due to [127, Lemma 1] and by using the hypothesis on the function g, the operator B' maps $\mathcal{C}_b(I)$ into itself and is continuous.

(iii) Concerning the operator D, in view of hypothesis $(\mathcal{R}_{34})$, D is a k-contraction.

(iv) Next, let us show that $C(S) \subset (I - D)(S')$. To do so, let $x \in S$ be fixed and let us define a mapping

$$\begin{aligned} \varphi_x : \mathcal{C}_b(I) &\longrightarrow \mathcal{C}_b(I) \\ y &\longrightarrow Cx + Dy. \end{aligned}$$

Notice that φ_x is a k-contraction. Hence, by the Banach fixed point theorem, we deduce that there exists a unique fixed point $y \in \mathcal{C}_b(I)$ such that $Cx+Dy = y$, which means that $C(S) \subset (I - D)(\mathcal{C}_b(I))$. Clearly, we have

$$\|y\| = \|Cx + Dy\| \leq \theta r + \|f_2(.,0)\| + k\|y\| + \|g_2(.,0)\|,$$

which leads to

$$\begin{aligned} \|y\| &\leq \frac{\theta r + \|f_2(.,0)\| + \|g_2(.,0)\|}{1-k} \\ &\leq p. \end{aligned}$$

(v) In order to achieve the proof, it is sufficient to verify that $Ax+Tx.T'x \in S$

for all $x \in S$. To do so, let us consider $x \in S$,

$$\begin{aligned}\|Ax + Tx.T'x\| &\leq \|Ax\| + \|Tx\|.\|T'x\| \\ &\leq \|Ax\| + \|Bx_1\|.\|B'x_1\| \\ &\leq \alpha\|x\| + \|h(0)\| + (\eta\|x_1\| + \|f_1(.,0)\|).\|B'x_1\| \\ &\leq \alpha r + \|h(0)\| + (\eta p + \|f_1(.,0)\|).\frac{(1-\alpha)r - \|h(0)\|}{\eta p + \|f_1(.,0)\|} \\ &\leq r,\end{aligned}$$

where $x_1 = (I-D)^{-1}Cx$. Hence, $Ax + Tx.T'x \in S$ for all $x \in S$. This ends the proof since all the hypotheses of Theorem 7.5.2 are fulfilled. Q.E.D.

7.6 Differential Inclusions

The results of this section can be found in [102].

7.6.1 Multi-valued initial value problems IVP

Given a closed and bounded interval $J = [0, a]$ in $\mathbb{R}$ for some $a \in \mathbb{R}_+^*$, let us consider the system describing the initial value problem (in short IVP).

$$\begin{cases} \left(\dfrac{x(t) - k(t, x(t))}{f(t, y(t))}\right)' \in G(t, y(t)) \\ \\ y(t) = \dfrac{1}{1 + b(t)|x(\theta(t))|} - p(t, \dfrac{1}{1 + b(t)|x(\theta(t))|}) + p(t, y(t)) \\ \\ (x(0), y(0)) = (x_0, y_0) \in \mathbb{R}^2, \end{cases} \tag{7.42}$$

where $t \in J$, and the functions b, θ, k, f, and $G : J \times \mathbb{R} \longrightarrow \mathcal{P}_{cp,\ cv}(\mathbb{R})$ are given, whereas $x = x(t)$, and $y = y(t)$ are unknown functions that satisfy:

(i) The function $t \longrightarrow \dfrac{x(t) - k(t, x(t))}{f(t, y(t))}$ is differentiable, and

(ii) $\left(\dfrac{x(t) - k(t, x(t))}{f(t, y(t))}\right)' = v(t),\ t \in J$ for some $v \in L^1(J, \mathbb{R})$ such that $v(t) \in G(t, x(t))$, a. e. $t \in J$ satisfying $(x(0), y(0)) = (x_0, y_0)$.

We will seek the solution of IVP (7.42) in the space $\mathcal{C}(J,\mathbb{R})$ of continuous and real-valued functions on J. Consider the norm $\|.\|$ and the multiplication "$\cdot$" in the Banach algebra $\mathcal{C}(J,\mathbb{R})$ of continuous functions on J by $\|x\| = \sup_{t\in J}|x(t)|$ and

$$(x \cdot y)(t) = x(t) \cdot y(t) \ ; \ t \in J$$

for all $x,\ y \in \mathcal{C}(J,\mathbb{R})$.

First, let us recall the following definitions.

Definition 7.6.1 *A multi-valued map $Q : J \longrightarrow \mathcal{P}_{cp}(\mathbb{R})$ is said to be measurable if, for any $y \in X$, the function $t \longrightarrow d(y, Q(t)) = \inf\{|y - x| \ ; \ x \in Q(t)\}$ is measurable.*

Definition 7.6.2 *A multi-valued map $Q : J \longrightarrow \mathcal{P}_{cp}(\mathbb{R})$ is called Carathéodory, if*

(i) $t \longrightarrow Q(t, x)$ is measurable for each $x \in \mathbb{R}$, and

(ii) $x \longrightarrow Q(t, x)$ is an upper semi-continuous almost everywhere for $t \in J$.

Again, a Carathéodory multi-valued function Q is called L^1-Carathéodory, if

(iii) for each real number $r > 0$, there exists a function $h_r \in L^1(J,\mathbb{R})$ such that

$$\|Q(t, x)\|_{\mathcal{P}} \le h_r(t) \quad a.e\ t \in J$$

for all $x \in \mathbb{R}$, with $|x| \le r$.

Moreover, a Carathéodory multi-valued function Q is called $L^1_{\mathbb{R}}$-Carathéodory, if

(iv) there exists a function $h \in L^1(J,\mathbb{R})$ such that

$$\|Q(t, x)\|_{\mathcal{P}} \le h(t) \quad a.e\ t \in J$$

for all $x \in \mathbb{R}$. The function h is called a growth function of Q on $J \times \mathbb{R}$.

For any multi-valued function $Q : J \times \mathbb{R} \longrightarrow \mathcal{P}_{cp}(\mathbb{R})$, we denote

$$S^1_Q(x) = \{v \in L^1(J,\mathbb{R}) \ ; \ v(t) \in Q(t, x(t)) \text{ for all } t \in J\}$$

for some $x \in \mathcal{C}(J,\mathbb{R})$. The integral of the multi-valued function Q is defined as

$$\int_0^t Q(s, x(s))ds = \left\{ \int_0^t v(s)ds \ ; \ v \in S^1_Q(x) \right\}.$$

The following lemmas can be found in [120].

Lemma 7.6.1 *Let X be a Banach space. If $dim(X) < \infty$ and $Q : J \times X \longrightarrow \mathcal{P}_{cp}(X)$ is L^1-Carathéodory, then $S^1_Q(x) \neq \emptyset$ for each $x \in X$.*

Lemma 7.6.2 *Let $Q : J \times X \longrightarrow \mathcal{P}_{cp}(X)$ be a Carathéodory multi-valued operator with $S^1_Q(x) \neq \emptyset$ and let $\mathcal{L} : L^1(J, X) \longrightarrow \mathcal{C}(J, X)$ be a linear continuous mapping. Then, the operator*

$$\mathcal{L} \circ S^1_Q : \mathcal{C}(J, X) \longrightarrow \mathcal{P}_{cp,\ cv}(\mathcal{C}(J, X))$$

has a closed graph.

Remark 7.6.1 *It is known that a multi-valued map Q is upper semi-continuous if, and only if, it has a closed graph.*

In what follows, suppose that:

($\mathcal{R}_{38}$) The single-valued mapping $k : J \times \mathbb{R} \longrightarrow \mathbb{R}$ is continuous and there is a bounded function $l_1 : J \longrightarrow \mathbb{R}$ with a bound $\|l_1\| \leq \frac{1}{4}$ satisfying:

$$|k(t, x) - k(t, y)| \leq l_1(t)|x - y| \ ; \quad \text{for all } x, y \in \mathbb{R}.$$

($\mathcal{R}_{39}$) The single-valued mapping $f : J \times \mathbb{R} \longrightarrow \mathbb{R}\backslash\{0\}$ is continuous and there is a bounded function $l_2 : J \longrightarrow \mathbb{R}$ with a bound $\|l_2\|$ satisfying:

$$|f(t, x) - f(t, y)| \leq l_2(t)|x - y| \ ; \quad \text{for all } x, y \in \mathbb{R}.$$

($\mathcal{R}_{40}$) The multi-valued operator $G : J \times \mathbb{R} \longrightarrow \mathcal{P}_{cp,\ cv}(\mathbb{R})$ is $L^1_{\mathbb{R}}$-Carathéodory with growth function h.

($\mathcal{R}_{41}$) The single-valued mapping $p : J \times \mathbb{R} \longrightarrow \mathbb{R}$ is such that:

- (a) The function $x \longrightarrow p(t, x)$ is a q-contraction,
- (b) The function $t \longrightarrow p(t, x)$ is continuous on J, for all $x \in \mathcal{C}(J, \mathbb{R})$, and
- (c) $|p(t, x)| \leq q|x|$, for all $t \in J$ and $x \in \mathbb{R}$.

($\mathcal{R}_{42}$) The single-valued mapping $b : J \longrightarrow \mathbb{R}_+$ is continuous.

Theorem 7.6.1 *Assume that the hypotheses* $(\mathcal{R}_{38})$–$(\mathcal{R}_{42})$ *hold. If*

$$\begin{cases} 4\|b\|_\infty\|l_2\|\left(\left|\dfrac{x_0-k(0,x_0)}{f(0,y_0)}\right|+\|h\|_{L^1}\right)\leq 1 \\ \\ 1-\sqrt{2}\sqrt{1-P}\leq q<\dfrac{1-4\|l_2\|\|b\|_\infty\left(\left|\dfrac{x_0-k(0,x_0)}{f(0,y_0)}\right|+\|h\|_{L^1}\right)}{1+4\|l_2\|\left(\left|\dfrac{x_0-k(0,x_0)}{f(0,y_0)}\right|+\|h\|_{L^1}\right)} \\ \\ \max\left\{\|l_1\|\ ;\ \|l_2\|\Big(M(x_0,y_0)+\|h\|_{L^1}\Big)\ ;\ K+F\Big((M(x_0,y_0)+\|h\|_{L^1}\Big)\right\}\leq\dfrac{1}{4} \end{cases} \tag{7.43}$$

where $M(x_0,y_0):=\left|\dfrac{x_0-k(0,x_0)}{f(0,y_0)}\right|$, $\frac{1}{2}<P=\sup\{|p(t,0)|\ ;\ t\in J\}<1$, $K=\sup\limits_{t\in J}\{|k(t,0)|\}$, *and* $F=\sup\limits_{t\in J}\{|f(t,0)|\}$. *Then, the system* (7.42) *has a solution.*

Proof. Consider the mappings A, B, C, D, and B' on $\mathcal{C}(J,\mathbb{R})$ by:

- $Ax(t)=k(t,x(t)),\quad Bx(t)=f(t,x(t))$, and $Dx(t)=p(t,x(t))$
- $Cx(t)=\dfrac{1}{1+b(t)|x(\theta(t))|}-p\big(t,\dfrac{1}{1+b(t)|x(\theta(t))|}\big)$, and
- $B'x=\left\{u\in X \text{ such that } u(t)=\dfrac{x_0-k(0,x_0)}{f(0,y_0)}+\displaystyle\int_0^t v(s)ds\ ;\ v\in S^1_G(x)\right\}.$

For all $t\in J$, the problem IVP (7.42) may be abstractly written in the form

$$\begin{cases} x(t)\in Ax(t)+By(t)\cdot B'y(t) \\ \\ y(t)=Cx(t)+Dy(t) \end{cases}$$

We will show that A, B, C, D, and B' meet all the conditions of Theorem 4.5.8. Let us define the subsets S and S' on $\mathcal{C}(J,\mathbb{R})$ by:

$$S=\{y\in\mathcal{C}(J,\mathbb{R})\ ;\ \|y\|_\infty\leq 1+q\}\ \text{ and }\ S'=\{y\in\mathcal{C}(J,\mathbb{R})\ ;\ \|y\|_\infty\leq 2+q\}.$$

It is obvious that S and S' are nonempty, bounded, convex, and closed subsets of $\mathcal{C}(J,\mathbb{R})$, and similarly it is clear that the operator B' is well defined since $S^1_G(x)\neq\emptyset$, for each $x\in\mathcal{C}(J,\mathbb{R})$.

Step 1: We start by showing that the operators A, B, C, and D define single-valued operators $A,\ B,\ C,\ D : \mathcal{C}(J,\mathbb{R}) \longrightarrow \mathcal{C}(J,\mathbb{R})$ and $B' : \mathcal{C}(J,\mathbb{R}) \longrightarrow \mathcal{P}_{cp,\ cv}(\mathcal{C}(J,\mathbb{R}))$. The claim regarding A, B, C, and D is clear, since the functions f and k are continuous on $J \times \mathbb{R}$. We only have to prove the claim for the multi-valued operator B' on $\mathcal{C}(J,\mathbb{R})$. First, we show that B' has compact values on $\mathcal{C}(J,\mathbb{R})$. Notice that the operator B' is equivalent to the composition $\mathcal{L} \circ \mathcal{K}$ of two operators on $L^1(J,\mathbb{R})$, where $\mathcal{K} : L^1(J,\mathbb{R}) \longrightarrow \mathcal{C}(J,\mathbb{R})$ is the continuous operator defined by:

$$\mathcal{K}v(t) = \frac{x_0 - k(0,x_0)}{f(0,y_0)} + \int_0^t v(s)ds.$$

In order to demonstrate that B' has compact values, it is sufficient to prove that the composition operator $\mathcal{L} \circ \mathcal{K}$ has compact values on $\mathcal{C}(J,\mathbb{R})$. Let $x \in \mathcal{C}(J,\mathbb{R})$ be arbitrary and let $\{v_n\}$ be a sequence in $S^1_G(x)$. Then, by using the definition of $S^1_G(x)$, we get $v_n(t) \in G(t,x(t))$ a.e. for $t \in J$. Since $G(t,x(t))$ is compact, then there is a convergent subsequence of $v_n(t)$ (for simplicity, call it $v_n(t)$ itself) that converges in measure to some $v(t) \in G(t,x(t))$ for $t \in J$. From the continuity of $\mathcal{L}$, it follows that $\mathcal{K}v_n(t) \to \mathcal{K}v(t)$ pointwise on J as $n \to \infty$. We need to show that $\{\mathcal{K}v_n\}$ is an equicontinuous sequence in order to demonstrate the uniform convergence. Let $t_1,\ t_2 \in J$. Then, we have

$$|\mathcal{K}v_n(t_1) - \mathcal{K}v_n(t_2)| \le \int_{t_1}^{t_2} |v(s)|\, ds. \tag{7.44}$$

As $v_n \in L^1(J,\mathbb{R})$, the right-hand side of Eq. (7.44) tends to 0 as $t_1 \to t_2$. Hence, the sequence $\{Kv_n\}$ is equicontinuous, and when applying the Ascoli theorem, we deduce that that there is a uniformly convergent subsequence. Therefore, $\left(\mathcal{K} \circ S^1_G\right)(x)$ is a compact set for all $x \in \mathcal{C}(J,\mathbb{R})$. Consequently, B' is a compact multi-valued operator on $\mathcal{C}(J,\mathbb{R})$. Again, let $u_1,\ u_2 \in B'x$. Then, there are $v_1,\ v_2 \in S^1_G(x)$ such that

$$u_1(t) = \frac{x_0 - k(0,x_0)}{f(0,y_0)} + \int_0^t v_1(s)ds \ ;\ t \in J,$$

and

$$u_2(t) = \frac{x_0 - k(0,x_0)}{f(0,y_0)} + \int_0^t v_2(s)ds \ ;\ t \in J.$$

Now, for any $\lambda \in [0,1]$, we have

$$\begin{aligned}\lambda u_1(t)+(1-\lambda)u_2(t) &= \frac{x_0-k(0,x_0)}{f(0,y_0)}+\int_0^t(\lambda v_1(s)+(1-\lambda)v_2(s))\,ds\\ &= \frac{x_0-k(0,x_0)}{f(0,y_0)}+\int_0^t v(s)ds\end{aligned}$$

where $v(s) = \lambda v_1(s) + (1-\lambda)v_2(s) \in G(s,x(s))$, for all $s \in J$. Hence, $\lambda u_1+(1-\lambda)u_2 \in B'x$ and consequently, $B'x$ is convex for each $x \in \mathcal{C}(J,\mathbb{R})$. As a result, B' defines a multi-valued operator $B' : \mathcal{C}(J,\mathbb{R}) \longrightarrow \mathcal{P}_{cp,\ cv}(\mathcal{C}(J,\mathbb{R}))$.

Step 2: We will show that A, B, and C are single-valued Lipschitzian operators on $\mathcal{C}(J,\mathbb{R})$. Let $x, y \in \mathcal{C}(J,\mathbb{R})$. Then,

$$\begin{aligned}\|Ax-Ay\| &= \sup_{t\in J}|k(t,x(t))-k(t,y(t))|\\ &\leq \sup_{t\in J}|l_1(t)||x(t)-y(t)| \leq \|l_1\|_\infty\|x-y\|,\end{aligned}$$

which demonstrates that A is a multi-valued Lipschitzian operator on $\mathcal{C}(J,\mathbb{R})$ with the constant $\|l_1\|$. In a similar way, it can be proved that B and C are also two Lipschitzian operators on $\mathcal{C}(J,\mathbb{R})$ with the constants $\|l_2\|$ and $(\|b\|_\infty+k)$, respectively. Now, let us prove that $C(S) \subset (I-D)(S')$. To do it, let $x \in S$ be a fixed point. Let us define a mapping

$$\begin{cases}\varphi_x : \mathcal{C}(J,\mathbb{R}) \longrightarrow \mathcal{C}(J,\mathbb{R})\\ \qquad y \longrightarrow Cx+Dy.\end{cases}$$

By using hypothesis $(\mathcal{R}_{41})(a)$, we reach the result that the operator φ_x is a contraction with a constant k. Then, an application of the Banach's theorem implies that there is a unique point $y \in \mathcal{C}(J,\mathbb{R})$ such that $Cx+Dy = y$. Hence,

$$C(S) \subset (I-D)(\mathcal{C}(J,\mathbb{R})).$$

Since $y \in \mathcal{C}(J,\mathbb{R})$, we deduce the existence of $t^* \in J$, such that

$$\begin{aligned}\|y\|_\infty &= |y(t^*)| = |Cx(t^*)+Dy(t^*)|\\ &\leq \frac{1}{1+b(t^*)|x(\theta(t^*))|}+\left|p(t^*,\frac{1}{1+b(t^*)|x(\theta(t^*))|})\right|+|p(t^*,y(t^*))|\\ &\leq 1+q+2|p(t,0)|+q|y(t^*)|.\end{aligned}$$

This implies that

$$|y(t^*)| \leq \frac{1+q+2P}{1-q}.$$

Now, in view of the last inequality in Eq. (7.43), it follows that

$$q^2 + 2q + 2P - 1 \leq 0.$$

Accordingly, we have

$$1 + q + 2P \leq (2+q)(1-q)$$

or equivalently, $\|y\|_\infty \leq 2 + q$. We conclude that $C(S) \subset (I - D)(S')$.

Step 3: In this step, we have to show that B' is completely continuous on S'. First, we need to prove that B' is a compact operator on S'. To do this, it is sufficient to prove that $B'(S')$ is a uniformly bounded and equicontinuous set. For this purpose, let $u \in B'(S)$ be arbitrary. Then, there is a $v \in S^1_G(x)$, such that

$$u(t) = \frac{x_0 - k(0, x_0)}{f(0, y_0)} + \int_0^t v(s)ds,$$

for some $x \in S'$. Hence, by using assumption $(\mathcal{R}_{40})$, we have

$$\begin{aligned} u(t) &\leq \left|\frac{x_0 - k(0, x_0)}{f(0, y_0)}\right| + \int_0^t |v(s)|\, ds \\ &\leq \left|\frac{x_0 - k(0, x_0)}{f(0, y_0)}\right| + \int_0^t \|G(s, x(s))\|\, ds \\ &\leq \left|\frac{x_0 - k(0, x_0)}{f(0, y_0)}\right| + \int_0^t h(s)ds \\ &\leq \left|\frac{x_0 - k(0, x_0)}{f(0, y_0)}\right| + \|h\|_{L^1} \end{aligned}$$

for all $x \in S'$ and so, $B'(S')$ is a uniformly bounded set. Again, when we proceed with the same arguments as in Step 1, we notice that $B'(S')$ is an equicontinuous set in $\mathcal{C}(J, \mathbb{R})$. Next, we demonstrate that B' is an upper semi-continuous multi-valued mapping on $\mathcal{C}(J, \mathbb{R})$. Let $\{x_n\}$ be a sequence in $\mathcal{C}(J, \mathbb{R})$ such that $x_n \to x$. Let $\{y_n\}$ be a sequence such that $y_n \in B'x_n$ and $y_n \to y$. We will prove that $y \in B'x$. Since $y_n \in B'x_n$, then there exists a $v_n \in S^1_G(x_n)$, such that

$$y_n(t) = \frac{x_0 - k(0, x_0)}{f(0, y_0)} + \int_0^t v_n(s)ds \ ; \ t \in J.$$

We must prove that there is a $v \in S^1_G(x)$, such that

$$y(t) = \frac{x_0 - k(0, x_0)}{f(0, y_0)} + \int_0^t v(s)ds \ ; \ t \in J.$$

Consider the continuous linear operator $\mathcal{K} : L^1(J, \mathbb{R}) \longrightarrow \mathcal{C}(J, \mathbb{R})$ defined by:

$$\mathcal{K}v(t) = \int_0^t v(s)ds \ ; \ t \in J.$$

Then, $\mathcal{K} \circ S^1_G(x)$ is a closed graph operator, in view of Lemma 5.4 in [74]. Also, from the definition of $\mathcal{K}$, we have

$$y_n(t) - \frac{x_0 - k(0, x_0)}{f(0, y_0)} \in \mathcal{K} \circ S^1_G(x).$$

Since $y_n \to y$, there is a point $v \in S^1_G(x)$, such that

$$y(t) = \frac{x_0 - k(0, x_0)}{f(0, y_0)} + \int_0^t v(s)ds \ ; \ t \in J.$$

This demonstrates that B' is a u.s.c. operator on $\mathcal{C}(J, \mathbb{R})$. Moreover, B' is compact and hence, it is completely continuous multi-valued operator on $\mathcal{C}(J, \mathbb{R})$. Now, from the hypothesis $(\mathcal{R}_{40})$, it follows that

$$\begin{aligned} M &= \| \bigcup B'(I - D)^{-1} C(S) \|_{\mathcal{P}} \\ &= \sup\{ \| B'(I - D)^{-1} Cx \|_{\mathcal{P}} \ ; \ x \in S \} \\ &\leq \sup\{ \| B'x \|_{\mathcal{P}} \ ; \ x \in S \} \\ &\leq \left| \frac{x_0 - k(0, x_0)}{f(0, y_0)} \right| + \|h\|_{L^1}. \end{aligned}$$

In order to simplify the calculation, it is assumed that $x_0 = k(0, x_0)$. From the second inequality of Eq. (7.43), it follows that

$$q\left(1 + \frac{1}{4\|h\|_{L^1}\|l_2\|}\right) + \|b\|_\infty < \frac{1}{4\|h\|_{L^1}\|l_2\|}.$$

Then,

$$\frac{\|l_2\| \left(\|b\|_\infty + q\right)}{1 - q} \|h\|_{L^1} < \frac{1}{4}.$$

Consequently,

$$\|l_1\| + \frac{\|l_2\| \left(\|b\|_\infty + q\right)}{1 - q} \|h\|_{L^1} < \frac{1}{2}.$$

Finally, it remains to verify the hypothesis (iv) of Theorem 4.5.8. Let $x \in S$ and

$$y \in Ax + B(I-D)^{-1}Cx \cdot B'(I-D)^{-1}Cx.$$

Then, there is $z \in S'$ and $v \in S^1_G(z)$, such that

$$y(t) = k(t, x(t) + f(t, z(t)) \cdot \int_0^t v(s)ds \ ; \ t \in J.$$

Therefore, we have

$$\begin{aligned}
|y(t)| &\leq |k(t, x(t))| + |f(t, z(t))| \int_0^t |v(s)|ds \\
&\leq |k(t, x(t)) - k(t, 0)| + |k(t, 0)| + \Big(|f(t, z(t)) - f(t, 0)| \\
&\quad + |f(t, 0)| \Big) \|h\|_{L^1} \\
&\leq \|l_1\| |x(t)| + K + (\|l_2\| |z(t)| + F) \|h\|_{L^1} \\
&\leq (1+q)(\|l_1\| + \|l_2\| \|h\|_{L^1}) + K + \|l_2\| \|h\|_{L^1} + F\|h\|_{L^1} \\
&\leq 1 + q.
\end{aligned}$$

So, $\|y\|_\infty \leq 1 + q$ and consequently,

$$Ax + B(I-D)^{-1}Cx \cdot B'(I-D)^{-1}Cx \subset S, \quad \text{for all} \ \ x \in S.$$

We deduce that A, B, C, D, and B' meet all the requirements of Theorem 4.5.8. Now, the result follows from Theorem 4.5.8. Q.E.D.

7.6.2 Multi-valued periodic boundary value problem of first order

Let $J = [0, \mathcal{T}]$ be the closed and bounded interval in $\mathbb{R}$. Let $\mathcal{C}(J, \mathbb{R})$ be the Banach algebra of all continuous functions from J to $\mathbb{R}$ endowed with the sup-norm $\|.\|_\infty$ and defined by $\|f\|_\infty = \sup_{t \in J} |f(t)|$, for each $f \in \mathcal{C}(J, \mathbb{R})$. Consider the periodic boundary value problem for the first-order ordinary differential inclusion

$$\begin{cases} \left(\dfrac{x(t)-k(t,x(t))}{f(t,y(t))}\right)' + h(t)\left(\dfrac{x(t)-k(t,x(t))}{f(t,y(t))}\right) \in G_h(t,x(t),y(t)) \\ y(t) = \dfrac{1}{1+b(t)|x(\theta(t))|} - p(t, \dfrac{1}{1+b(t)|x(\theta(t))|}) + p(t,y(t)) \\ (x(0),y(0)) = (x(\mathcal{T}),y(\mathcal{T})) \in \mathbb{R}^2, \end{cases} \tag{7.45}$$

where $h \in L^1(J,\mathbb{R})$ is bounded and the multi-valued function $G_h : J \times \mathbb{R} \times \mathbb{R} \longrightarrow \mathcal{P}_{cp,\ cv}(\mathbb{R})$ is defined by:

$$G_h(t,x,y) = G(t,y) + h(t)\left(\frac{x-k(t,x)}{f(t,y)}\right).$$

A solution of the system (7.45) stands for two functions $x,\ y \in \mathcal{AC}(J,\mathbb{R})$ that satisfies:

(i) The function $t \longrightarrow \dfrac{x(t)-k(t,x(t))}{f(t,y(t))}$ is absolutely continuous, and

(ii) there exists a function $v \in L^1(J,\mathbb{R})$ such that $v(t) \in G(t,y(t))$ satisfying the following equality

$$\left(\frac{x(t)-k(t,x(t))}{f(t,y(t))}\right)' = v(t)\ ;\ (x(0),y(0)) = (x(\mathcal{T}),y(\mathcal{T})),$$

where $\mathcal{AC}(J,\mathbb{R})$ is the space of all absolutely continuous real-valued functions on J. The following useful lemma can be found in [132].

Lemma 7.6.3 *For any $h \in L^1(J,\mathbb{R}^+)$ and $\sigma \in L^1(J,\mathbb{R})$, x is a solution of the differential equation*

$$\begin{cases} x' + h(t)x = \sigma(t) \\ x(0) = x(\mathcal{T}) \end{cases}$$

if, and only if, x is a solution of the integral equation

$$x(t) = \int_0^{\mathcal{T}} g_h(t,s)\sigma(s)ds, \text{ where}$$

$$g_h(t,s) = \begin{cases} \dfrac{e^{H(s)-H(t)+H(\mathcal{T})}}{e^{H(\mathcal{T})}-1} & \text{if } 0 \le s \le t \le \mathcal{T} \\ \dfrac{e^{H(s)-H(t)}}{e^{H(\mathcal{T})}-1} & \text{if } 0 \le t < s \le \mathcal{T} \end{cases} \tag{7.46}$$

with $H(t) = \int_0^t h(s)ds.$

Let us consider the following hypotheses needed in the sequel.

($\mathcal{R}_{43}$) The functions $t \longrightarrow f(t,x)$ and $t \longrightarrow k(t,x)$ are periodic of period T for all $x \in \mathbb{R}$.

($\mathcal{R}_{44}$) The function $(x,y) \longrightarrow \dfrac{x - k(0,x)}{f(0,y)}$ is injective on $\mathbb{R}^2$.

Lemma 7.6.4 *Assume that the hypotheses* ($\mathcal{R}_{43}$) *and* ($\mathcal{R}_{44}$) *hold. Then, for any bounded integrable function* h *on* J, (x,y) *is a solution of the differential inclusion* (7.45) *if, and only if, it is a solution of the integral equation*

$$\begin{cases} x(t) \in k(t,x(t)) + f(t,y(t)) \cdot \displaystyle\int_0^{\mathcal{T}} g_h(t,s)G_h(s,x(s),y(s))ds \\ y(t) = \dfrac{1}{1+b(t)|x(\theta(t))|} - p(t, \dfrac{1}{1+b(t)|x(\theta(t))|}) + p(t,y(t)) \\ (x(0),y(0)) = (x(\mathcal{T}),y(\mathcal{T})) \in \mathbb{R}^2 \end{cases},$$

where the Greens function g_h *is defined in Eq.* (7.46).

In the sequel, we will also use the following assumptions.

($\mathcal{R}_{45}$) The single-valued mapping $k : J \times \mathbb{R} \longrightarrow \mathbb{R}$ is continuous and there is a bounded function $l_1 : J \longrightarrow \mathbb{R}$ with a bound $0 < \|l_1\|_\infty \leq \frac{1}{6}$ satisfying:

$$|k(t,x) - k(t,y)| \leq l_1(t)|x-y| \; ; \quad \text{for all } x,y \in \mathbb{R}.$$

($\mathcal{R}_{46}$) The single-valued mapping $f : J \times \mathbb{R} \longrightarrow \mathbb{R}\backslash\{0\}$ is continuous and there is a bounded function $l_2 : J \longrightarrow \mathbb{R}$ with a bound $0 < \|l_2\|$ satisfying:

$$|f(t,x) - f(t,y)| \leq l_2(t)|x-y| \; ; \quad \text{for all } x, \; y \in \mathbb{R}.$$

($\mathcal{R}_{47}$) The single-valued mapping $p : J \times \mathbb{R} \longrightarrow \mathbb{R}$ is such that:

(*a*) p is Lipschitzian with a constant q in $]0, \frac{1}{4}[$ with respect to the first variable,

(*b*) p is continuous with respect to the second variable, and

(*c*) $|p(t,x)| \leq q|x|$, for all $t \in J$ and $x \in \mathbb{R}$.

($\mathcal{R}_{48}$) The single-valued mapping $b : J \longrightarrow \mathbb{R}_+$ is continuous with $4\|b\|_\infty \leq 1$.

($\mathcal{R}_{49}$) The multi-valued operator $G : J \times \mathbb{R} \longrightarrow \mathcal{P}_{cp,\ cv}(\mathbb{R})$ is Carathéodory.

($\mathcal{R}_{50}$) There is a function $\varrho \in L^1(J, \mathbb{R}_+^*)$, and a nondecreasing continuous function $\varphi : \mathbb{R}_+ \longrightarrow \mathbb{R}_+$ such that

$$\|G_h(t, x, y)\|_{\mathcal{P}} \leq \varrho(t)\varphi(|y|)$$

for each $(x, y) \in \mathbb{R}^2$.

Theorem 7.6.2 *Assume that the hypotheses* ($\mathcal{R}_{43}$)–($\mathcal{R}_{50}$) *hold. Moreover, if there exists a real number* $r > 0$ *such that:*

$$\begin{cases} \|l_1\|_\infty + \|l_2\|_\infty M_h \|\varrho\|_{L^1}\varphi(r) < \frac{1}{2} \\ \\ \max\left\{K + FM_h\|\varrho\|_{L^1}\varphi(r)\ ;\ \dfrac{3}{4} + P\right\} \leq \dfrac{r}{2}, \end{cases} \tag{7.47}$$

where $\frac{1}{2} < P = \sup\{|p(t, 0)|\ ;\ t \in J\} < 1$, $K = \sup\limits_{t\in J}\{|k(t, 0)|\}$, *and* $F = \sup\limits_{t\in J}\{|f(t, 0)|\}$, *then, Eq.* (7.45) *has, at least, a solution.*

Proof. Let us define an open ball $\overline{B}_r(0)$ in $\mathcal{C}(J, \mathbb{R})$, centered at the origin and with radius r, where the real number r satisfies the inequalities of Eq. (7.47). Consider the mapping A, B, C, D, and B' on $\mathcal{C}(J, \mathbb{R})$ by:

(*i*) $Ax(t) = k(t, x(t))$, $\quad Bx(t) = f(t, x(t))$, $Dx(t) = p(t, x(t))$,

(*ii*) $Cx(t) = \dfrac{1}{1 + b(t)|x(\theta(t))|} - p(t, \dfrac{1}{1 + b(t)|x(\theta(t))|})$, and

(*iii*) $B'x = \left\{u \in \mathcal{C}(J, \mathbb{R}) : u(t) = R(x_0, y_0) + \displaystyle\int_0^t g_h(t, s)v(s)ds\ ;\ v \in S^1_{G_h}(x)\right\}$

for all $t \in J$, where $R(x_0, y_0) := \dfrac{x_0 - k(0, x_0)}{f(0, y_0)}$. Then, the system IVP (7.42) is equivalent to the operator inclusion

$$\begin{cases} x(t) \in Ax(t) + By(t) \cdot B'y(t) \\ \\ y(t) = Cx(t) + Dy(t). \end{cases}$$

We will show that A, B, C, D, and B' meet all the conditions of Theorem 4.5.8 on $\overline{B}_r(0)$. Since $S^1_{G_h}(x) \neq \emptyset$ for each $x \in \overline{B}_r(0)$, it follows that B' is well defined. We start by showing that B' defines a multi-valued operator B' :

$\overline{B}_r(0) \longrightarrow \mathcal{P}_{cp,\ cv}(\mathcal{C}(J,\mathbb{R}))$. Let $u_1,\ u_2 \in B'x$, Then, there are $v_1, v_2 \in S^1_{G_h}(x)$ such that

$$u_1(t) = \frac{x_0 - k(0,x_0)}{f(0,y_0)} + \int_0^t g_h(t,s)v_1(s)ds \ ; \ t \in J,$$

and

$$u_2(t) = \frac{x_0 - k(0,x_0)}{f(0,y_0)} + \int_0^t g_h(t,s)v_2(s)ds \ ; \ t \in J.$$

Now, for any $\lambda \in [0,1]$, we have

$$\begin{aligned}\lambda u_1(t) + (1-\lambda)u_2(t) &= R(x_0,y_0) + \int_0^t g_h(t,s)\left(\lambda v_1(s) + (1-\lambda)v_2(s)\right)ds \\ &= \frac{x_0 - k(0,x_0)}{f(0,y_0)} + \int_0^t v(s)ds,\end{aligned}$$

where $v(s) = \lambda v_1(s) + (1-\lambda)v_2(s) \in G_h(s,x(s),y(s))$ for all $s \in J$. Hence, $\lambda u_1 + (1-\lambda)u_2 \in B'x$ and consequently, $B'x$ is convex for each $x \in \mathcal{C}(J,\mathbb{R})$. Then, B' defines a multi-valued operator $B' : \mathcal{C}(J,\mathbb{R}) \longrightarrow \mathcal{P}_{cp,\ cv}(\mathcal{C}(J,\mathbb{R}))$. Proceeding as in the proof of Theorem 7.6.1, we deduce that the single-valued operators A, B, and C are Lipschitzian with the constants $\|l_1\|_\infty$, $\|l_2\|_\infty$ and $\|b\|_\infty + q$, respectively. Under the assumption $(\mathcal{R}_{47})(a)$, we get $C(\overline{B}_r(0)) \subset (I-D)(\mathcal{C}(J,\mathbb{R}))$. Since $y \in \mathcal{C}(J,\mathbb{R})$, then there exists $t^* \in J$ such that

$$\|y\|_\infty = |Cx(t^*) + Dy(t^*)| \leq \frac{1+q+2P}{1-q} \leq r.$$

As a result, $y \in \overline{B}_r(0)$ and consequently, $C(\overline{B}_r(0)) \subset (I-D)(\overline{B}_r(0))$. Next, we show that B' is completely continuous on $\overline{B}_r(0)$. First, let us prove that $B'(\overline{B}_r(0))$ is a totally bounded subset of $\mathcal{C}(J,\mathbb{R})$. To do this, it is enough to prove that $B'(\overline{B}_r(0))$ is a uniformly bounded and an equicontinuous set in $\mathcal{C}(J,\mathbb{R})$. For this purpose, let $u \in B'(\overline{B}_r(0))$ be arbitrary. Then, there is a $v \in S^1_{G_h}(x)$ such that

$$u(t) = \frac{x_0 - k(0,x_0)}{f(0,y_0)} + \int_0^t g_h(t,s)v(s)ds \ ; \ t \in J.$$

Hence,

$$\begin{aligned}
|u(t)| &\leq \left|\frac{x_0 - k(0,x_0)}{f(0,y_0)}\right| + \int_0^t g_h(t,s)\,\|G_h(s,x(s),y(s))\|_{\mathcal{P}}\,ds \ ; \ t \in J, \\
&\leq \left|\frac{x_0 - k(0,x_0)}{f(0,y_0)}\right| + \int_0^t g_h(t,s)\varrho(s)\varphi(|y(s)|)ds \ ; \ t \in J, \\
&\leq \left|\frac{x_0 - k(0,x_0)}{f(0,y_0)}\right| + \int_0^t g_h(t,s)\|\varrho\|_{L^1}\varphi(r)ds \ ; \ t \in J, \\
&\leq \left|\frac{x_0 - k(0,x_0)}{f(0,y_0)}\right| + M_h\|\varrho\|_{L^1}\varphi(r) \ ; \ t \in J.
\end{aligned}$$

Consequently, $B'(\overline{B}_r(0))$ is a uniformly bounded set in $\mathcal{C}(J,\mathbb{R})$. Finally, it is sufficient to show that $B'(\overline{B}_r(0))$ is an equicontinuous set. Indeed, for any t_1, $t_2 \in [0,\mathcal{T}]$, we have

$$\begin{aligned}
|B'x(t_1) - B'x(t_2)| &\leq \left(\int_0^t \frac{\partial}{\partial t} g_h(t,s)\,\|G_h(s,x(s),y(s))\|_{\mathcal{P}}\,ds\right)|t_1 - t_2| \\
&\leq \left(\int_0^t (-h(t))\|\varrho\|_{L^1}\varphi(r)ds\right)|t_1 - t_2| \\
&\leq \left(\max_{t\in J}(h(t))M_h\,\|\varrho\|_{L^1}\varphi(r)\right)|t_1 - t_2|.
\end{aligned}$$

This shows with the Arzelà–Ascoli theorem, that $B'(\overline{B}_r(0))$ is totally bounded. Next, we demonstrate that B' is an upper semi-continuous multi-valued mapping on $\mathcal{C}(J,\mathbb{R})$. Let $\{x_n\}$ be a sequence in $\mathcal{C}(J,\mathbb{R})$ such that $x_n \to x$. Let $\{y_n\}$ be a sequence such that $y_n \in B'x_n$ and $y_n \to y$. We will prove that $y \in B'x$. Since $y_n \in B'x_n$, there exists a $v_n \in S^1_G(x_n)$ such that

$$y_n(t) = \frac{x_0 - k(0,x_0)}{f(0,y_0)} + \int_0^t g_h(t,s)v_n(s)ds \ ; \ t \in J.$$

We must prove that there exists a $v \in S^1_{G_h}(x)$ such that

$$y(t) = \frac{x_0 - k(0,x_0)}{f(0,y_0)} + \int_0^t g_h(t,s)v(s)ds \ ; \ t \in J.$$

Consider the continuous linear operator $\mathcal{K} : L^1(J,\mathbb{R}) \longrightarrow \mathcal{C}(J,\mathbb{R})$ defined by:

$$\mathcal{K}v(t) = \int_0^t g_h(t,s)v(s)ds \ ; \ t \in J.$$

Then, $\mathcal{K} \circ S^1_{G_h}(x)$ is a closed graph operator, in view of Lemma 5.4 in [74]. Also from the definition of $\mathcal{K}$ we have

$$y_n(t) - \frac{x_0 - k(0,x_0)}{f(0,y_0)} \in \mathcal{K} \circ S^1_{G_h}(x).$$

Since $y_n \to y$, there is a point $v \in S^1_{G_h}(x)$ such that

$$y(t) = \frac{x_0 - k(0,x_0)}{f(0,y_0)} + \int_0^t g_h(t,s)v(s)ds \ ; \ t \in J.$$

This demonstrates that B' is a u.s.c. operator on $\mathcal{C}(J,\mathbb{R})$. Hence, B' is a completely continuous multi-valued operator on $\mathcal{C}(J,\mathbb{R})$. Now, it remains to verify the hypothesis (iv) of Theorem 4.5.8 on $\overline{B}_r(0)$. it follows that

$$\begin{aligned} M &= \sup\{\|B'(I-D)^{-1}Cx\|_{\mathcal{P}} \ ; \ x \in S\} \\ &\leq \sup\{\|B'x\|_{\mathcal{P}} \ ; \ x \in S\} \\ &\leq \left|\frac{x_0 - k(0,x_0)}{f(0,y_0)}\right| + \int_0^t |g_h(t,s)|\|G_h(s,x(s),y(s))\|_{\mathcal{P}} ds \\ &\leq \left|\frac{x_0 - k(0,x_0)}{f(0,y_0)}\right| + M_h\|\varrho\|_{L^1}\varphi(r). \end{aligned}$$

In order to simplify the calculation, it is assumed that $x_0 = k(0,x_0)$. From the first inequality of Eq. (7.47), it follows that

$$\frac{\|l_2\|\left(\|b\|_\infty + q\right)}{1-q} M + \|l_1\|_\infty < \frac{\|b\|_\infty + q}{1-q}\left(\frac{1}{2} - \|l_1\|_\infty\right) + \|l_1\|_\infty < \frac{1}{2}.$$

Finally, it remains to verify the hypothesis (iv) of Theorem 4.5.8. Let $x \in S$ and

$$y \in Ax + B(I-D)^{-1}Cx \cdot B'(I-D)^{-1}Cx.$$

Then, there is $z \in \overline{B}_r(0)$ and $v \in S^1_{G_h}(z)$ such that

$$y(t) = k(t,x(t) + f(t,z(t)) \cdot \int_0^t g(h,s)v(s)ds \ ; \ t \in J.$$

Therefore, we have

$$\begin{aligned}
|y(t)| &\leq |k(t,x(t))| + |f(t,z(t))| \int_0^t g(h,s)|v(s)|ds \\
&\leq |k(t,x(t)) - k(t,0)| + |k(t,0)| + \Big(|f(t,z(t)) - f(t,0)| \\
&\quad + |f(t,0)|\Big) M_h \|\varrho\|_{L^1}\varphi(r) \\
&\leq \|l_1\||x(t)| + K + (\|l_2\|_\infty |z(t)| + F) M_h \|\varrho\|_{L^1}\varphi(r) \\
&\leq (\|l_1\|_\infty + \|l_2\|_\infty M_h \|\varrho\|_{L^1}\varphi(r))\, r + K + F M_h \|\varrho\|_{L^1}\varphi(r) \\
&< r.
\end{aligned}$$

So, $\|y\|_\infty \leq r$, and consequently,

$$Ax + B(I-D)^{-1}Cx \cdot B'(I-D)^{-1}Cx \subset \overline{B}_r(0), \quad \text{for all } \; x \in \overline{B}_r(0).$$

We deduce that A, B, C, D, and B' satisfy the requirements of Theorem 4.5.8. Now, the result follows from Theorem 4.5.8. Q.E.D.

Bibliography

[1] Agarwal, R. P., Hussain, N., Taoudi, M. A.: Fixed point theorems in ordered Banach spaces and applications to nonlinear integral equations. *Abstr. Appl. Anal.,* Art. ID 245872, 15 pp. MR2947757 (2012).

[2] Agarwal, R. P., Meehan, M., O'Regan, D.: *Fixed Point Theory and Applications.* Cambridge University Press, Cambridge (2001).

[3] Agarwal, R. P., O'Regan, D., Sahu, D. R.: *Fixed Point Theory for Lipschitzian type Mappings with Applications.* Springer, New York (2009).

[4] Agarwal, R. P., O'Regan, D., Taoudi, M. A.: Fixed point theorems for (ws)-compact mappings in Banach spaces. *Fixed Point Theory Appl.* 1–13 (2010).

[5] Agarwal, R. P., O'Regan, D., Taoudi, M. A.: Fixed point theorems for convex-power condensing operators relative to the weak topology and applications to Volterra integral equations. *J. Integral Equations Appl.* 24: 167–181 (2012).

[6] Akhmerov, P. P., Kamenskii, M. I., Potapov, A. S., Sadovskii, B. N.: *Measures of Noncompactness and Condensing Operators,* Birkhäuser (1992).

[7] Alexiewcz, A.: *Analiza funkcjionala.* PWN, Warszawa (2001).

[8] Appell, J.: Measures of noncompactness, condensing operators and fixed points: An application-oriented survey. *Fixed Point Theory* 6: 157–229 (2005).

[9] Appell, J., De Pascale, E.: Su alcuni parametri connessi con la misura di non compattezza di Haussdorff in spazi di funzioni misurabili. *Boll. Unione Mat. Ital. Sez. B* 6: 497–515 (1984).

[10] Arino, O., Gautier, S., Penot, J. P.: A fixed point theorem for sequentially continuous mapping with application to ordinary differential equations. *Functional. Ekvac.* 27: 273–279 (1984).

[11] Atkinson, F. V., Langer, H., Mennicken, R., Shkalikov, A. A.: The essential spectrum of some matrix operators. *Math. Nachr.* 167: 5–20 (1994).

[12] Averna, D., Marano, S. A.: Existence of solutions for operator inclusions: A unified approach. *Rendiconti del Seminario Matematico della Universit di Padova* 102: 285–303 (1999).

[13] Ball, J. M.: Weak continuity properties of mapping and semi-groups. *Proc. Royal Soc. Edinburgh Sect. A* 72: 275–280 (1979).

[14] Ball, G.: Diffusion approximation of radiative transfer equations in a channel. *Tran. Theo. Stat. Phys.* 30: 269–293 (2001).

[15] Banas, J., Lecko, L.: Fixed points of the product of operators in Banach algebras. *Panamer. Math. J.* 12: 101–109 (2002).

[16] Banas, J., Olszowy, L.: On a class of measures of non-compactness in Banach algebras and their application to nonlinear integral equations. *Z. Anal. Anwend.* 28: 475–498 (2009).

[17] Banas, J., O'Regan, D.: On existence and local attractivity of solutions of a quadratic integral equation of fractional order. *J. Math. Anal. Appl.* 345: 573–582 (2008).

[18] Banas, J., Rivero, J.: On measures of weak noncompactness. *Ann. Mat. Pura Appl.* 151: 213–224 (1988).

[19] Banas, J., Rzepka, B.: Monotonic solutions of a quadratic integral equation of fractional order. *J. Math. Anal. Appl.* 332: 1371–1379 (2007).

[20] Banas, J., Sadarangani, K.: Solutions of some functional-integral equations in Banach algebras. *Math. Comput. Modelling* 38: 245–250 (2003).

[21] Banas, J., Taoudi, M. A.: Fixed points and solutions of operator equations for the weak topology in Banach algebras. *Taiwanese Journal of Mathematics* 18: 871–893 (2014).

[22] Barroso, C. S.: Krasnoselskii's fixed point theorem for weakly continuous maps. *Nonlinear Anal.* 55: 25–31 (2003).

[23] Barroso, C. S., Teixeira, E. V.: A topological and geometric approach to fixed point results for sum of operators and applications. *Nonlinear Anal.* 60: 625–660 (2005).

[24] Bátkai, A., Binding, P., Dijksma, A., Hryniv, R., Langer, H.: Spectral problems for operator matrices. *Math. Nachr.* 278: 1408–1429 (2005).

[25] Ben Ali, N., Jeribi, A., Moalla, N.: On some subsets of Schechter's essential spectrum of a matrix operator and application to transport operator. *Math. Nachr.* 283: 1245–1256 (2010).

[26] Ben Amar, A., Chouayekh, S., Jeribi, A.: New fixed point theorems in Banach algebras under weak topology features and applications to nonlinear integral equations. *J. Funct. Anal.* 259: 2215–2237 (2010).

[27] Ben Amar, A., Chouayekh, S., Jeribi, A.: Fixed-point theory in a new class of Banach algebras and applications. *Afr. Mat.* 24: 705–724 (2013).

[28] Ben Amar, A., Chouayekh, S., Jeribi, A.: Fixed point results in Banach algebras for weakly convex-power condensing operators and applications. Preprint (2015).

[29] Ben Amar, A., Feki, I., Jeribi, A.: Krasnoselskii–Schaefer type fixed point theorems for weakly sequentially cotinuous mappings. *Fixed Point Theory* (2015).

[30] Ben Amar, A., Jeribi, A., Krichen, B.: Essential spectra of a 3×3 operator matrix and an application to three-group transport equations. *Integr. Equ. Oper. Theory* 68: 1–21 (2010).

[31] Ben Amar, A., Jeribi, A., Krichen, B.: Fixed point theorems for block operator matrix and an application to a structured problem under boundary conditions of Rotenberg's model type. *Math. Slovaca* 64: 155–174 (2014).

[32] Ben Amar, A., Jeribi, A., Mnif, M.: On generalization of the Schauder and Krasnosel'skii fixed point theorems on Dunford–Pettis space and applications. *Math. Meth. Appl. Sci.* 28: 1737–1756 (2005).

[33] Ben Amar, A., Jeribi, A., Mnif, M.: Some fixed point theorems and application to biological model. *Numer. Funct. Anal. Optim.* 29: 1–23 (2008).

[34] Ben Amar, A., Jeribi, A., Moalla, R.: Leray–Schauder alternatives in Banach algebra involving three operators with applications. *Fixed Point Theory* 2: 359–372 (2014).

[35] Ben Amar, A., Mnif, M.: Leray–Schauder alternatives for weakly sequentially continuous mappings and application to transport equation. *Math. Methods Appl. Sci.* 33: 80–90 (2010).

[36] Ben Amar, A., Sikorska-Nowak. A.: On some fixed point theorems for 1-set weakly contractive multi-valued mappings with weakly sequentially closed graph. *Adv. Pure. Math.* 1: 163–169 (2011).

[37] Bonanno, G., Marano, S. A.: Positive solutions of elliptic equations with discontinuous nonlinearities. *Topological Methods in Nonlinear Analysis* 8: 263–273 (1996).

[38] Boulanouar, M., Leboucher, L.: Une équation de transport dans la dynamique de population cellulaires. *C. R. Acad. Sci. Paris, t. 321: Série I,* 305–308 (1995).

[39] Boyd, D.W., Wong, J.S.W.: On nonlinear contractions. *Proc. Amer. Math. Soc.* 20: 458–464 (1969).

[40] Brezis, H.: *Analyse fonctionnelle.* Tho. et Appl., Masson, Paris (1983).

[41] Brooks, R. M., Schmitt, K.: The contraction mapping principle and some applications. *Electronic Journal of Differential Equations,* Monograph 09 (2009).

[42] Brouwer, L. E. J.: Uber Abbildungen von Mannigfaltigkeiten, *Math. Ann.* 71: 97–115 (1912).

[43] Browder, F. E.: *Problèmes Non-Linéaires.* Séminaire Math. Supérieures 1965, Presse de l'université de Montreal (1966).

[44] Browder, F. E.: Nonlinear operators and nonlinear equations of evolution in Banach spaces. Nonlinear functional analysis (*Proc. Sympos. Pure Math.,* Vol. XVIII, Part 2, Chicago, Ill., 1968). Amer. Math. Soc., Providence, R. I, 1–308 (1976).

[45] Brown, R. F., Furi, M., Gorniewicz, L., Jiang, B.: *Handbook of Topological Fixed Point Theory: Fixed Point Theory.* Springer, Dordrecht (2005).

[46] Burton, T. A.: *Volterra Integral and Differential Equations.* Academic Press, New York (1983).

[47] Burton, T. A.: Integral equations, implicit functions, and fixed points. *Proc. Amer. Math. Soc.* 124: 2383–2390 (1996).

[48] Burton, T. A.: A fixed point theorem of Krasnoselskii. *Appl. Math. Lett.* 11: 85–88 (1998).

[49] Caballero, J., Lopez, B., Sadarangani, K.: On solutions of an integral equation related to traffic flow on unbounded domains. *Arch. Math.* (Basel) 82: 551–563 (2004).

[50] Caballero, J., Lopez, B., Sadarangani, K.: Existence of non decreasing and continuous solutions of an integral equation with linear modification of the argument. *Acta Math. Sin.* (Engl. Ser.) 23: 1719–1728 (2007).

[51] Cain, Jr., G. L., Nashed, M. Z.: Fixed points and stability for a sum of two operators in locally convex spaces. *Pacific J. Math.* 39: 581–592 (1971).

[52] Cessenat, M.: Théorèmes de trace L_p pour des espaces de fonctions de la neutronique. *C. R. Acad. Sci. Paris, t. 299, Série I* 831–834 (1984).

[53] Cessenat, M.: Théorèmes de trace pour des espaces de fonctions de la neutronique. *C. R. Acad. Sci. Paris, t. 300, Série I* 89–92 (1985).

[54] Chang, S. S., Cho, Y. J., Huang, N. J.: Coupled fixed point theorems with applications. *J. Korean Math. Soc.* 33: 575–585 (1996).

[55] Charfi, S., Jeribi, A.: On a characterization of the essential spectra of some matrix operators and applications to two-group transport operators. *Math. Z.* 262: 775–794 (2009).

[56] Charfi, S., Jeribi, A., Walha, I.: Essential spectra, matrix operator and applications. *Acta Appl. Math.* 111: 319–337 (2010).

[57] Chow, S. N., Hale, J. K.: *Methods of Bifurcations Theory, Grundlehren der Mathematischen Wissenschaften* 251, Springer-Verlag, New York-Berlin, (1982).

[58] Covitz, H., Nadler jr., S. B.: Multivalued contraction mappings in generalized metric spaces. *Israel J. Math.* 8: 5-11 (1970).

[59] Damak, M., Jeribi, A.: On the essential spectra of some matrix operators and applications. *Electron. J. Differential Equations* 11: 1–16 (2007).

[60] Darbo, G.: Punti uniti in transformazioni a condominio non compatto, *Rend. Sem. Math. Univ. Padova* 24: 84-92 (1955).

[61] Darwish, M. A.: On existence and asymptotic behaviour of solutions of a fractional integral equation with linear modification of the argument, arXiv: 0805.1422v1 (2008).

[62] Darwish, M. A., Henderson, J.: Existence and asymptotic stability of solutions of a perturbed quadratic fractional integral equation. *Fract. Calc. Appl. Anal.* 12: 71–86 (2009).

[63] Dautray, R., Lions, J. L.: *Analyse Mathématiques et Calcul Numérique.* Masson, Paris, Tome 9 (1988).

[64] De Blasi, F. S.: On a property of the unit sphere in a Banach space. *Bull. Math. Soc. Sci. Math. R. S. Roumanie* (N.S.) 21: 259–262 (1977).

[65] Deimling, K.: *Multi-Valued Differential Equations.* De Gruyter, Berlin, (1998).

[66] Dhage, B. C.: On some variants of Schauder's fixed point principle and applications to nonlinear integral equations. *J. Math. Phys. Sci.* 22: 603–611 (1988).

[67] Dhage, B. C.: On a fixed point theorem of Krasnoselskii–Schaefer type. *Electronic J. Qualitative Theory of Differential Equations* 6: 1–9 (2002).

[68] Dhage, B. C.: Local fixed point theory for sum of two operators. *Fixed Point Theory* 4: 49–60 (2003).

[69] Dhage, B. C.: Multivalued operators and fixed-point theorems in Banach algebras. *Comput. Math. Appl.* 48: 1461–1476 (2004).

[70] Dhage, B. C.: A fixed point theorem in Banach algebras involving three operators with applications. *Kyungpook Math. J.* 44: 145–155 (2004).

[71] Dhage, B. C.: On a fixed point theorem in Banach algebras with applications. *Appl. Math. Lett.* 18: 273–280 (2005).

[72] Dhage, B. C.: Existence of extremal solutions for discontinuous functional integral equations. *Appl. Math. Lett.* 19: 881–886 (2006).

[73] Dhage, B. C.: On some nonlinear alternatives of Leray–Schauder type and functional integral equations. *Arch. Math.* (Brno) 42: 11–23 (2006).

[74] Dhage, B. C.: On solvability of operator inclusions $x \in AxBx + Cx$ in Banach algebras and differential inclusions. *Commun. Appl. Anal.* 14: 567–596 (2010).

[75] Dhage, B. C.: Some generalizations of multi-valued version of Schauder's fixed point theorem with applications. [Some generalizations of multi-valued version of Schauder's fixed point theorem with applications]. *Cubo* 12: 139–151 (2010).

[76] Dhage, B. C., Ntouyas, S. K.: An existence theorem for nonlinear functional integral equations via a fixed point theorem of Krasnoselskii–Schaefer type. *Nonlinear Digest* 9: 307–317 (2003).

[77] Dhage, B. C., O'Regan, D.: A fixed point theorem in Banach algebras with applications to functional integral equations. *Funct. Differ. Equ.* 7: 259–267 (2000).

[78] Diestel, J.: A survey of results related to Dunford–Pettis property. In *Cont. Math.* 2, Amer. Math. Soc. Conf. on Integration, Topology and Geometry in Linear Spaces. 15–60 (1980).

[79] Djebali, S., Górniewicz, L., Ouahab, A.: *Solution Sets for Differential Equations and Inclusions.* De Gruyter Series in Nonlinear Analysis and Applications, 18. Walter de Gruyter, Berlin. (2013).

[80] Djebali, S., Sahnoun, Z.: Nonlinear alternatives of Schauder and Krasnosel'skij types with applications to Hammerstein integral equations in L^1 spaces, *J. Differential Equations* 249: 2061–2075 (2010).

[81] Dobrakov, I.: On representation of linear operators on $C_0(T, X)$. *Czechoslovak Math. J.* 21: 13–30 (1971).

[82] Dodds, P., Fremlin, J.: Compact operator in Banach lattices. *Isra. J. Math.* 34: 287–320 (1979).

[83] Dunfort, N., Schwartz, J. T.: *Linear Operators:* Part I. Intersciences. (1958).

[84] Engel, Klaus. J.: Operator matrices and systems of evolution equations. Manuscript, Tübingen (1996).

[85] Furi, M., Pera, P.: A continuation method on locally convex spaces and applications to ordinary differential equations on noncompact intervals. *Ann. Polon. Math.* 47: 331–346 (1987).

[86] Garcia-Falset, J: Existence of fixed points and measure of weak noncompactness. *Nonlinear Anal.* 71: 2625–2633 (2009).

[87] Garcia-Falset, J., Latrach, K.: Krasnoselskii-type fixed-point theorems for weakly sequentially continuous mappings *Bull. Lond. Math. Soc.* 44: 25–38 (2012).

[88] Granas, A., Dugundji, J.: *Fixed Point Theory* Springer-Verlag, New York (2003).

[89] Gorniewicz, L.: *Topological Fixed Point Theory of Multivalued Mappings.* Springer, Berlin (2009).

[90] Gowda, M. S., Isac, G.: Operators of class $(S)^1_+$, Altman's condition and the complementarity problem. *J. Fac. Sci. Univ. Tokyo, Sect. IA, Math.* 40: 1–16 (1993).

[91] Grothendieck, A.: Sur les applications linéaires faiblement compactes d'espaces du type $C(K)$. *Canad. J. Math.* 5: 129–173 (1953).

[92] Guo, D., Lakshmikantham, V.: *Nonlinear Problems in Abstract Cones.* Notes and Reports in Mathematics in Science and Engineering, 5. Academic Press, Inc., Boston, MA (1988).

[93] Guo, D., Lakshmikantham, V., Liu, X.: *Nonlinear Integral Equations in Abstract Cones.* Kluwer Academic Publishers (1996).

[94] Heikkilä, S., Lakshmikantham, V.: *Monotone Iterative Techniques for Discontinuous Nonlinear Differential Equations.* Monographs and Textbooks in Pure and Applied Mathematics, 181. Marcel Dekker, Inc., New York (1994).

[95] Himmelberg, J.: Fixed points of multifunctions. *Journal of Mathematical Analysis and Applications* 38: 205–207 (1972).

[96] Hong-Bo, S.: The fixed point theorem of convex-power condensing and its applications in locally convex spaces. *J. Southwest Univ.* 29: 1–5 (2007).

[97] Hu, S., Khavanin, M., Zhuang, W.: Integral equations arising in the kinetic theory of gases. *Applicable Analysis* 34: 261–266 (1989).

[98] Hu, S., Papageorgiou, N. S.: *Handbook of Multivalued Analysis.* Vol. I, *Mathematics and Its Applications,* Kluwer Academic Publishers, Dordrecht, The Netherlands (1997).

[99] Jeribi, A.: A nonlinear problem arising in the theory of growing cell populations. *Nonl. Anal.: Real World Applications* 3: 85–105 (2002).

[100] Jeribi, A.: *Spectral Theory and Applications of Linear Operators and Block Operator Matrices.* Springer-Verlag, New-York (2015).

[101] Jeribi, A., Kaddachi, N., Krichen, B.: Existence results for a system of nonlinear functional integral equations in Banach algebras under weak topology. To appear in *Fixed point Theory* (2015).

[102] Jeribi, A., Kaddachi, N., Krichen, B.: Existence Results for a coupled system of perturbed functional differential inclusions in Banach algebras. Preprint (2015).

[103] Jeribi, A., Krichen, B., Mefteh, B.: Existence of solutions of a two-dimensional boundary value problem for a system of nonlinear equations arising in growing cell populations. *J. Biol. Dyn.* 7: 218–232 (2013).

[104] Jeribi, A., Krichen, B., Mefteh, B.: Existence of solutions of a nonlinear Hammerstein integral equation. *Numer. Funct. Anal. Optim.* 35: 1328–1339 (2014).

[105] Jeribi, A., Krichen, B., Mefteh, B.: Fixed point theory in WC–Banach algebras. Preprint (2015).

[106] Jeribi, A., Moalla, N., Walha, I.: Spectra of some block operator matrices and application to transport operators. *J. Math. Anal. Appl.* 351: 315–325 (2009).

[107] Jeribi, A., Walha, I.: Gustafson, Weidmann, Kato, Wolf, Schechter and Browder essential spectra of some matrix operator and application to two-group transport equation. *Math. Nachr.* 284: 67–86 (2011).

[108] Kaddachi, N., Jeribi, A., Krichen, B.: Fixed point theorems of block operator matrices On Banach algebras and an application to functional integral equations. *Math. Meth. Appl. Sci.* 36: 659–673 (2013).

[109] Kakutani, S.: A generalization of Brouwer's fixed point theorem. *Duke Math. J.* 8: 457–459 (1941).

[110] Krasnosel'skii, M. A., Zabreyko, P. P., Pustylnik, E. L., Sobolevski, P. E.: *Integral Operators in Space of Summable Functions.* Leyden: Noordhoff International Publishing (1976).

[111] Krichen, B.: *Spectral Properties and Fixed Point Theory for Block Operator Matrix: Fixed Point Theory, Spectral Properties for Block Operator Matrix and Applications to Transport Equation.* Scholars' Press (2014).

[112] Kubiaczyk, I.: On a fixed point theorem for weakly sequentially continuous mapping. *Discuss. Math., Differ. Incl.* 15: 15–20 (1995).

[113] Kubiaczyk, I., Szufla, S.: Kneser's theorem for weak solutions of ordinary differential equations in Banach spaces. *Publ. Inst. Math.* (Beograd) (N.S.) 46: 99–103 (1982).

[114] Kuratowski, K.: Sur les espaces complets. *Fund. Math.* 15: 301-309 (1930).

[115] Latrach, K.: On a nonlinear stationary problem arising in transport theory. *J. Math. Phys.* 37: 1336–1348 (1996).

[116] Latrach, K.: Compactness results for transport equations and applications. *Math. Mode. Meth. Appl. Sci.* 11: 1182–1202 (2001).

[117] Latrach, K., Jeribi, A.: Sur une équation de transport intervenant en dynamique des populations. *C. R. Acad. Sci. Paris, t. 325 Série I* 1087–1090 (1997).

[118] Latrach, K., Jeribi, A.: A nonlinear boundary value problem arising in growing cell populations. *Nonlinear Analysis, T.M.A.* 36: 843–862 (1999).

[119] Latrach, K., Taoudi, M. A., Zeghal, Z.: Some fixed point theorems of the Schauder and the Krasnosel'skii type and application to nonlinear transport equations. *Differential Equations* 221: 256–271 (2006).

[120] Lasota, A., Opial, Z.: An application of the Kakutani–Ky Fan theorem in the theory of ordinary differential equations, *Bull. Acad. Pol. Sci. Ser. Sci. Math. Astronom. Phy.* 13: 781–786 (1965).

[121] Lebowitz, J. L., Rubinow, S. I.: A theory for the age and generation time distribution of microbial population. *J. Math. Bio.* 1: 17–36 (1974).

[122] Legget, R. W.: On certain nonlinear integral equations. *J. Math. Anal. Appl.* 57: 462–468 (1977).

[123] Leray, J., Schauder, J.: Topologie et équations fonctionnelles. *Annales Scientifiques de l'Ecole Normale Supérieure* 51: 45–78 (1934).

[124] Liu, Y. C., Li, Z. X.: Schaefer type theorem and periodic solutions of evolution equations. *J. Math. Anal. Appl.* 316: 237–255 (2006).

[125] Liu, Y. C., Li, Z. X.: Krasnoselskii type fixed point theorems and applications. *Proc. Amer. Math. Soc.* 136: 1213–1220 (2008).

[126] Lods, B.: On linear kinetic equations involving unbounded cross-sections. *Math. Mode. Meth. Appl. Sci.* 27: 1049–1075 (2004).

[127] Majorana, A., Marano, S. A.: Continuous solutions of a nonlinear integral equation on an unbounded domain. *J. Integral Equations Appl.* 6: 119–128 (1994).

[128] Megginson, Robert E.: *An Introduction to Banach Space Theory.* Graduate Texts in Mathematics, Springer Verlag (1988).

[129] Moalla, N., Damak, M., Jeribi, A.: Essential spectra of some matrix operators and application to two-group transport operators with general boundary conditions. *J. Math. Anal. Appl.* 323: 1071–1090 (2006).

[130] Moreira, D. R., Teixeira, E. V. O.: Weak convergence under nonlinearities. *Ann. Braz. Acad. Sci.* 75: 9–19 (2003).

[131] Nashed, M. Z., Wong, J. S. W.: Some variants of a fixed point theorem of Krasnoselskii and applications to nonlinear integral equations. *J. Math. Mech.* 18: 767–777 (1969).

[132] Nieto, J. J.: Periodic boundary value problems for first order impulsive ODEs. *Nonlinear Anal.* 51: 1223–1232 (2002).

[133] O'Regan, D.: Existence results for nonlinear Volterra and Hammerstein integral equations. Dynamical systems and applications. (Edited by R. P. Agarwal) World Scientific Series in Applicable Analysis, River Edge, New Jersey 4: 601–615 (1995).

[134] O'Regan, D.: Fixed point theory for weakly sequentially continuous mappings. *Math. Comput. Modelling.* 27: 1–14 (1998).

[135] O'Regan, D.: Fixed point theorems for weakly sequentially closed maps. *Archivum Mathematicum* (Brno) *Tomus*, 36: 61–70 (2000).

[136] O'Regan, D., Meehan, M.: *Existence Theory for Nonlinear Integral and Integrodifferential Equations.* Kluwer Academic Publishers, Dordrecht (1998).

[137] Palmucci, M., Papalini, F.: Periodic and boundary value problems for second order differential inclusions. *Journal of Applied Mathematics and Stochastic Analysis.* 14: 161–182 (2001).

[138] Perthame, B.: *Transport Equations in Biology.* Frontiers in Mathematics. Birkhäuser Verlag, Basel (2007).

[139] Petryshyn, W. V.: Structure of the fixed points sets of k-set-contractions. *Archive for Rational Mechanics and Analysis.* 40: 312–328 (1971).

[140] Petryshyn, W. V.: Fixed point theorems for various classes of 1-set-contractive and 1-ball-contractive mappings in Banach spaces. *Transactions of the American Mathematical Society.* 182: 323–352 (1973).

[141] Potter, A. J. B.: An elementary version of the Leray–Schauder theorem. *Journal of the London Mathematical Society.* 5: 414–416 (1972).

[142] Rotenberg, M.: Transport theory for growing cell populations. *J. Theo. Biol.* 103: 181–199 (1983).

[143] Rybinski, L.: An application of the continuous selection theorem to the study of the fixed points of multivalued mappings. *J. Math. Anal. Appl.* 153: 391–396 (1990).

[144] Schaefer, H. H.: Neue Existenzstze in der theorie nichtlinearer Integragleichungen. *Sitzungsbh. Sächs. Akademie Wiss. Leipzig Matematisch-Naturwissenschaftliche* 101: 7–40 (1955).

[145] Schaefer, H. H.: Uber Methode der a priori-Schranken. *Math. Ann.* 129: 415–416 (1955).

[146] Schaefer, H. H.: *Topological Vector Spaces.* Macmillan Company: New York, Dordrecht (1966).

[147] Schauder, J.: Der Fixpunktsatz in Funktionalriiumen. *Studia Math.* 2: 171–180 (1930).

[148] Showalter, R. E.: *Monotone Operators in Banach Space and Nonlinear Partial Differential Equations.* Mathematical Surveys and Monographs. American Mathematical Society, Providence, RI (1997).

[149] Smart, D.R.: *Fixed Point Theorems.* Cambridge University Press, Cambridge (1980).

[150] Taoudi, M. A.: Krasnosel'skii type fixed point theorems under weak topology features. *Nonlinear Anal.* 72: 478–482 (2010).

[151] Tretter, C.: *Spectral Theory of Block Operator Matrices and Applications.* Impe. Coll. Press, London (2008).

[152] Van der Mee, C., Zweifel, P.: A Fokker–Plank equation for growing cell population. *J. Math. Biol.* 25: 61–72 (1987).

[153] Vrabie, I. I., Cuza, Al. I., Mayer, O.: *C_0-Semigroups and Applications.* Elseiver, New york (2003).

[154] Zeidler, E.: *Nonlinear Functional Analysis and Its Applications.* Vol I, Springer, New York (1986).

Index